AF682283

The Science of AI in Water Management

The Science of AI in Water Management offers a comprehensive exploration of how artificial intelligence is transforming the management of one of Earth's most vital resources. Written in an engaging and accessible style, the book examines AI's applications across various water management domains, including real-time water quality monitoring, leak detection, flood risk assessment, and wastewater treatment. Readers will gain insights into predictive maintenance, AI-powered irrigation, groundwater management, and the intricate relationship between AI and sustainable water practices. Addressing technical, economic, and social aspects, this book serves as a valuable resource for water managers, engineers, policymakers, students, and industry professionals.

- It offers insights into AI-driven strategies that promote sustainable water use, enhance water conservation efforts, and address the impacts of climate change on water resources.
- It includes practical case studies, real-world applications, and actionable solutions to modern water management challenges.
- It combines expertise from environmental science, engineering, AI technology, and policy to provide a well-rounded understanding of water management challenges and solutions.

The Science of AI in Water Management

Frank R. Spellman

CRC Press
Taylor & Francis Group
Boca Raton London New York

CRC Press is an imprint of the
Taylor & Francis Group, an **informa** business

Designed cover image: Shutterstock

First edition published 2026
by CRC Press
2385 NW Executive Center Drive, Suite 320, Boca Raton FL 33431

and by CRC Press
4 Park Square, Milton Park, Abingdon, Oxon, OX14 4RN

CRC Press is an imprint of Taylor & Francis Group, LLC

© 2026 Frank R. Spellman

Reasonable efforts have been made to publish reliable data and information, but the author and publisher cannot assume responsibility for the validity of all materials or the consequences of their use. The authors and publishers have attempted to trace the copyright holders of all material reproduced in this publication and apologize to copyright holders if permission to publish in this form has not been obtained. If any copyright material has not been acknowledged please write and let us know so we may rectify in any future reprint.

Except as permitted under U.S. Copyright Law, no part of this book may be reprinted, reproduced, transmitted, or utilized in any form by any electronic, mechanical, or other means, now known or hereafter invented, including photocopying, microfilming, and recording, or in any information storage or retrieval system, without written permission from the publishers.

For permission to photocopy or use material electronically from this work, access www.copyright.com or contact the Copyright Clearance Center, Inc. (CCC), 222 Rosewood Drive, Danvers, MA 01923, 978-750-8400. For works that are not available on CCC please contact mpkbookspermissions@tandf.co.uk

Trademark notice: Product or corporate names may be trademarks or registered trademarks and are used only for identification and explanation without intent to infringe.

ISBN: 978-1-041-10590-9 (hbk)
ISBN: 978-1-041-10586-2 (pbk)
ISBN: 978-1-003-65584-8 (ebk)

DOI: 10.1201/9781003655848

Typeset in Times
by Apex CoVantage, LLC

Contents

PART 4 Drinking Water Supplies, Pollution, and Climate Change

PART 5 Water Management Applications

Preface

This book, *The Science of AI in Water Management,* is the eighteenth volume in the acclaimed series that includes *The Science of AI in Energy Storage* (in production), *The Science of AI in Environmental Engineering, The Science of Energy: Principles, Concepts, and Applications, The Science of Fluid Mechanics: Concepts and Applications in Water and Wastewater (in production), The Science of Water Reuse, The Science of Land Subsidence, The Science of Green Energy, The Science of Carbon Capture and Storage (CCS), The Science of Ocean Pollution, The Science of Lithium (Li), The Science of Electric Vehicles (EVs): Concepts and Applications, The Science of Rare Earth Elements: Concepts and Applications, The Science of Water, The Science of Air, The Science of Environmental Pollution, The Science of Renewable Energy, The Science of Waste, and The Science of Wind Power,* all of which bring this highly successful series fully into the twenty-first century. This book continues the series' mantra based on sound rather than feel-good science. It also continues to be presented in the Author's trademark conversational style—making sure communication is certain—not a failure . . . I like to ensure that the reader sees both the forest and the trees, and maybe because I want to converse with the reader. I aim to be comprehensive in coverage, clear in my delivery, and clear in my language.

All living things on Earth need certain essential elements to survive. First, we must be in the right galaxy. We need the life-giving and supporting star within the right galaxy—the Sun. Then, we need Earth to be at the perfect or correct distance from the Sun. Next is the essential atmosphere. We need a hospitable environment. We need the atmosphere to hold oxygen, a few other gases, and safe breathing. We also need nutrients; we need safe, healthy food. Water is an absolute must; we cannot survive for long without water. This book is all about our absolute need for secure, healthy water. In this book, all references to water refer to freshwater. This book specifically discusses

- AI in real-time water quality monitoring.
- AI in drinking water demand management.
- AI in wastewater treatment.
- AI for water conservation.
- AI-enabled flood risk assessment and management.
- AI-powered leak detection and repair.
- AI-assisted groundwater management.
- predictive maintenance of water infrastructure.
- the technical, economic, and social aspects of managing water resources.
- water fundamentals involving physical, chemical, and biological systems provide a foundation.
- observed and predicted climate change impacts on water resources.
- the management of water supply and demand in the United States.
- fundamentals of flood and drought management are covered.

- fundamentals of AI-directed irrigation.
- industrial uses of water.
- AI in water treatment plants.
- Lean and Water Toolkit.
- Lean Six Sigma and AI.

To ensure the sustainability of our water supply, we must implement effective water management practices rather than superficial or ineffective measures. To ensure good water management, we must have water management experts and informed individuals who are well-rounded in knowledge. In addition to good management skills and common-sense leadership, water managers must be generalists. Effective water managers require expertise in fields such as environmental lab operations, math, geology, biology, hydrology, geography, meteorology, economics, political science, psychology, and practical common sense. This book introduces water sources, development, recycling, and management practices.

This book follows the successful format of the other editions in this series to achieve its goals. This unrestrained book points to artificial intelligence's present and future applications in efficient water management procedures and processes, such as water and wastewater treatment technology.

Population and AI can coexist in conflict or harmony, benefiting humanity as with all other life forms on Earth. The point is that effective water management must include a comprehensive understanding of how these forces intermingle, connect, and exist in the critical water resource management that is so critical to us all.

Artificial intelligence (AI) can play an essential role in ensuring that water used for consumption is safe for public health and the environment by reducing the impact of pollutants discharged into water bodies. The challenges associated with achieving this goal are becoming more complex due to rapid population growth, urban development, sprawl, and industrialization. These challenges include predicting the removal of organic and inorganic pollutants, monitoring water quality, and detecting faults in purification and machinery equipment. In this work, the reader is introduced to "Lean," "5S," and the Water Professional Manager's Toolkit.

In addition, this book introduces the science of water management in a manner that assures its correlation to modern practice and design.

The bottom line: This book is designed for and is accessible to water managers, water practitioners, engineers, designers, policy and decision-makers, corporate planners, water industries, students, and anyone else interested in water, water technologies, lesser-known technologies, and AI (artificial intelligence) as it is employed in water resource management, production, usage, and the goal of reaching sustainability. We are on the cusp of entering an AI-powered future. This points out that we need to board the AI train and not ride the caboose—we need to be the engine pulling the rest of the world into the AI Age. Eventually, growth in the global population and the material economy will be confronted by humanity. When this occurs (and it will), we will either adjust and survive or join the ranks of the dinosaurs, dodo birds, passenger pigeons, golden toads, or several other species presently experiencing the Sixth Extinction.

The combination of artificial intelligence and its subsections is creating massive new opportunities for water and public health practitioners and companies to improve efficiency, optimize performance, drive innovation, and accelerate growth. Improving efficiency and performance, driving innovation, and accelerating growth are critical because AI offers unlimited possibilities.

Many people feel fear and anxiety about AI interfering with their lives. Several others and I have said that mindset is everything, and hopefully, humans and AI have connected minds.

Note to Reader: In this book, "Water Manager" is deliberately capitalized to emphasize the importance of this position.

Frank R. Spellman
Norfolk, Virginia

About the Author

Frank R. Spellman, PhD, is a retired assistant professor of environmental health at Old Dominion University, Norfolk, Virginia, and the author of more than 165 books covering topics ranging from concentrated animal feeding operations (CAFOs) to all areas of environmental science and occupational health. Many of his texts are readily available online, and several have been adopted for classroom use at major universities throughout the United States, Canada, China, Europe, South Korea, and Russia; two have been translated into Spanish for South American markets. Dr. Spellman has been cited in more than 1,250 publications. He is a professional expert witness for three law groups and an incident/accident investigator for the U.S. Department of Justice and a Northern Virginia law firm. Additionally, he consults on homeland security vulnerability assessments for critical infrastructures, including water/wastewater facilities nationwide. He conducts nationwide pre-Occupational Safety and Health Administration (OSHA)/Environmental Protection Agency (EPA) audits. Dr. Spellman receives frequent requests to co-author with well-recognized experts in several scientific fields; for example, he is a contributing author of the prestigious text *The Engineering Handbook*, 2nd ed. (CRC Press). Dr. Spellman lectures on sewage treatment, water treatment, homeland security, and safety topics throughout the country and teaches water/wastewater operator short courses at Virginia Tech (Blacksburg, Virginia). He holds a BA in public administration and a BS in business management, both Summa Cum Laude, an MBA, and an MS and PhD in environmental engineering. In 2011, Dr. Spellman traced the ancient water system at Machu Picchu, Peru, and also two different drinking water systems in the Galapagos Islands.

Acronyms/Abbreviations

°C	Degrees Centigrade or Celsius
°F	Degrees Fahrenheit
μ	Micron
μg	Microgram
μm	Micrometer
A-C	Alternating Current
ACEE	American Council for an Energy-Efficient Economy
AI	Artificial Intelligence
Amp	Ampere
APPA	American Public Power Association
ASCE	American Society of Civil Engineers
BEP	Best Efficiency Point
bhp	Brake Horsepower
CCS	Carbon Capture and Storage
CEC	California Energy Commission
CEE	Consortium for Energy Efficiency
DC	Direct Current
DOE	Department of Energy
ECM	Energy Conservation Measure
EIS	Environmental Impact Statement
ENR	*Engineering News-Record*
EPA	Environmental Protection Agency
EPACT	Energy Policy Act
EPC	Energy Performance Contracting
EPRI	Electric Power Research Institute
ESCO	Energy Services Company
EV	Electric Vehicle
GHG	Greenhouse Gas
GIS	Geographic Information System
GPD	Gallons per Day
GPM	Gallons per Minute
GW	Gigawatt
HVAC	Heating, Ventilation, and Air Conditioning
Hz	Hertz
I&C	Instrumentation and Control
IoT	Internet of Things
kW	Kilowatt-hour
kWh/year	Kilowatt-hours per year
M	Mega
ML	Machine Learning
MPN	Most Probable Number
MW	Molecular Weight

NEPA	National Environmental Policy Act
NREL	National Renewable Energy Laboratory
O&M	Operation and Maintenance
Pa	Pascal
PG&E	Pacific Gas and Electric
psi	Pounds per Square Inch
psig	Pounds per Square Inch Gauge
rpm	Revolutions per Minute
SCADA	Supervisory Control and Data Acquisition
SCFM	Standard Cubic Feet per Minute
TVA	Tennessee Valley Authority
VFD	Variable Frequency Drive
VSS	Volatile Suspended Solids
W	Watt

Units/Conversions

Most calculations made in energy storage operations/equations involve using *units.* While the number tells us how many, the units tell us what we have. Examples of units include inches, feet, square feet, cubic feet, gallons, pounds, watts, kilowatts, liters, pounds per square inch, and miles per hour. *Conversions* are a process of changing the units of a number to make the number usable in a specific instance. Multiplying or dividing by another number to change the units of the number accomplishes conversions. Common conversions are as follows:

Note: In many instances, the conversion factor cannot be derived—it must be known. Therefore, we use tables like the one below to determine the common conversions.

Note: Conversion factors change measurements or calculated values from one unit of measure to another. In making the conversion from one unit to another, you must know two things:

(1) The exact number that relates the two units
(2) Whether to multiply or divide by that number

Common Conversions

Linear Measurements	Weight
1 inch = 2.54 cm	1 ft^3 of water = 62.4 lbs
1 foot = 30.5 cm	1 gal = 8.34 lbs
1 meter = 100 cm 3.281 ft = 39.4 in.	1 lb = 453.6 grams
1 acre = 43,560 ft^2	1 kg = 1,000 g = 2.2 lbs
1 yard = 3 feet	1 % = 10,000 mg/l
Volume	**Pressure**
1 gal = 3.78 liters	1 ft of head = 0.433 psi
1 ft^3 = 7.48 gal	One psi = 2.31 ft of head
1 l = 1,000 ml	**Flow**
1 acre foot = 43,560 cubic feet	1 cfs = 448 gpm
1 gal = 32 cups	1 gpm = 1,440 gpd
One pound = 16 oz dry weight	

Temperature Conversions

$$°C = 5/9(°F - 32°)$$

$$°F = 5/9(°C + 32°)$$

These conversions are easy to perform. The difficulty arises when we must recall these formulas from memory. The easiest way to recall these crucial formulas is to remember three basic steps for both Fahrenheit and Celsius conversions:

(1) Add 40°.
(2) Multiply by the appropriate fraction (5/9 or 9/5).
(3) Subtract 40°.

The only variable in this method is the choice of 5/9 or 9/5 in the multiplication step. To make the proper choice, you must be familiar with two scales. On the Fahrenheit scale, the freezing point of water is 32°F, and 0°C on the Celsius scale. The boiling point of water is 212°F on the Fahrenheit scale and 100°C on the Celsius scale.

Why is it important? Note, for example, that higher numbers are associated with the Fahrenheit scale at the same temperature and lower numbers with the Celsius scale. This vital relationship helps you decide whether to multiply by 5/9 or 9/5. Let us look at a few conversion problems to see how the three-step process works.

Problem #1: Convert 220°F to Celsius. Using the three-step process, we proceed as follows:

Solution:

(1) Step 1: Add 40°F.

$$\begin{array}{r} 220^\circ \\ \underline{+\ 40^\circ} \\ 260^\circ \end{array}$$

(2) Step 2: 260°F must be multiplied by either 5/9 or 9/5. Since the conversion is to the *Celsius* scale, you will move to a number *smaller* than 260. Through reason and observation, we see that multiplying 260 by 9/5 would be the same as multiplying by 2, which would double 260 rather than make it smaller. On the other hand, multiplying by 5/9 is approximately the same as multiplying by 1/2, which would cut 260 in half. Since, in this problem, you wish to move to a smaller number, you should multiply by 5/9:

$$(5/9)(260^\circ) = 144.4^\circ\text{C}$$

(3) Step 3: Now subtract 40°.

$$\begin{array}{r} 144.4^\circ\text{C} \\ \underline{-\ 40.0^\circ\text{C}} \\ 104.4^\circ \end{array}$$

Thus, 220°F = 104.4°C.

Problem #2: Convert 22°C to Fahrenheit.

(1) Step 1: Add 40°.

$$\begin{array}{r} 22^\circ\text{F} \\ \underline{+\ 40^\circ} \\ 62^\circ\text{F} \end{array}$$

Because we are converting from Celsius to Fahrenheit, you are moving from a smaller to a larger number and should use 9/5 in the multiplication:

(2) Step 2:

$$(9/5)(62^\circ) = 112^\circ$$

(3) Step 3: Subtract 40°.

$$\begin{array}{r} 112^\circ \\ \underline{-\ 40^\circ} \\ 72^\circ \end{array}$$

Thus, 22°C = 72°F.

Knowing how to make these temperature conversion calculations is useful. However, in practical (real-world) operations, you may wish to use a temperature conversion table.

Note: Additional units are used in energy and are listed in the Energy Storage Technician's Toolbox below.

THE WATER PROFESSIONAL'S CONVENTIONAL TOOLBOX

In many cases, when a water professional's toolbox is mentioned or thought about by domestic and even some industrial users, it is likely that the tools contained in the toolbox are plumbers' tools. In this regard, the tools assumed to be in the plumber's toolbox include the following:

- Pipe wrenches
- Plungers and drain snakes
- Teflon tape and Plumber's Putty
- Basin wrench
- Pipe cutters
- Plumber's torch
- Adjustable wrenches (**note**: based on personal experience, I have had no problem with maintenance operators using a wide assortment of adjustable wrenches, but I always ensure that channel lock wrenches are not included in their toolboxes. The adjustable jaws of an adjustable channel lock plier are designed to be set to different widths, allowing them to grip a wide range of sizes. When properly used, whether straight, curved, or angled, channel locks are handy tools for experienced hands. The problem arises whenever a maintenance operator uses channel locks for any task. This practice can lead to more damage to various parts than desired.)
- Screwdrivers
- Pliers
- Safety gear.

As with all professionals, the water professional who is adequately educated, trained, and has learned through on-the-job experience has a toolbox.

Does he or she have the plumber's conventional toolkit?

The accessories in the toolbox include a practical suite of proven and cost-effective tools, measures, and activities to improve water efficiency, reduce waste and environmental damage, and, when needed, devise appropriate mitigation strategies. This section discusses different water sources, types, and computations related to the environment; these are elements contained in the toolbox.

Energy Units

Energy and its associated units cannot be ignored in water management activities simply because they are essential in the finding, processing, and delivery of safe drinking water for consumption and for other water uses. As an example of how vital energy is to the water management professional, consider the operation of a wastewater treatment plant. Wastewater managers must be acutely aware of energy and its use in the treatment process because energy in the form of electrical power is the number one cost involved in plant operations.

Watts (W) is the term we usually use when measuring energy. In practical usage, a single watt is too tiny, so to make energy numbers usable, scientists use multiples of thousands of watts, such as kilowatts, megawatts, gigawatts, and terawatts. **Kilowatts (kW)** are 1,000 watts, most used for domestic energy. **Megawatts (MW)** are 1,000 kW and are most widely used for power generation and distribution. **Gigawatts (GW)** is 1,000 megawatts, which is used to describe notional demand. **Terawatts (TW)** is 1,000 gigawatts, which describes country-level consumption.

MKS System (SI Units)

The *MKS system* (so named because it uses the meter, kilogram, and second as base units) comprises *SI units*. All other units are derived from the base units listed in Table TB. This is an entirely consistent system; there is only one recognized unit for each variable (physical quantity).

TABLE TB
SI Base Units

Quantity	Name	Symbol
Length	Meter	m
Mass	Kilogram	kg
Time	Seconds	
Electric current	Ampere	A
Temperature	Kelvin	K
Amount of substance	Mole	mol
Luminous intensity	Candela	cd

DID YOU KNOW?

U.S. water practitioners find identifying specific quantities used in the United States confusing. For example, if the question was asked, how many gallons are contained in a barrel of U.S. liquid compared to a barrel of petroleum, how many would know the correct answer? Along with answering this question and to clarify the air on other quantities, several are listed as follows:

- atmosphere standard atmosphere
- barrel 31.5 gallons (U.S., liquid)
- barrel 42 gallons (petroleum)
- Btu tradition (thermochemical) value
- calories thermochemical
- gallons U.S., liquid
- horsepower U.S. or mechanical
- ounces avoirdupois
- pints U.S., liquid
- pounds avoirdupois
- quarts U.S., liquid
- tons short tons of 2,000 pounds

DID YOU KNOW?

An abampere, in electricity, is a centimeter-gram-second unit of electromagnetic current, equivalent to 10 amperes. A statampere is the electric unit of current equal to the current produced by an electromotive force of one statvolt acting through a resistance of one statohm. An abampere is multiplied by 2.99793×10^{10} to obtain a statampere.

Electricity Reductions (Kilowatt-Hours)

USEPA's Greenhouse Gas Equivalencies Calculator uses the Emissions & Generation Resource Integrated Database (eGRID) U.S. annual non-baseload CO_2 output emission rate to convert reductions of kilowatt-hours into avoided units of carbon dioxide emissions. Most users of the Equivalencies Calculator who seek equivalencies for electricity-related emissions want to know the equivalencies for emissions reductions from energy efficiency or renewable energy programs that are responsible for water operations. These programs are not assumed to affect baseload emissions (the emissions from power plants that run constantly) but rather non-baseload generation (power plants brought online as necessary to meet demand). Therefore, the Equivalencies Calculator uses a non-baseload emission rate (USEPA 2012a).

Emission Factor

$$7.0555 \times 10^{-4} \text{ metric tons } CO_2 \text{ / kWh}$$

Notes:

- This calculation does not include any greenhouse gases other than CO_2.
- This calculation does not include line losses.

Tons of Waste Recycled Instead of Landfilled

To develop the conversion factor for recycling rather than landfilling waste, emission factors from EPA's WAste Reduction Model (WARM) were used (USEPA, 2012e). These emission factors were developed following a life-cycle assessment methodology using estimation techniques developed for national inventories of greenhouse gas emissions. According to WARM, the net emission reduction from recycling mixed recyclables (e.g., paper, metals, plastics), compared with a baseline in which the materials are landfilled, is 0.73 metric tons of carbon equivalent per short ton. This factor was converted to metric tons of carbon dioxide equivalent by multiplying it by 44/12, which is the molecular weight ratio of carbon dioxide to carbon.

Calculation

Note: Due to rounding, the calculation given in the equation below may not return the exact results shown.

0.73 metric tons of carbon equivalent/ton × 44 g CO_2/12 g C = **2.67 metric tons of CO_2 equivalent/ton of waste recycled instead of landfilled**

Water Management by Other Terms[1]

When talking about managing water resources and usage, it seems like straightforward lingo, easily understood with little explanation required, right?

Wrong. Well, sort of, anyway. So, when we discuss management of water resources and usage, are we discussing and thinking about this vital management of the resource and its use in conventional, out-of-date terms, or are we focusing on the transformation from the old to the new, the new being the digital, AI, AI subfields, and the lean thinking age?

Understanding water management science and its connection to artificial intelligence requires knowing the terminology related to this notable change, this transition, this transformation. This shift moves from the traditional "batch and queue" mass production to a product-aligned "one-piece flow" pull production (aka a Lean One Piece flow system), as described by the USEPA (2024): "The conventional *batch and queue* involves mass production of large lots of products in advance based on potential or predicted customer demands, a *one-piece flow* systems rearranges production activities in a way that processing steps of different types are conducted immediately adjacent to each other in a continuous flow" (p. 1).

One of the most essential tools in a Water Professional Manager is to know, understand, and use properly the language (the jargon or lingo), terms, and definitions associated with this vital service. Thus, the following jargon and explanations related to water resource management cost and performance metrics are presented. Note that the terms and definitions provided below are essential tools, a significant part of the Water Managers' toolbox. In some cases, they are deliberately repetitive for retention and for explaining the broad scope of their meaning, but later, their meaning may differ slightly depending on their application.

Note to Reader: You may wonder why the following terms and definitions are presented here early in this work and not listed until presented and defined in the glossary. After years of teaching college students and on-site operators the material within this text, I have found that my former students appreciated an early introduction to the key terms used in this text.

- **3P**: Production Preparation Process or Pre-Production Planning
- **5S**: Sort, Set in order, Shine, Standardize, Sustain
- **6S:** Sort, Set in order, Shine, Standardize, Sustain, Safety
- **C/O**: Changeover time
- **E3:** Economy, Energy, and Environment
- **A3**: A3 refers to the paper size used by manufacturers for creating reports using the A3 method, which is applied to solve problems, report project status, and propose policy changes.

- **5 Whys Approach**: This is a root cause analysis method that involves asking "why" five times to identify the cause/effect relationships underlying a specific problem. It is used to determine the root cause of a defect or problem and improve manufacturing processes by eliminating waste and enhancing quality. An example is shown below.
  1. *Why did the aeration blower stop?*
     There was an overload, and the circuit breaker tripped.
  2. *Why was there an overload?*
     The bearings were not properly lubricated.
  3. *Why was it not lubricated sufficiently*?
     Preventive maintenance was skipped.
  4. *Why was it not aerating sufficiently?*
     The shaft of the blower motor was worn and rattling.
  5. *Why was the shaft worn out?*
     There was no filter attached, and fugitive metal scraps got in.
- **Batch and Queue**: The mass production process of making large lots of a part and then sending the batch to wait in the queue until the next operation in the production process begins. This process is the contrast of one-piece flow.
- **Cellular Manufacturing**: An approach in which manufacturing work centers (cells) have the total capabilities needed to produce an item or group of similar items; contrasts with setting up work centers based on similar equipment or capabilities, in which case items must move among multiple work centers before they are completed.
- **Chaku-Chaka (Japanese for "load, load")**: This is a method of conducting one-piece flow, where the operator proceeds from machine to machine, taking the part from one machine and loading it into the next.
- **DMAIC**: This is the Six Sigma five-step process (Define, Measure, Analyze, Improve, and Control) used to implement Six Sigma tools and identify process inefficiencies and variations.
- **Design for Lean Six Sigma**: This method involves designing processes that align with Lean Six Sigma objectives, including reducing variability, improving yield, minimizing waste, and accelerating time-to-market.
- **Failure Mode and Effects Analysis (FMEA):** This is a technique used to identify potential failure modes or causes of failures due to design or process deficiencies. Failure Mode and Effects Analysis (FMEA) evaluates the severity of potential failures and determines necessary design or process modifications to mitigate risks.
- **Footprint (square feet)**: The physical area a water utility requires.
- **Kaizen (Japanese, "to take apart" and "make good")**: This is the incremental and continual improvement of production activities aimed at reducing waste, and designed around planned, structured worker-oriented events.
- **Lean Management**: This is a business improvement method that eliminates waste using practical methods, often combined with Six Sigma to reduce defects and variation.

- **Six Sigma**: This refers to a set of well-established statistical control techniques and data analysis methods used to identify and reduce variation in products and processes.
- **Non-Value-Added**: Activities or actions taken that add no real value to the product or service from a customer's perspective, making such activities or actions a form of waste.
- **One-Piece Flow (single-piece flow)**: A method where each product moves through design, order-taking, and production without interruptions, backflows, or scrap.
- **Queue Time**: The time a material stays in a queue waiting to be used.
- **Rate Base**: The portion of total assets (principally investments in plant and equipment) for regulated utilities, as defined by a regulatory body, on which a utility can earn a return.
- **Return on Debt**: A component of the carrying charge, return on debt is the revenue required to pay for using borrowed money. It is usually stated as a percentage and applies to unearned capital in a particular year. Numerically, it equals the cost of debt money times the debt ratio.
- **Return on Equity**: A component of the carrying charge, return on equity is the revenue required to pay for using equity money. It is commonly stated as a percentage and is applied to unrecovered capital in a particular year. Numerically, it equals the cost of equity money times the equity ratio.
- **Security Analysis**: The evaluation of a project, using several different assumptions, on the values of one or more uncertain variables.
- **Service Territory**: The area where a utility currently provides service to customers and specific areas adjacent to the utility's water distribution lines or where the water utility holds a franchise.
- **Six Sigma**: This methodology emphasizes continuous improvement and statistical analysis, aiming to identify and reduce variation in products and processes.
- **Supply Chain**: A group of all suppliers involved in the manufacture of a product, beginning with the simplest part and ending with the production of the final product.
- **Time-of-Use Pricing**: A rate design imposing higher charges to customers during periods when higher demand is experienced.
- **Toyota Productive Maintenance (TPM):** The manufacturing strategy of Toyota, upon which the terms "Lean Production" and "Lean Manufacturing" are based.
- **Value Stream:** All activities (value added and non-value added) involved in producing a product or delivering a service to a customer, from the receipt of raw materials to the delivery of finished products to the customer.
- **Water Utilities:** All enterprises involved in the production and/or distribution of electricity for public use, including investor-owned water utility companies and government-owned water utilities and purveyors (such as municipal systems, federal agencies, and state systems).

- **Use Management**: The application of measures to influence customers' use of water to modify the demand and load factor.
- **Value Stream Mapping**: A process mapping method used to document the current and future states of the information and material flows in a value stream.
- **Weight (kg)**: How heavy the water storage and conveyance system is.

Prelude: During a Recent Rabbit and Grasshopper Conversation

At a recent meeting at their favorite location in the pristine wilderness of western Washington State— a place humans call Olympic National Park, Mr. Rabbit and his friend Mr. Grasshopper were hidden together in the Hoh Rain Forest beneath their favorite fall of moss standing on a bed of soft ferns next to a popular hiking trail just across from a shared camping spot for humans. They were quiet as seven hikers stood on the trail, observing the moss and Oregon oxalis painting the understory green amid Sitka spruce. The hikers were soaked despite their rain gear—staying dry is impossible in a place with over 200 inches of rain annually. Then, the hikers turned off to their right and entered the campsite. While removing their soaked clothing during a brief pause in the rain, Mr. Grasshopper and his friend Mr. Rabbit remained hidden. They were about to beat feet, literally and otherwise, toward the deep, dark surroundings when they paused as one of the male hikers stated: "Wow, there certainly is no shortage of water here . . . we are one with Nature and her natural washing machine, so to speak."

This statement was like listening to something foreign to Mr. Rabbit and Mr. Grasshopper.

Mr. Grasshopper asked Mr. Rabbit, "What are those humans talking about?"

Mr. Rabbit, still listening to the human talk, thumped a leg against the wet ferns, shrugged his shoulders, and did not answer.

Another female individual stated, "At least the rain is warm right now and kind of feels good. Though I am drenched, it is of no concern. We are embracing nature and liberating ourselves from our everyday lives in the real world. We are free from the adverse impacts of various forms of this or that polluting our usual environment, making life unbearable at times . . . and, as already said around here, we do not have to worry about our water supply . . . like we have to worry about quantity and quality of our drinking water and other freshwater uses."

For a moment, humans and their hidden observers could only hear the pit-pat, pat, pit-pat of raindrops striking against the lance-shaped blades of the copse of blue Piper's bellflowers. Mr. Grasshopper and Mr. Rabbit took shelter beneath the blades high on the stems. With time and lots of rain, they learned that these natural umbrellas help them from getting overwatered by the wetness. They both looked at each other, shrugged their shoulders, and turned their attention back to the camping area where the humans were pitching their tents. Mr. Grasshopper and Mr. Rabbit were surprised by the news about the water supply and quality, knowing that their water from the pristine forest was always pure due to minimal pollution.

Then they heard another older lady state, "I am so happy to be away from it all . . . this place is paradise, a wet one, for sure, but away from all the bull we put up out there in la-la land."

Mr. Rabbit and Mr. Grasshopper looked at each other. They had heard these comments many times before from humans who were new to the surroundings, but they still wondered what the humans were discussing, what La-La Land was, and where it was. They were unaware, and to be candid, they preferred to remain uninformed.

One of the young male campers said, "Well, we certainly don't lack freshwater out here, and I am not worried about paying a horrendous water bill out here . . . but I am sure one of them terrible bills will be waiting for me . . . waiting for us all when we return to what Melody just called la-la land . . . which I agree with that."

Again, Mr. Grasshopper and Mr. Rabbit were all ears, so to speak.

A familiar wailing sound rang through the dense forest. Mr. Rabbit and Mr. Grasshopper paid little attention to the bugling of the Roosevelt Elk. In this place, this is a familiar sound. And for these two, it is a sound that is so reassuring to them . . . in their case, there is no need for CBD chill gummies like those just consumed by the hikers.

"Yup, I like no need for worry about water and its purity and it's costs to worry about here," says another male camper. He says this before trying to power up his iPhone, but the phone battery is not cooperating and, in the dense forest, getting a signal from a distant tower is not possible.

There are just some things humans do not want to be without, even in one of the most pristine and beautiful surroundings like the park.

"This is a real bummer, my friends. I like to stay informed about what is happening, but with a dead phone, this is impossible."

All the campers agreed, and one stated, "Let us get inside the tents, dry off, eat, and drink good wine."

Meanwhile, Mr. Rabbit and Mr. Grasshopper moved closer to the plant stems to seek more cover under the flower blades from the now-pouring rain.

"Electricity . . . you know I often hear our camping visitors talk about electricity, especially when they try to use their flat-plate things or their phones," says Mr. Rabbit while peering through the rain at the six colorful wet tents in the camping area.

"Well, Mr. Rabbit, I have heard, over the season, you understand, the campers talking about electricity, water, TV, computers, AI (whatever that is?), and many other items that they do not initially miss. However, after a few days out here in Elk country, they began talking about getting back to their 'la-la land'."

Mr. Grasshopper said. "Them human types sure are a strange bunch."

Mr. Rabbit says, "You know you are correct . . . one time when I was eavesdropping on a couple who were sharing a tent mention water—clean water—and one of the campers went so far as to state what she was missing being out here in our grand neighborhood . . . that my long-legged friend was a whole laundry list of missing items."

During their conversation, Mr. Rabbit told his friend that modern human life relies on water and other essential items.

Mr. Rabbit paused momentarily and listened to the rain . . . so soothing, and licking the wet leaves was refreshing—no problem with their water supply here.

Then, being one of those rabbits in the know, so to speak, he explained to his friend what humans need water for. He stated that not only did they need to drink water, as the two of them do, but also, they water lawns, wash vehicles, take baths, cook with water, and use water for other purposes.

It is essential to point out that Mr. Rabbit had no idea about the things he mentioned, but he had heard about them by overhearing the many visitors during the summer. He also listened to a few of them say they wished they could store more energy and take it camping with them.

Grasshopper reminded his friend Rabbit, "To fix human water needs, what humankind needs right now, my friend, is innovation, innovation, innovation."

After a deliberate and well-practiced thumping of his foot, Rabbit replied to his friend, Grasshopper: "No, my long-legged friend . . . to fix all of humankind's water and other problems, what they need first is discovery, discovery, discovery and then invention, invention, invention . . . followed up by innovation, innovation, innovation . . . hmmmm, human leadership, brain power, common sense, (hmmm, common sense is not that common with them folks) anyway accountability would also help."

Grasshopper: hmmmmmmmmmm, buzzzzzzzz, etc., "Well, sometimes they ain't too smart . . . them humans. . . . All they need to do is ask us. We know how to economize . . . how to appreciate our adequate water supply and how to be efficient and use it—well, at least for our use."

Rabbit replied: "Right on, Grasshopper!"

—A friend of all grasshoppers and rabbits, 2025

NOTE

1 Based on information in USEPA's (2024). *Environmental Professional's Guide to Lean and Six Sigma.* Washington, DC: United States Environmental Protection Agency.

Part 1

Digital Basics

1 AI in Water Management

Ignorance, misuse, and mismanagement of water can have profound effects on life on Earth. The good news is that several public utilities are using AI to improve their water management and preserve water, our most valuable resource.

1.1 THE PARADIGM SHIFT

Scientific research is undergoing a paradigm shift, a change in thinking that is also impacting water management. In this transformation, this paradigm of science has evolved through four principal scientific paradigm shifts. In the beginning, science relied solely on the first paradigm of science: **Empirical Evidence**. During this period, scientists relied on observable evidence, experimentation, and measurement to verify and certify hypotheses, or, in some cases, guesses as to their validity. The second scientific paradigm shift was the transition from relying only on empirical evidence to incorporating **Scientific Theory**. During this shift, we will try to explain why things happened the way they did. Supported theories are kept; unsupported ones are discarded. Note that empirical evidence makes the theory a scientific theory. Then we moved on, advanced, and entered the era of **Computational Science,** the third scientific paradigm shift, the change in thinking. This is the modern approach in the sciences, facilitated by the computer. I often wonder and continue to ask myself what the great geniuses of all time, such as Einstein, Newton, Da Vinci, Tesla, Aristotle, Archimedes, Fleming, Marie Curie, and Pasteur, could have discovered, found, and accomplished using computer science. Would we have a cure for cancer because of them and others who had access to computers during their time on Earth? Would we have a better understanding of science and its potentially great benefits to all living things and to Earth itself? We can only wonder. The fourth scientific paradigm is Data-intensive Scientific Discovery: This is an evolving shift powered by data science resulting from the emergence of massive amounts of data. We have only scratched the surface in scientific research using data-intensive scientific discovery. Those presently involved in the practical and positive applications of this fourth scientific paradigm generated by AI are like a horde of children in a candy store with an inexhaustible supply.

Water and wastewater utilities, purveyors of water, industrial users of water, and others who use substantial amounts of water (e.g., AI data centers) have, over time, collected vast amounts of data about water usage. Water managers who use artificial intelligence (AI) to process the trove of data can plan for and address the need to preserve water supplies, reduce contaminants, ensure wiser and more equitable distribution, enhance water quality, and better protect people and property from droughts and floods.

A transformative change in water management is ongoing. It means doing things differently. Actual transformative change is the change that becomes sweeping.

DOI: 10.1201/9781003655848-2

Transformative change is like a railroad locomotive transiting an endless track at high speed. As the locomotive wheels along the endless track, it picks up passengers and boxcars filled with new technological developments. All these advancements are coupled with the locomotive as it continues along its infinite track. One of the boxcars recently added is the digital metamorphosis—digital progress that continues to ride the endless rails. This technology is moving up the train, passing the other cars in the train, leaving those cars with older technology and moving toward dominance in management practices in several different enterprises. Water managers today must choose whether to lead improvements in efficiency, productivity, and management or remain passive followers—riders of the caboose, if they can catch it.

AI enables water managers to manage water resources and operations better. Water managers are better able to predict when and how predictive maintenance can be accomplished to promote the repair or replacement of key components before they fail. AI, as used in predictive maintenance, is revolutionizing water management by optimizing water utilization and creating a more resilient water infrastructure.

Another benefit of employing AI in water management is related to water distribution networks. Large datasets contain numerous trends that may not immediately be apparent to water managers. AI can help sort through the data (the maze) and help water managers gain understanding and forecast water demand to enhance supply throughout any given day. Machine learning (ML), a subset of AI, assists managers in reducing water wastage and efficiently meeting water demand. ML achieves this by helping to maintain a constant, consistent supply of water even with some equipment failures and when repairs are in progress. ML algorithms optimize water flow and pressure in real time, reducing losses and improving network performance.

AI allows Water Managers to monitor quality, trace usage, and recognize potential infrastructure problems in water systems. Moreover, AI in water management helps managers identify and address issues before they become problems. Water Managers should train employees in computer operations, AI, and IoT devices to interpret necessary data, rather than hiring outside experts (at present, it is difficult to find trained experts). Employee training on AI and its subsets, edge computing, IoT, cloud computing, and associated technologies allows them to improve water operation capacities. What this employee training accomplishes, along with employee buy-in (very important), is providing a workforce that can use AI and its subsets in real time to monitor pH, turbidity, contaminants, and other quality parameters. In this monitoring activity, operators can analyze IoT sensor data to trigger alerts about anomalies and guide corrective actions to maintain clean and safe water supplies.

We usually do not think about floods until they occur. However, meteorologists use AI and other technologies to predict flooding in their area of concern, allowing for advanced warnings about the potential for flooding events. Water managers use weather data, river levels, and historical flood patterns to predict and mitigate flood damage.

Before AI, flood experts predicted flooding and damage based on personal experience and intuition. AI has transformed weather prediction, improving accuracy for events like flooding. This ability to leverage data from generative AI, IoT, and other combinations of AI-enabled water management technologies not only enables experts to predict and mitigate flooding risks but also allows water managers to respond to the critical information provided by the technology.

The real bottom line in water management is to ensure efficiency through AI-driven algorithms. For instance, the efficiency of water operations in managing water/wastewater treatment plants can be improved by optimizing energy-intensive processes such as aeration used in activated sludge treatment. This can be achieved by adjusting the aeration blowers used for activated sludge operations to enhance operational performance while reducing electricity consumption.

AI allows water managers to access vast amounts of information and insights. AI can combine operational data to accurately show the real-time status of water utilities, industrial users, treatment plants, and other water management entities. For example, drought cannot be avoided entirely, but AI now enables meteorologists to analyze extensive datasets on weather, soil moisture, water tables, vegetation health, and other factors for more accurate drought predictions. One great example of an AI subfield usage in water/wastewater treatment operations is the employment of the Internet of Things (IoT). IoT is used for sensors to measure pH, temperature, turbidity, dissolved oxygen content, chemical concentration, flow rate, pressure, and chemical leak detection sensors. All of these sensors play a crucial role in real-time monitoring and management of water and, where applicable, in the wastewater treatment process, ensuring optimal performance and compliance with regulatory standards.

1.2 ALL ABOARD!

We must adapt as we ride the digital wave (aka the digital train). In this adaptation process, we need to learn the nuts and bolts and the applications of AI and its subfields, including ML, Deep Learning (DL), Natural Language Processing, Expert Systems, Computer Vision, Edge Computing, and the IoT. While traditional web technologies (i.e., programming languages, databases, server administration, servers, markup languages, style rules, and cloud-side code) remain central to current digital services, cloud-based solutions are leading the way. In addition to the digital locomotive, technologies such as mobile Internet, Big Data, design thinking, AI, robotics, advanced neural ML, augmented reality (AR), and additive manufacturing are advancing along the tracks of widespread adoption and innovation.

The present call of technology: All aboard the digital train.

DID YOU KNOW?

Natural Language Processing (NLP) focuses on understanding and generating human language. Expert Systems use rules and knowledge to make decisions in specific domains. Computer vision involves analyzing visual information to understand and interpret images and videos.

1.3 THE BIG QUESTION?

Note: Consider the real-time, real-world personal experience related to the following:

In teaching my college-level classes in environmental engineering and associated disciplines, I always spend considerable time at the start of my lectures on AI. I do this in classes related not only to the science of energy, energy storage, energy applications, water, air, soil, pollution, waste, climate change, assorted environmental health classes, with a few digressions thrown in here and there (commonly called "Sea Stories"). Again, I always begin my lectures on topics, subjects, disciplines, or whatever you want to call them, with a brief discussion on AI. For example, at the beginning of each semester, at the start of the first class, with a new group of students, undergrads, and grad students in combined classes (ranging from class size of 45 to 150 students), I ask each student to take a scrap of paper and with a pencil to write down no more than three words that, in their opinion best, describe AI—what it is, what it does, and so forth. I always receive a whole bunch of "and so forth" answers. For example, consider the following:

"It's Magic" (had a number of these responses)
"It's Today's World"
"We've been digitized."
"My job replacement"
"A communist plot" (more than one respondent provided this answer)
"The great repeater"
"AI break up"
"More data, please"
"Ability to learn"

The last response shown here, "ability to learn," was the only answer I sought, because learning answers the big question and many more. The other hundreds of responses are not printable in a work like this one. Anyway, because the nexus between the science of water management and AI is what this book is about, we need to look at the current AI components that make up the digital train and list their function:

- **Cloud-based services:** Traditionally, organizations had to purchase and maintain their physical servers. Now, cloud-based servers run in the cloud on infrastructure owned by the cloud service provider; they remove the need to buy, run, and manage physical servers.
- **Mobile Internet:** Also known as the *mobile web* for handheld devices, such as smartphones or feature phones (i.e., pre-wireless smartphones) through a wireless mobile network.
- **Big Data Architecture:** This is a data management system's systemic design and structure.
- **The IoT:** A network of physical objects embedded with sensors, software, and other technologies that connect and exchange data with other devices and systems via the Internet. IoT sensors are necessary for many applications.

Note: A few of these essential sensors were mentioned earlier, but it is vital to follow up with an explanation of their purpose or function:

- **Pressure Sensors**: These sensors detect changes in gas or liquid.
- **Chemical and Gas Sensors**: Gas and chemical sensors are key devices for detecting and measuring unwanted ambient gases and chemicals, and generating warnings.

- **Infrared Sensors**: These devices measure heat emissions from an object in applications such as remote controls and detecting environmental changes (temperature, motion, and proximity).
- **Water Quality Sensors**: These devices collect real-time data on water quality. They measure pH, temperature, turbidity, and contaminant levels.
- **Proximity Sensors**: These devices detect the presence or absence of objects near the sensor and operate without physical contact.
- **Temperature Sensors**: These devices measure the amount of heat generated from an area or an object.
- **Smoke Sensors**: Most of us are familiar with smoke detectors, as they have protected our homes and businesses for a long time. However, these IoT sensors are much more sensitive than the standard smoke detector.
- **Motion Sensors**: These devices are essential for ensuring security. Their applications include automated sinks and toilet flushers, automatic door controls, and energy management systems.
- **Level Sensors**: These sensors detect the level of various substances, including energy operations. They measure light and convert it into electrical signals. They are used in the automotive, energy, healthcare, and aerospace industries.
- **Digital Thinking:** This is a nonlinear, iterative aspect or process that challenges assumptions, redefines complications, and creates ground-breaking models to be evaluated.
- **AI technology:** This enables computers and machines to simulate human learning, understanding, problem-solving, decision-making, originality, and independence.
- **Robotics:** This follows the digital train. Within environmental engineering and science, robotics focuses on robotic automation algorithms used in ecological functions such as performing vital tasks related to air pollution monitoring, water, soil, wildlife monitoring, and waste cleanup operations.
- **Advanced Neural ML:** The "name neural" is based on a model of how the human brain functions and comprises a collection of processing units called *nodes*. Advanced Neural ML involves learning without definite instructions and is primarily used in DL.
- **Augmented-Reality Technology:** AR technology is following in the digital train. AR is popular because it presents an enhanced real-world version through computer-generated digital information, including visual, sound, and other sensory elements.
- **Additive Manufacturing:** The last section of the digital train, not the caboose presented in this discussion, is known as additive manufacturing (aka 3D printing). In operation, additive manufacturing constructs a three-dimensional object from a CAD or digital 3D model.

Note: To be successful in AI-based water management, you must know the terminology, live it, write it, sleep with it, and use it in speech. Therefore, the keywords and abbreviations related to AI used in this book are presented and defined in the following section.

1.3.1 Key AI Terms and Abbreviations

An algorithm: a specification of how to solve a class of problems by performing calculations, data processing, and automated reasoning tasks.

AI: intelligence produced by machines. AI is applied when a machine mimics cognitive functions that humans associate with learning and problem-solving. EO 14110 and the United States Code define AI as

> a machine-based system that can, for a given set of human-defined objectives, make predictions, recommendations, or decisions influencing real or virtual environments. Artificial Intelligence systems use machine- and human-based inputs to perceive real and virtual environments, abstract such perceptions into models through analysis in an automated manner, and use model inference to formulate options for information or action.
>
> (USDOE, 2008)

The reality is that there are various categories of AI; these categories overlap with each other to a point and include the following:

Artificial General Intelligence: AI, or human-level intelligence, aims to replicate human cognitive abilities across various tasks. It involves learning material or information and applying it to other domains.

- **Limited-memory Artificial Intelligence:** This refers to a subset of AI narrow intelligence in which algorithms and networks can manage data exceeding the machine's memory capacity.
- **Artificial Narrow Intelligence**: This is AI, also known as **weak AI**, which performs only one task. ANI includes image recognition, autonomous vehicles, chatbots, spam filters, car engine diagnostics, handwriting recognition, and other applications.
- **Theory of Mind Artificial Intelligence**: This type of AI interacts socially and can discern and respond appropriately to people's beliefs, thoughts, emotions, expectations, and even facial expressions.
- **Reactive Artificial Intelligence**: Reactive AI is best known for implementing if-then rules to work through a set of inputs to determine what action to take or recommend. It can only react to current situations without memory of past experiences.
- **Self-aware Artificial Intelligence: This purely theoretical system can be aware of its existence and understand its** thoughts.
- **Artificial Superintelligence**: This is a purely theoretical system in which the scope of intelligence surpasses that of humans.

Chatbot: An AI that conducts conversation via auditory or word-based methods.

Generative AI: This refers to models or algorithms that generate new output, such as text, photos, videos, data code, and 3D representations, from vast amounts of data they were trained on. Understanding the differences between AI and generative AI is crucial in today's technology-driven world, as these two concepts, while closely related, offer unique capabilities and applications in several different fields. Traditional AI systems are designed to understand, interpret, and make decisions based on the data they have been supplied. They can answer questions, recommend purchasing products, and drive motor vehicles autonomously. This is accomplished through ML algorithms that allow these systems to gain knowledge from experience and improve over time. On the other hand, generative AI goes beyond understanding and interpreting data. It generates entirely new content from the very beginning. It can create human-like text through advanced DL, compose music, and create images indistinguishable from those made by humans.

Generative Pre-trained Transformers: This denotes a family of AI models. These models, primarily larger language models such as those developed by OpenAI (founded in 2015 in San Francisco by Elon Musk and others), are based on a transformer architecture capable of understanding and generating human-like language.

Edge Computing: This is a distributed computing model that brings data processing and storage closer to the source of data generation—especially vital for applications where real-time decision-making is critical, such as in water management.

ML: The scientific study of statistical models and algorithms that computer systems use to perform specific tasks.

AE—auto recorder
AI—artificial intelligence
ANN—artificial neural network
CFD—computational fluid dynamics
CUSUM—Shewhart cumulative sum
DBN—Deep Belief Network
DL—Deep Learning
DRL—Deep Reinforcement Learning
GPU—Graphical Processing Unit
GRU—Gated Recurrent Unit
LSTM—Long Short-Term Memory
MLP—Multi-Layer Perception
ML—Machine Learning
NN—Neural Network
TL—Transfer Learning

1.4 AI AND THE CONNECTED HUMAN MINDSET

For several years, the technology used to produce, generate, and develop effective systems to collect, treat, and distribute freshwater has been the primary goal of Water Managers. Historically, the prioritization of water production and distribution has often overshadowed the importance of safe freshwater storage for immediate use.

Competent water management has become an essential element in water supply, as well as in water production and distribution. This is the case because competent water management is the key to unlocking the future of the Earth's most important substance, just behind breathable air, clean, safe, consumable freshwater.

Advancement of AI with human actions or activities comes down to an issue of attitude, outlook, approach, belief, and conviction that humans and AI can meld into a workable enterprise or mission that can accomplish great things for our benefit that we are still discovering, imagining, and hoping for. This certainly applies to water management.

Advancing water and wastewater technology with AI requires a comprehensive approach that integrates diverse data sources such as research, numerical models, emerging data, and environmental factors. It all concerns the need for AI technologies to implement domain-aware algorithms. These domain-aware algorithms leverage context-specific procedures and constraints to achieve advanced accuracy and efficiency. This provides critical insights that guide the ML process and ensure that the models developed are applicable and effective in their intended application.

Presented in simple, understandable language, the human/AI mindset, the connection, and concerns about applying AI to accelerate the development and deployment of water technology and usage are what we are talking about. The capabilities of multimodal AI foundation models should not be ignored. The possibility of enabling the discovery of new patterns and the prediction of overall impacts, physically and environmentally, should not be understated.

It is interesting to note the parallel between the 1970s environmental movement, the incorporation of personal computers, and the programs that run them, which have blossomed to the present extent. The movement toward combating climate change and the growing demand for clean, safe drinking water has provided an impetus for developing water operations that can incorporate AI to accelerate the development of new water technologies. AI-supported water management technologies can reliably control and efficiently optimize multi-scale treatment, storage, and distribution networks. AI is the tool needed to quickly incorporate and distribute data to a wide range of recipients. A key problem is the slow transfer of knowledge and requirements between different areas. Proper implementation of AI and its subsections can address and mitigate these issues.

1.5 THE BOTTOM LINE

The bottom line on AI in water management is that we have only touched the surface of what AI and its further development and application are potentially able to produce or accommodate in future operations. AI tools and approaches can optimize manufacturing processes that use water. The adoption of AI in automation technologies can improve production efficiency. AI has the potential to optimize every step of various water management activities. AI can also help develop modular and flexible manufacturing systems responsive to evolving technologies and market demands.

Part 2

Water Basics

When color photographs of the Earth as it appears from space were first published, it was a revelation: They showed our planet to be astonishingly beautiful. We were taken by surprise. What makes the Earth so beautiful is its abundant water. The great expanses of vivid blue ocean with swirling, sunlit clouds above them should not have caused surprise, but the reality exceeded everybody's expectations. The pictures must have brought home to all who saw them the importance of water to our planet.

—E. C. Pielou, Preface, 1998[1]

NOTE

1 From the Preface of E. C. Pielou (1998). *Fresh Water*. Chicago: University of Chicago Press.

DOI: 10.1201/9781003655848-3

2 Earth's Blood[1]

The watery environment in which single-cell organisms live provides them with food and removes their wastes, a function that the human circulatory system offers for the 60–100 trillion cells in a human body. The circulatory system brings each cell its daily supply of nutritive amino acids and glucose, and carries away waste carbon dioxide and ammonia, which will be filtered out of our systems and flushed away through micturition and excretory functions. The heart, the center of our circulatory system, keeps blood moving on its predetermined circular path, so if the pump fails, we quickly fail as well, and we die.

As single-celled organisms no longer, humans sometimes assume they no longer need a watery environment in which to live—but they are not paying close attention to the world around them. Those of us who live on Earth are as dependent upon the Earth's circulatory system as we are on our own circulatory system. Just as our human hearts pump blood, circulating it through a series of vessels, and just as our lives are dependent upon that flow of blood, life on Earth is dependent on the Earth's water cycle.

This cycle is so automatic that we ignore it until we are slapped in the face by it. Just as we don't control or pay attention to the beating of our hearts unless the beat skips or falters, until we are confronted by flood or drought or until our plans are disrupted by rain, we ignore the water cycle, preferring to believe that the water we drink comes out of the faucet, not from deep within the belly of the Earth, placed there by a process we only dimly comprehend. But water is as essential to us and the Earth as blood is in our bodies, and the constant cycle of water makes our lives possible.

Earth's blood, water, is pumped, not by a heart, but by the hydrological cycle, the water cycle. A titanic force of nature, the water cycle is beyond our control—a fact that we ignore until weather patterns shift and suddenly inundated rivers flow where they will and not within human-engineered banks, floodwalls, dikes, or levees. In the water cycle, water evaporates from the oceans; falls as rain, hail, sleet, or snow; and strikes the Earth again . . . thus, the cycle continues.

In cities, in summer, rain strikes hot cement and asphalt and evaporates or runs into storm drains, swiftly rejoining the cycle. In fields, rain brings essential moisture to crops and, sinking deeper into the Earth, ends up as groundwater. If water strikes a forested area, the forest canopy breaks the force of the falling drops. The forest floor, carpeted in twigs, leaves, moss, dead and decaying vegetation, keeps the soil from splashing away in erosion as the water returns to the depths of the earth or runs over the land to join a stream.

Whenever water strikes the earth, it flows along four pathways, which carry water through the cycle as our veins, arteries, and capillaries carry our blood to our cells. Water may evaporate directly into the air. It may flow overland into a stream as run-off. It may soak into the ground and be taken up by plants for evapotranspiration. Or

DOI: 10.1201/9781003655848-4

water may seep down to the groundwater. Whatever pathway water takes, one fact is certain: Water is dynamic, vital, constantly on the move. And like human blood, which sustains our lives, Earth's blood, to sustain us as well, must continue to flow.

—Frank R. Spellman (2007)

2.1 WATER USE

Have you ever considered the intricacies of fresh water, advanced water treatment systems, or drinking water distribution networks? No, or yes? **No** comes into play whenever we use water for our personal or workplace use and turn on the water faucet, push a handle or push a button, or uncork a bottle of water, and the water pours. Water is readily available for use, whether from a pond, river, lake, or swimming pool. (Avoid drinking pool water unless essential.) Turning a faucet, pressing a button, or seeing an empty bottle with no water may make us wonder about the absence of water.

The truth is that there are so many uses of water that we do not often think about until they are not operable or available, or the water is disgustingly putrid. So, what are some of the uses of water? Okay, well, there are numerous potential applications for water, known as the universal solvent. The following provides a partial list of several current uses.

- Agriculture—water is used in farming operations for irrigation, machinery, and crop processing.
- Environmental—water is vital for nature and environmental conservation.
- Industrial manufacturing—energy is used for operating machinery, powering assembly lines.
- Mining and extraction—water is used in mining operations for drilling and excavation.
- Cleaning—water is essential for washing applications.
- Transportation infrastructure—water is used in transportation equipment.
- See alsoWater supply and treatment—water for residential and industrial use.
- Automotive—water is used in the production and use of vehicles.
- Chemical industry—water is required for chemical manufacturing processes.
- Firefighting—water is used in firefighting.
- Forest industry—water is used in sawmills, paper mills, and wood processing.
- Marine transportation—water is a conveyor of ships, boats, and marine vessels for cargo hauling and fishing.
- Oil and gas industry—water is required for the oil and gas sector's drilling, extraction, and refining processes.
- Biofuels production—water is used in the production of biofuels.
- HVAC systems—water is used in heating, ventilation, and air conditioning.

- Renewable energy production—water is used for hydropower via dams and ocean waves.
- Artificial Intelligence (AI)—water is used in AI data centers.

2.2 WATER AND LIFE

Life on Earth cannot exist without water. The paradox is that if you have never given water a first, second, third, or any thought at all, you need to recognize, as shown in Figure 2.1, that water does a lot for the human body—water make up a lot of the human body.

Remember, while water is essential for all life on Earth, it also serves many other purposes, which will be discussed later in this book.

2.2.1 Survival Rule of 5

The Survival Rule of 5 provides a simple yet crucial framework for prioritizing your actions during a survival situation. When facing adversity, knowing this rule increases your chances of survival. Although some sources mention a two- or four-minute Survival Rule, depending on the circumstances and the person's health, this text refers to the Survival Rule of 5 as another approach. The point is that when you understand and apply this Rule (any of them), you increase your chances of surviving in the face of adversity.

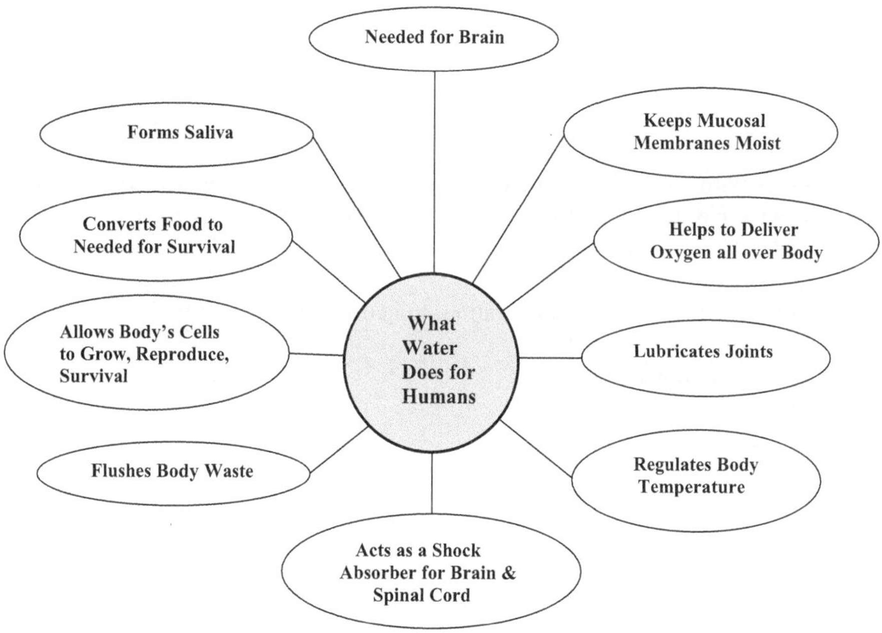

FIGURE 2.1 What water does for humans.

The "Survival Rule of 5" is a simple guideline outlining key survival principles, all based on the number five. The Rule states that you can survive:

- See alsofive minutes without air
- See alsofive hours without shelter
- See alsofive days without water
- See alsofive weeks without food.

The order of these determines what you should prioritize in your survival situation. You need to be able to breathe. Without breathable air, the others simply do not matter. If you cannot breathe, what good is finding shelter, water, and/or food?

Note that each rule assumes you meet the requirement immediately above it.

2.3 GOING WITH THE FLOW

The goal in this book is to take a prosaic subject and make it endlessly fascinating . . . and appreciated. Many might ask how anyone can make water fascinating? After all, water is a contradiction, a riddle. How? Consider the Chinese proverb that states, "[W]ater can both float and sink a boat." Water's presence everywhere feeds these contradictions. Lewis (1996) points out that "water is the key ingredient of mother's milk and snake venom, honey and tears" (p. 90). Leonardo da Vinci gave us insight into more of water's apparent contradictions:

> Water is sometimes sharp and sometimes strong, sometimes acid and sometimes bitter.
> Water is sometimes sweet and sometimes thick or thin.
> Water sometimes brings hurt or pestilence, health-giving, sometimes poisonous.
> Water suffers changes into as many natures as are the various places through which it passes.
> Water, as with the mirror that changes with the color of its object, so it alters with the nature of the place, becoming noisome, laxative, astringent, sulfurous, salt, incarnadined, mournful, raging, angry, red, yellow, green, black, blue, greasy, fat or slim.
> Water sometimes starts a conflagration, sometimes it extinguishes one.
> Water can be warm or cold.
> Water carries away or sets down.
> Water hollows out or builds up.
> Water tears down or establishes.
> Water empties or fills.
> Water rises or sinks.
> Water spreads or is still.
> Water is the cause at times of life or death, or the increase of privation, nourishes at times and at others does the contrary.
> Water, at times, has a tang; at times, it is without savor.
> Water sometimes submerges in the valleys during a great flood.
> In time and with water, everything changes."

Although water is abundant on Earth, it is not a single entity but rather a fundamental element that plays a significant role in shaping our environment and lives. Da Vinci's last contradiction, "In time and with water, everything changes," concerns us most in this text.

Why?

It is a basic fact that humans are born of water and require it to sustain life.

The development of safe drinking water supplies is a significant concern today. This may seem strange to the average person (depending, of course, on where this individual resides), some of whom might be surrounded by various water bodies. Bodies of water encompass us. Drinking water practitioners—those responsible for finding a source, certifying its safety, and providing it to the consumer—know better. The drinking water practitioner knows, for example, that two key concerns drive the development of safe drinking water supplies: quantity and quality. Herein lies the problem: Quantity (or water stress) may indeed be a significant issue (a limiting factor) for a particular location, often simply because water suitable for consumption is not evenly distributed throughout the world. Places with plenty of surface or groundwater may not face shortages if their supply meets all consumer needs. Some areas lack enough water to meet residents' needs. This is one of the primary reasons, of course, that significant portions of the globe are either uninhabited or sparsely populated at best.

However, not all the news on this topic is bad. Extreme variations in water quantity do not automatically indicate that a region will experience scarcity. Water quantity issues can be managed. Given the right information and possessing a high degree of common sense, locations facing extremely high quantity issues can implement management and conservation strategies to secure their water supplies. Take Singapore, for example. It has one of the highest water quantity concerns in the world. Yet, the area is highly populated with booming economic enterprise capacity and output, even though the location does not possess surface freshwater or groundwater sources. Moreover, Singapore's demand for water far exceeds its naturally occurring supply. The point is that if one needs to look at an example of water management, Singapore is the place. In what way, you may ask? Well, to meet its freshwater needs, Singapore invests heavily in technology, responsible management, and international agreements. For instance, forward-thinking and innovative management practices, including the use of advanced rainwater capture systems, contribute at least 20% of Singapore's water supply (total annual precipitation averages 2,150 mm (84.6 in.), equivalent to 2,150 l/m^2 or 52.73 gal/ft^2). In addition, 40% of the water supply is imported from Malaysia. The use of gray water or sullage (i.e., wastewater from homes and offices excluding toilet water) provides 30% of the supply, while desalination accounts for the remaining 10% to meet total demand. Unfortunately, Singapore is the exception and not the trend. Because Singapore's example and other contributing factors are not the trend, this is the primary reason, of course, why much of the globe is either uninhabited or sparsely populated at best.

The priority is water quality. Having enough freshwater available does little good if the water is unsafe for consumption or for other uses. Another issue is that it is easy to determine the quantity of a substance such as water. We can say there is too little, enough, or too much. A quantity is a metric that indicates or signifies a number. In the case of water, quantity can be the number of gallons or acre-feet of water available or not available. Trying to quantify the quality of water is an entirely different matter. Quality is a characteristic that is left up to judgment call: a judgment that is not a quantity. The problem with judging quality is that it can be subjective—a glass that holds water that is dim or has an odor will turn many off from drinking it; however, one dying of thirst will consume the water with one gulp.

Again, while quantity can usually be explained or understood by simple metrics (a gallon or ounce or teaspoon of this or that, etc.), quality is an entirely different issue. Indeed, the appearance of anything we are about to or intend to put into our mouths is a quality factor based on appearance. But keep in mind that quality judgments can be based on other factors as well. In the case of drinking water, we expect (and make a judgment) that our glass of tap water contains water that is odorless, tasteless, and transparent. We also expect our glass of tap water to be safe. We hope it is pure and free from biological and/or chemical contamination.

The bottom line: Obviously, having enough freshwater available does little good if the water is unsafe for consumption or for other uses.

Another contradiction—one for which humans bear considerable responsibility—is best illustrated by considering that most of the early settlements of the world began along waterways. Waterways were important primarily because of the ease of transportation they afforded and because of their industrial value (waterpower, etc.), as well as because of the food supply they provided to the settlers (fish and other wildlife). These waterways offer a natural, clean, and safe source of drinking water. However, these early waterways soon became polluted. When you get right down to it, this should not surprise humans; we can say that pollution is a natural by-product of civilization. We eat, we work, we do whatever is necessary to sustain our existence, and in doing so, we pollute.

DID YOU KNOW?

Washington Irving is known to have been one of the greatest and most interesting writers who ever lived. In addition to having written *Rip Van Winkle* and *The Legend of Sleepy Hollow*, he also wrote *The Life and Voyage of Christopher Columbus*, in which Irving profiled many famous people who were Columbus' passengers or crew members during his epic voyages. One of his passengers was the famous explorer, former governor of Puerto Rico, and the person who named Florida, Ponce de Leon. Of course, whenever the name Ponce de Leon is mentioned, many people instantly think about his quest for the so-called Fountain of Youth—that magic water source that, if you drank from it, you would be bestowed eternal youth and the prevention of loss of vigor and health. Of course, today we know the facts about Ponce de Leon and his quest for the magic elixir, and that Irving employed a fountain of literary license when he wrote. But why let facts interfere with a good story, right? Anyway, what Ponce de Leon was really searching for was not a fountain of magical water, but instead he was looking for that elusive pot of gold. One must wonder if Ponce would have chosen the pot of gold over the fountain of youth—that is, if he had found both. Which leads us to the bottom line: If you had the choice of either, which would you choose? Simply because we know at our advanced ages that youth is not all it is counted on being; instead, we will settle for that pot of gold, thank you very much!

Are we the only freshwater polluter? No, not exactly. Natural occurrences also pollute our water sources, especially our surface waters. For example, a stream that flows in a heavily wooded area—in and through a deciduous forest, for instance—suffers from the effects of natural pollution each year during leaf-fall. Leaves drop into streams and float along with the current until they either become saturated and sink or get caught on something in the water.

Leaves are organic and eventually degrade. During this process, the microbes degrading the leaves accept and use dissolved oxygen (DO) in the water. In some cases, the amount of oxygen used during degradation is of such quantity (especially in slow-moving or stagnant water areas) that the natural biota in the stream suffers from a lack of oxygen and, therefore, either moves on to healthier parts of the stream or succumbs because of a lack of oxygen. Note that other natural water polluters also affect water quality, including forest fires, earthquakes, and floods.

Remember that Nature understands the contradictions of water . . . and because she understands, she is also well-suited and equipped to deal with such problems. When a stream is polluted (for whatever reason), Nature immediately goes to work and sets in motion natural processes—known as self-purification—designed to restore the stream to its normal, healthy state. Only when such streams become overloaded with pollution or with non-biodegradable manufactured ingredients (contaminants) does Nature have difficulty in restoring the stream to its normal quality.

DID YOU KNOW?

Contrary to popular belief, raindrops are not shaped like teardrops. Small raindrops (with a diameter of less than 0.1 inch [0.25 cm]) are shaped like a sphere; larger ones are shaped more like a hamburger bun.

Oh, if only myth were real! Legend states that Alexander the Great discovered Adam's will. The will relates that God had created a spring beyond the mountains surrounding the world, in the Land of Darkness. This spring was unique—"whiter than milk, colder than ice, sweeter than honey, softer than butter, and sweeter smelling than musk" (Varner, 2005). Those who drank it would be granted eternal life. Will this book lead you to that spring and eternal life? The simple answer is no. The compound answer is also no, but with the caveat that this text, instead, is intended to serve as a convenient, quick reference and to provide technical support in the hands of the public, sanitary engineers, public health administrators, public works engineers, water treatment operators, and college students in environmental health or public health engineering. The purpose of this handbook is to evaluate and emphasize drinking water quality control, from the source to the treatment plant, and from the distribution system to the consumer. The goal is to look at the factors that affect water quality, including trihalomethanes (THMs), *Cryptosporidium*, viruses, carcinogens, and polychlorinated biphenyls, in addition to the traditional physical, chemical, and bacteriological parameters. We address challenges faced by drinking water practitioners striving to provide the best drinking water quality to the consumer. Specifically, we deal with the nature of these challenges (unsafe drinking water) and their solutions (how to make it safe), in addition to exploring techniques that can be employed to mitigate the problem of unsafe drinking water through technology. We focus on science and technology instead of "feel-good" approaches or myths simply because we have a clear understanding that technology, politics, and fantasy are seldom a rational mix.

Take a moment and perform an action that most people never think about. Hold a glass of water (like the one shown in Figure 2.2) and think about the substance within the glass—about the substance you are getting ready to drink. Also, think about the ice cubes floating in the glass, unlike those shown in Figure 2.2. Typically, people do not focus on the water or ice cubes in a drinking glass unless they are responsible for supplying it.

Let us think about water now. Earlier, we stated that water is special, strange, and different. We find water fascinating—a subject worthy of endless interest because of

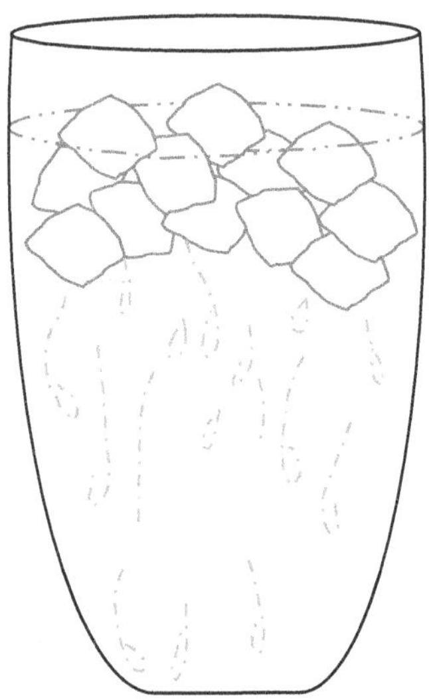

FIGURE 2.2 Glass of iced water.

its unique behavior, endless utility, and ultimate and intimate connection with our existence. We agree with Tom Robbins (1976, pp. 1–2), whose description of water follows:

> Stylishly composed in any situation—solid, gas or liquid—speaking in penetrating dialects understood by all things—animal, vegetable or mineral—water travels intrepidly through four dimensions, *sustaining* (Kick a lettuce in the field and it will yell "Water!") *destroying* (The Dutch boy's finger remembered the view from Ararat) and *creating* (It has even been said that human beings were invented by water as a device for transporting itself from one place to another, but that's another story). Always in motion, ever flowing (whether at stream rate or glacier speed), rhythmic, dynamic, ubiquitous, changing and working its changes, mathematics turned wrong side out, a philosophy in reverse, the ongoing odyssey of water is irresistible.

Let's review a few basic facts about the water in the glass you're holding: Water is liquid between 0°C and 100°C (32°F and 212°F), solid at or below 0°C (32°F), and gaseous at or above 100°C (212°F). One gallon of water weighs 8.33 pounds (3.778 kilograms). One gallon of water equals 3.785 liters. One cubic foot of water equals 7.50 gallons (28.35 liters). One ton of water equals 240 gallons. One acre-foot of water equals 43,560 cubic feet (325,900 gallons). Earth's rate of rainfall equals 340 cubic miles per day (16 million tons per second).

As Robbins observed, water is always in motion. The one most essential characteristic of water is that it is dynamic: Water constantly evaporates from the sea, lakes, and soil and transpires from foliage. It is transported through the atmosphere and falls to Earth, where it runs across the land and filters down to flow along rock strata into aquifers. Eventually, water finds its way to the sea again—indeed, water never stops moving.

A thought that might not have occurred to most people as we look at our glass of water is, "Has someone tasted this same water before us?" Absolutely. Remember, water is a finite entity. What we have now is what we have had in the past. We are drinking the same water consumed by Cleopatra, Aristotle, da Vinci, Napoleon, Joan of Arc, Stonewall Jackson, General Patton (and several billion other folks who preceded us) because water is dynamic (never at rest), and because water constantly cycles and recycles, as we discuss in the next section.

DID YOU KNOW?

A person must consume 2.5 quarts of water per day from all sources (drinking and eating) to maintain health. However, we wonder if it is true or an urban myth that we must absolutely drink eight glasses of water to remain fit and healthy. According to the *Journal of the American Society of Nephrology,* "there is no clear evidence of benefit from drinking increased amounts of water" (Foreman, 1998).

2.4 THE WATER CYCLE

The natural *water cycle,* or *hydrological cycle* (as the journey water takes in its constant, inevitable motion), is how water in all three forms—solid, liquid, and vapor—circulates through the biosphere. Water lost from the Earth's surface to the atmosphere, either by evaporation from the surface of lakes, rivers, and oceans or through the transpiration of plants, forms clouds that condense to deposit moisture on the land and sea. A drop of water may travel thousands of miles between the time it evaporates and the time it falls to Earth again as rain, sleet, or snow. The water that collects on land flows to the ocean in streams and rivers or seeps into the earth, joining groundwater. Even groundwater eventually flows toward the ocean for recycling (see Figure 2.3).

Note: Only about 2% of the water absorbed into plant roots is used in photosynthesis. All of it travels through the plant to the leaves, where transpiration into the atmosphere begins the cycle again.

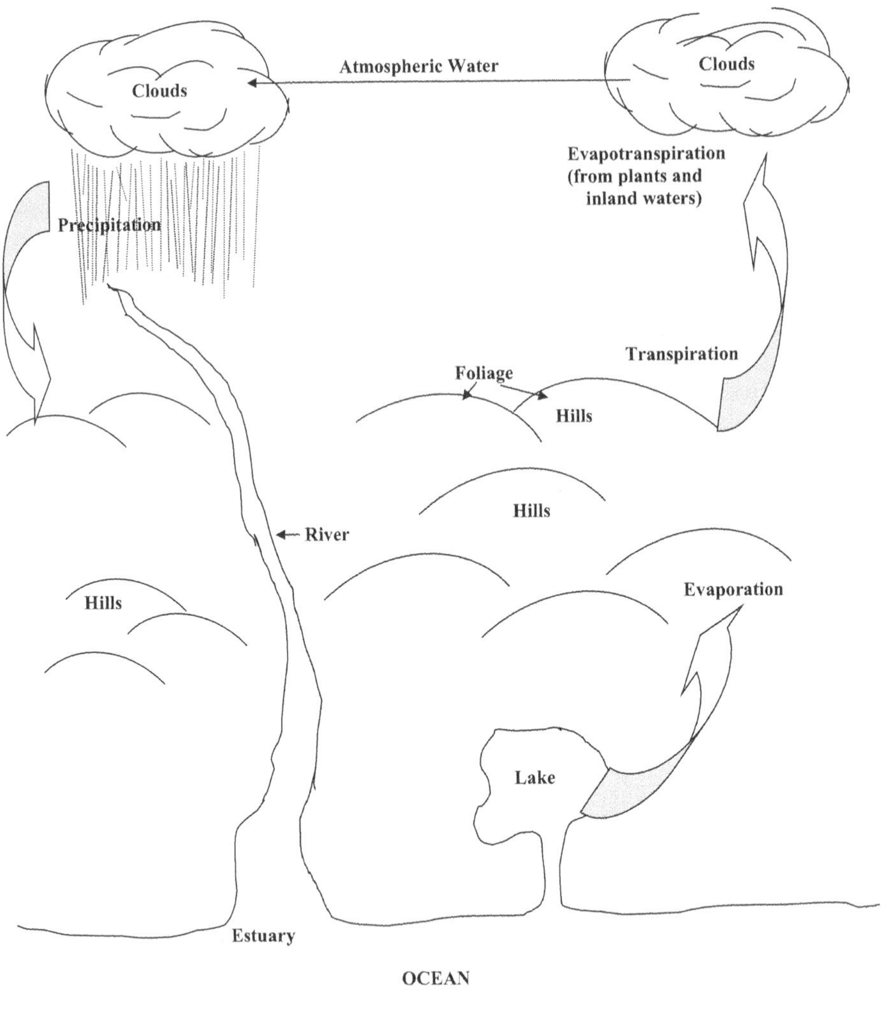

FIGURE 2.3 Water cycle.

***Key Concept*:** The hydrologic cycle describes water's circulation through the environment. Evaporation, transpiration, runoff, and precipitation describe specific water movements.

DID YOU KNOW?

If it rains one inch, the amount of water that falls on a quarter-acre lot is about 7,000 U.S. gallons (26,500 liters), or nearly 30 tons (27 metric tons) of water.

When humans intervene in the natural water cycle, they generate artificial water cycles or *urban water cycles* (local sub-systems of the water cycle—an integrated water cycle; see Figure 2.3). Although many communities withdraw groundwater for public supply, the majority rely on surface sources. After treatment, water is distributed to households and industries. Water that is wasted (wastewater) is collected in a sewer system and transported to a treatment plant for processing before disposal. Current processing technologies provide only partial recovery of the original water quality. The upstream community (the first water user, shown in Figure 2.3) can achieve additional quality improvement by dilution into a surface water body and natural purification. However (as also shown in Figure 2.4), the next community downstream is likely to withdraw the water for a drinking water supply before complete restoration. This practice is intensified and further complicated as existing communities continue to grow, and new communities spring up along the same watercourse. Increases in the number of users bring an additional need for more water. This withdrawal and return process by successive communities in a river basin results in *indirect water reuse* (the use of used toilet water via de facto water recycling).

Note: *Indirect water reuse* has been defined as potable reuse by incorporation of reclaimed wastewater into a raw water supply; the wastewater effluent is discharged to the water source, where it mixes with the source water and is assimilated with it, with the intent of reusing the water instead of as a means of disposal (Tchobanoglous, 1991, pp. 1139–1140). This type of potable reuse is becoming more common as water resources become less plentiful.

The indirect water reuse process (demonstrated in Figure 2.4) is a clear example of an artificial or de facto water cycle within the natural hydrologic scheme. It involves (1) surface-water withdrawal, processing, and distribution; (2) wastewater collection, treatment, and disposal back to surface water by dilution; (3) natural purification in a river; and (4) repetition of this scheme by communities downstream (Hammer and Hammer, 1996, p. 1).

DID YOU KNOW?

There is the same amount of water on Earth as there was when the Earth was formed. The water from your faucet could contain molecules that dinosaurs drank.

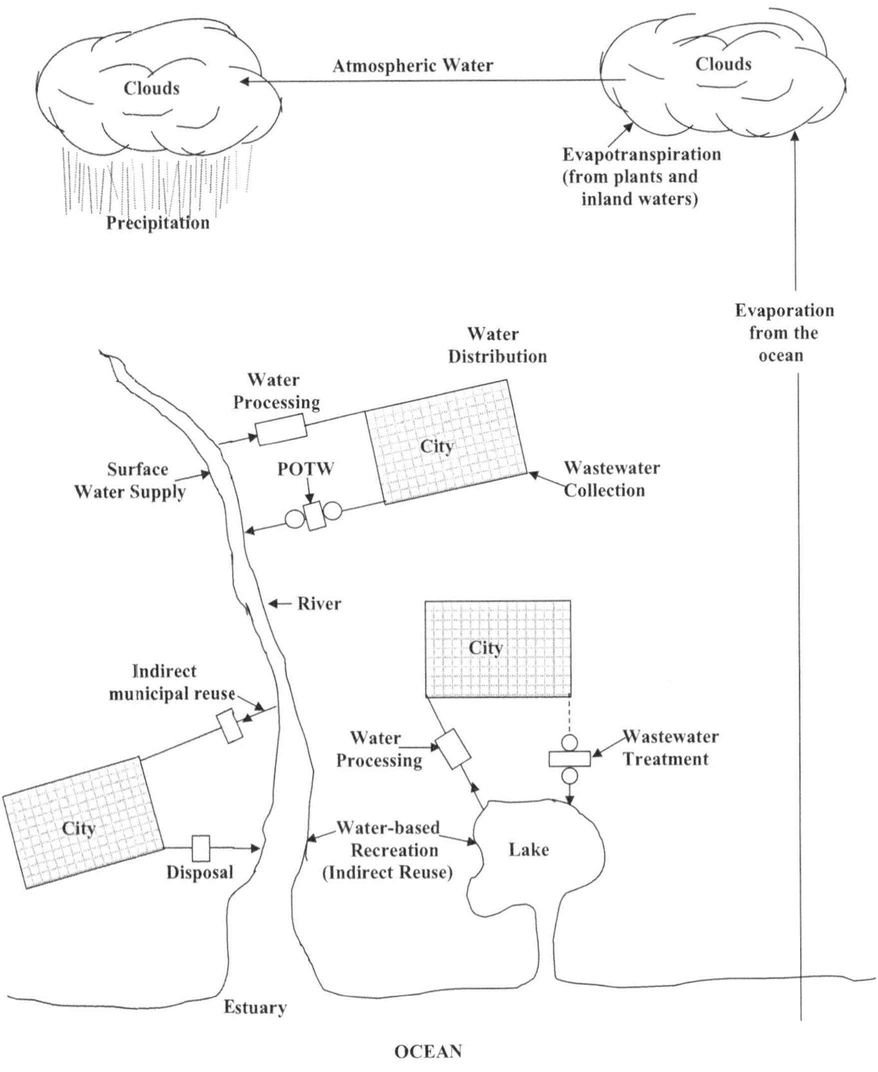

FIGURE 2.4 Urban or integrated water cycle.

2.5 WATER SUPPLY: THE Q AND Q FACTORS

Whereas drinking water practitioners must have a clear and complete understanding of the natural and manufactured water cycles, they must also factor in two primary considerations of *quantity and quality*, the Q and Q factors. They are responsible for (1) providing a *quality* potable water supply—one that is clean, wholesome, and safe to drink and (2) finding a water supply in adequate *quantities* to meet the anticipated demand.

Note: Two central facts important to our discussion of freshwater supplies are that (1) water is very much a local or regional resource and (2) problems of its shortage or pollution are equally local problems. Human activities affect the quantity of water available at a locale at any time by changing either the total volume that exists there or aspects of quality that restrict or devalue it for a particular use. Thus, the total human impact on water supplies is the sum of the separate human effects on the various drainage basins and groundwater aquifers. In the global system, the central, critical fact about water is the natural variation in its availability (Meyer, 1996). Simply put, not all lands are watered equally.

To meet the Q and Q requirements of a potential water supply, the drinking water managers and other practitioners (whether the design engineer, community planner, plant manager, plant administrator, plant engineer, or other responsible person in charge) must determine the answers to several questions, which include the following:

1. Is there a potable water supply with the capacity to always be distributed in sufficient quantity and pressure?
2. Will constructing a centralized treatment and distribution system for the entire community be best, or would using individual well supplies be better?
3. If a centralized water treatment facility is required, will the storage capacity at the source as well as at intermediate points of the distribution system maintain the water pressure and flow (*quantity*) within the conventional limits, particularly during loss-of-pressure events, such as major water main breaks, rehabilitation of the existing system, or major fires, for example?
4. Is a planned or preventive maintenance program in place (or anticipated) for the distribution system that can be adequately planned, implemented, and controlled at the optimum level possible?
5. Is the type of water treatment process selected in compliance with federal and state drinking water standards?
6. Are all employees trained along with the others who might be involved in one way or another with water and the nexus with artificial intelligence and its subsets, and how do they augment operations?

Note: Water from a river or a lake usually requires more extensive treatment than groundwater does to remove bacteria and suspended particles.

Note: The primary concern for the drinking water practitioner involved with securing an appropriate water supply, treatment process, and distribution system must be the protection of public health. Contaminants must be eliminated or reduced to a safe level to minimize menacing waterborne diseases (to prevent another Milwaukee *Cryptosporidium event*) and to avoid forming long-term or chronic injurious health effects.

6. When the source and treatment processes are selected, has the optimum hydraulic design of the storage, pumping, and distribution network been determined to ensure that enough water can be delivered to consumers at adequate pressures?
7. Have community leaders and the consumer (the public) received continuing and realistic information about the functioning of the proposed drinking water service?

Note: Drinking Water Managers are wise to follow the guidance given in #7, simply because public buy-in for any proposed drinking water project that involves new construction and/or retrofitting, expansion, or upgrading of an existing facility is essential to ensure that necessary financing is forthcoming. In addition to the finances required for any waterworks construction project, public and financial support is also essential to ensure the safe operation, maintenance, and control of the entire water supply system. The acronym POTW stands for "publicly owned treatment works," and the public foots the bills.

8. Does planning include steps to ensure the elimination of waste, leakages, and unauthorized consumption?

Note: Industry-wide operational experience has shown that the cost per cubic foot, cubic meter, liter, or gallon of water delivered to the customer has steadily increased because of workforce, automation, laboratory, and treatment costs. To counter these growing costs, treatment works must meter consumers, measure the water supply flow, and evaluate the entire system annually.

9. Does the waterworks or proposed waterworks physical plant include adequate laboratory facilities to ensure proper monitoring of water quality?

Note: Some waterworks facilities routinely perform laboratory work. However, water pollution control technologists must ensure that the waterworks laboratory or other laboratory used is approved by the appropriate health authority. Keep in mind that the laboratory selected to evaluate and analyze the waterworks samples must be able to analyze chemical, microbiological, and radionuclide parameters.

10. Are procedures in place to evaluate specific problems, such as the lead content in the distribution systems and at the consumer's faucet, or suspected contamination due to cross-connection potentials?
11. Is a cross-connection control program in place to make sure that the distribution system (in particular) is protected from plumbing errors and illegal connections that may lead to injection of non-potable water into public or private supplies of drinking water?
12. Are water utility operators and laboratory personnel adequately trained and licensed?
13. Are water utility managers properly trained and licensed?
14. Are proper operating records and budgetary records kept?

2.5.1 Drinking Water Q and Q: Key Definitions

As with any other technical presentation, understanding the information presented is difficult unless a shared vocabulary is established. Voltaire said it best: "If you wish to converse with me, please define your terms." We define many key terms used in the text in gloss form in this section. We define other terms as they appear in the text. Before we list and explain these key terms, however, we must first define the key term that this text is all about: *Drinking water.*

Potable water can be defined as the water delivered to the consumer that can be safely used for drinking, cooking, washing, and other household applications. In the past, drinking water suitable for use (i.e., safe) was certified as safe by a professional engineer specialized in the field, but times have changed. This practice is no longer accepted. Why? Because public health aspects have reached such a high level of importance and complexity, local licensed health officials are usually designated as those with the authority and jurisdiction in the community to review, inspect, sample, monitor, and evaluate the water supplied to a community continuously. This professional scrutiny is driven by updated drinking water standards, of course. When you factor in the importance of providing a safe, palatable product to the public, public health control is required to help ensure a continuous supply of safe drinking water, which makes real sense.

On many occasions, drinking water managers have attempted to explain to interested parties the complexities of providing safe drinking water to their household taps. Of course, we know the process is complex, but we continue to be surprised when listeners are astonished by its complexity. It becomes readily apparent that a commonly held view is that the provision of drinking water to the consumer tap involves nothing more than going down to the local river or stream, installing a suction pipe into the watercourse, and pumping the water out and into a distribution network that somehow delivers clean, safe drinking water to the household tap.

We then explain that, yes, water is often taken from a local river or stream and that it eventually reaches the household tap. However (isn't there always a "however" in any explanation?), we explain that the water drawn from any local surface water supply must pass through specific processes to ensure wholesomeness and safety. In addition to screening, filtering, and disinfecting water, these processes involve continuous inspection, sampling, monitoring, and evaluation of the supply.

Drinking water practitioners learn the ins and outs of providing drinking water through experience. The main lesson learned is that supplying drinking water to the household tap is a complex and demanding process. For example, water analysis is required (obviously), but it alone is not sufficient to maintain quality. It must be combined with the periodic review and acceptance of the facilities involved.

What does this mean? Stated, acceptance or approval consists of the evaluation and maintenance of proper protection of the water source; the qualifications of waterworks personnel; adequate monitoring procedures by water suppliers; and assessment of the quality and performance of laboratory work (DeZuane, 1997). Keep in mind that when we attempt to define "drinking" water, we must define it in all-encompassing terms. Potable water is a product from an approved source that falls within specific physical, chemical, bacteriological, and radionuclide parameters and is delivered to treatment works for processing and disinfecting. Such treatment works must be appropriately designed, constructed, and operated. Drinking water must be delivered to the consumer in sufficient quantity and pressure. Drinking water quality must meet stringent standards. It must be palatable, be within reasonable temperature limits, and be completely trustworthy to the consumer.

The bottom line is that drinking water is the substance available to the consumer at the household tap that can perform one essential function: It can satisfy thirst without threatening life and health.

2.5.1.1 Definitions

Absorption: A process where one substance penetrates the interior of another substance.

Acid Rain: Precipitation with higher-than-normal acidity, caused primarily by sulfur and nitrogen dioxide air pollution.

Activated carbon: A very porous material that can be used to adsorb pollutants from water after the material has been subjected to intense heat to drive off impurities.

Adsorption: The process by which one substance is attracted to and adheres to the surface of another substance, without penetrating its internal structure.

Aeration: A physical treatment method that promotes biological degradation of organic matter. The process may be passive (when waste is exposed to air) or active (when a mixing or bubbling device introduces the air).

Aerobic Bacteria: A type of bacteria that requires free oxygen to conduct metabolic function.

Biochemical Oxygen Demand: The amount of oxygen required by bacteria to stabilize decomposable organic matter under aerobic conditions.

Biological Treatment: A process that uses living organisms to bring about chemical changes.

Breakpoint chlorination: The addition of chlorine to water until the chlorine demand has been satisfied and free chlorine residual is available for disinfection.

Chemical Treatment: A process that results in the formation of a new substance or substances. The most common chemical water treatment processes include coagulation, disinfection, water softening, and filtration.

Chlorination: The process of adding chlorine to water to kill disease-causing organisms or to function as an oxidizing agent.

Chlorine Demand: A measure of the amount of chlorine that will combine with impurities and is therefore unavailable to function as a disinfectant.

Clean Water Act (CWA): Federal law dating back to 1972 (with several amendments) aimed at restoring and maintaining the chemical, physical, and biological integrity of the nation's waters. Its long-range goal is to eliminate the discharge of pollutants into navigable waters and to make national waters fishable and swimmable.

Coagulants: Chemicals that cause small particles to stick together to form larger particles.

Coagulation: A chemical water treatment method that causes tiny, suspended particles to attract one another and form larger particles. This is accomplished by the addition of a coagulant that neutralizes the electrostatic charges that cause particles to repel each other.

Coliform Bacteria: Bacteria typically found in human and animal intestines, but occasionally elsewhere. The presence of bacteria in water is used as an indication of fecal contamination (contamination by animal or human wastes).

Color: A physical characteristic of water. Color is most commonly tan or brown from oxidized iron, but contaminants may cause other colors, such as green or blue. Color differs from turbidity, which is water's cloudiness.

Communicable Diseases: Usually caused by *microbes*—microscopic organisms including bacteria, protozoa, and viruses. Most microbes are essential components of our environment and do not cause disease. Those that do are called pathogenic organisms, or simply *pathogens.*

Community Water System: A public water system that serves at least 15 service connections used by year-round residents or regularly serves at least 25 year-round residents.

Composite Sample: A series of individual or grab samples taken at various times from the same sampling point and mixed.

Contaminant: A toxic material found as an unwanted residue in or on a substance.

Cross Connection: Any connection between safe drinking water and nonpotable water or fluid.

CxT Value: The product of the residual disinfectant concentration **C**, in milligrams per liter, and the corresponding disinfectant contact time **T**, in minutes. The SWTR specifies minimum CxT values as a means of ensuring adequate kill or inactivation of pathogenic microorganisms in water.

Disinfectants-disinfection by-products: A term used in connection with state and federal regulations designed to protect public health by limiting the concentration of either disinfectants or the by-products formed by the reaction of disinfectants with other substances in the water (such as THMs).

Disinfection: A chemical treatment method. The addition of a substance (e.g., chlorine, ozone, or hydrogen peroxide) that destroys or inactivates harmful microorganisms or inhibits their activity.

Dissociate: The process of ion separation occurs when an ionic solid is dissolved in water.

Dissolved Oxygen: The oxygen dissolved in water is usually expressed in milligrams per liter, parts per million, or percent of saturation.

Dissolved Solids: Any material that can dissolve in water and be recovered by evaporating the water after filtering the suspended material.

Drinking Water Standards: Water quality standards are measured in terms of suspended solids, unpleasant taste, and microbes harmful to human health. Drinking water standards are included in state water quality rules.

Drinking Water Supply: Any raw or finished water source that is or may be used as a public water system or as drinking water by one or more individuals.

Effluent Limitations: Standards developed by the EPA to define the levels of pollutants that could be discharged into surface waters.

Electrodialysis: The process of separating substances in a solution by dialysis, using an electric field as the driving force.

Electronegativity: The tendency for atoms that do not have a complete octet of electrons in their outer shell to become negatively charged.

Enhanced Surface Water Treatment Rule: A revision of the original SWTR that includes modern technology and requirements to deal with newly identified problems.

Facultative Bacteria: A type of anaerobic bacteria that can metabolize its food either aerobically or anaerobically.

Federal Water Pollution Control Act (FWPCA) (1972): The Act outlines the objective "to restore and maintain the chemical, physical, and biological integrity of the nation's waters." The 1972 act and subsequent CWA amendments are the most far-reaching water pollution control legislation ever enacted. They provided comprehensive programs for water pollution control, uniform laws, and interstate cooperation. They provided grants for research, investigations, training, and information on national programs on surveillance, the effects of pollutants, pollution control, and the identification and measurement of pollutants. Additionally, they offer a lot of grants and loans for the construction of treatment works. The Act established national discharge standards with enforcement provisions.

The FWPCA established several milestone achievement dates. It required secondary treatment of domestic waste by publicly owned treatment works (POTWs) and the application of the "best practicable" water pollution control technology by industry by 1977. All industrial sources have achieved compliance (because of economic difficulties and cumbersome federal requirements, certain POTWs obtained an extension to July 1, 1988, for compliance). The Act also called for new levels of technology to be imposed during the 1980s and 1990s, particularly for controlling toxic pollutants. The Act mandates a strong pretreatment program to control toxic pollutants discharged by industry into POTWs. The 1987 amendments require that stormwater from industrial activity be regulated.

Filtration: A physical treatment method for removing solid (particulate) matter from water by passing the water through porous media such as sand or a manufactured filter.

Flocculation: The water treatment process following coagulation uses gentle stirring to bring suspended particles together so that they will form larger, more settleable clumps called floc.

Grab Sample: A single water sample collected at one time from a single point.

Groundwater: The fresh water found under the Earth's surface, usually in aquifers. Groundwater is a significant source of drinking water and a source of growing concern in areas where leaching agricultural or industrial pollutants, or substances from leaking underground storage tanks, are contaminating groundwater.

Hardness: A characteristic of water caused primarily by the salts of calcium and magnesium. It causes deposition of scale in boilers, damage in some industrial processes, and sometimes an objectionable taste. It may also decrease the soap's effectiveness.

Hydrogen Bonding: The term used to describe the weak but effective attraction that occurs between polar covalent molecules.

Hydrologic Cycle: The water-earth cycle. The movement of water in all three physical forms through the various environmental media (air, water, biota, and soil).

Hygroscopic: A substance that readily absorbs moisture.

Influent: Water flowing into a reservoir, basin, or treatment plant.

Inorganic Chemical: A chemical substance of mineral origin not having carbon in its molecular structure.

Ionic Bond: The attractive forces between oppositely charged ions—for example, the forces between the sodium and chloride ions in a sodium chloride crystal.

Maximum Contaminant Level (MCL): The maximum allowable concentration of a contaminant in drinking water, as established by state and/or federal regulations. Primary MCLs are health-related and mandatory. Secondary MCLs are related to the aesthetics of the water and are highly recommended but not required.

Membrane Filter Method: A laboratory method used for coliform testing. The procedure uses an ultrathin filter with a uniform pore size smaller than bacteria (less than a micron). After water is forced through the filter, the filter is incubated in a special medium that promotes the growth of coliform bacteria. Bacterial colonies with a green-gold sheen indicate the presence of coliform bacteria.

Modes of Transmission of Disease: How diseases spread from one person to another.

Multiple-tube Fermentation Method: A laboratory method used for coliform testing, which uses a nutrient broth placed in culture tubes. Gas production indicates the presence of coliform bacteria.

National Primary Drinking Water Regulations: Regulations developed under the SDWA, which establishes MCLs, monitoring requirements, and reporting procedures for contaminants in drinking water that endanger human health.

National Pollutant Discharge Elimination System: A requirement of the CWA that discharges meet specific requirements prior to discharging waste to any water body. It sets the highest permissible effluent limits, by permit, before making any discharge.

Near Coastal Water Initiative: This initiative was developed in 1985 to provide for the management of specific problems in waters near coastlines that are not dealt with in other programs.

Nonbiodegradable: Substances that do not break down easily in the environment.

Nonpolar Covalently Bonded: A molecule composed of atoms that share their electrons equally, resulting in a molecule that does not have polarity.

Organic Chemical: A chemical substance of animal or vegetable origin having carbon in its molecular structure.

Oxidation: When a substance either gains oxygen or loses hydrogen or electrons in a chemical reaction, it is a type of chemical treatment.

Oxidizer: A substance that oxidizes another substance.

Parts per million: The number of weight or volume units of a constituent present in each one million units of the solution or mixture. Formerly used to express the results of most water and wastewater analyses, **ppm** is being replaced by milligrams per liter, **m/l**. For drinking water analyses, concentrations in parts per million and milligrams per liter are equivalent. A single ppm can be compared to a shot glass full of water inside a swimming pool.

Pathogens: Types of microorganisms that can cause disease.

Physical Treatment: Any process that does not produce a new substance (e.g., screening, adsorption, aeration, sedimentation, and filtration).

Polar Covalent Bond: The shared pair of electrons between two atoms is not equally held. Thus, one of the atoms becomes slightly positively charged, and the other atom becomes slightly negatively charged.

Polar Covalent Molecule (water): One or more polar covalent bonds result in a molecule that is polar covalent. Polar covalent molecules exhibit partial positive and negative poles, causing them to behave like tiny magnets. Water is the most common polar covalent substance.

Pollutant: Any substance introduced into the environment that adversely affects the usefulness of the resource.

Pollution: The presence of matter or energy whose nature, location, or quantity produces undesired environmental effects. Under the Clean Water Act, for example, the term is defined as a manufactured or human-induced alteration of the physical, biological, and radiological integrity of water.

Pretreatment: Any physical, chemical, or mechanical process used before the main water treatment processes. It can include screening, presedimentation, and chemical addition.

Primary Drinking Water Standards: Regulations on drinking water quality (under SWDA) are considered essential for the preservation of public health.

Primary Treatment: The first step of treatment at a municipal wastewater treatment plant. It typically involves screening and sedimentation to remove materials that float or settle.

Public Water System: As defined by the Safe Drinking Water Act, any system, publicly or privately owned, that serves at least 15 service connections for 60 days out of the year or serves an average of 25 people for at least 60 days out of the year.

Publicly Owned Treatment Works (POTW): A waste treatment works owned by a state, local government unit, or Indian tribe, usually designed to treat domestic wastewaters.

Receiving Waters: A river, lake, ocean, stream, or other water source into which wastewater or treated effluent is discharged.

Recharge: The process by which water is added to a zone of saturation, usually by percolation from the soil surface.

Reference Dose: An estimate of the amount of a chemical that a person can be exposed to daily that is not anticipated to cause adverse systemic health effects over the person's lifetime.

Representative Sample: A sample containing all the constituents present in the water from which it was taken.

Reverse Osmosis (RO): A semipermeable membrane separates solutions with different ion concentrations. Typically, water flows from the chamber with a lesser ion concentration into the chamber with a greater ion concentration, resulting in hydrostatic or osmotic pressure. In RO, enough external pressure is applied to overcome this hydrostatic pressure, thus reversing the flow of water. This results in the water on the other side of the membrane becoming depleted in ions and demineralized.

SDWA: A federal law passed in 1974 to establish federal standards for drinking water quality, protect underground sources of water, and set up a system of state and federal cooperation to ensure compliance with the law.

Screening: A pretreatment method that uses coarse screens to remove large debris from the water to prevent clogging of pipes or channels leading to the treatment plant.

Secondary Drinking Water Standards: Regulations developed under the SDWA that established maximum levels of substances affecting the aesthetic characteristics (taste, color, or odor) of drinking water.

Secondary Treatment: The second step of treatment at a municipal wastewater treatment plant. This step uses a growing number of microorganisms to digest organic matter and reduce the amount of organic waste. Water leaving this process is chlorinated to destroy any disease-causing microorganisms before its release.

Sedimentation: A physical treatment method that involves reducing the velocity of water in basins so that the suspended material can settle out by gravity.

Solvated: When either a positive or negative ion becomes surrounded by polar solvent molecules.

Surface Tension: The attractive forces exerted by the molecules below the surface upon those at the surface, resulting in them crowding together and forming a higher density.

Surface Water: All water naturally open to the atmosphere and all springs, wells, or other collectors that are directly influenced by surface water.

Surface Water Treatment Rule (SWTR): A federal regulation established by the USEPA under the SDWA that imposes specific monitoring and treatment requirements on all public drinking water systems that draw water from a surface water source.

Synthetic Organic Chemicals: Applied to manufactured chemicals that are not as volatile as volatile organic chemicals. Included are herbicides, pesticides, and chemicals widely used in industry.

Total Suspended Solids: Solids present in wastewater.

THMs: A group of compounds formed when natural organic compounds from decaying vegetation and soil (such as humic and fulvic acids) react with chlorine.

Turbidity: A measure of the cloudiness of water caused by the presence of suspended matter, which shelters harmful microorganisms and reduces the effectiveness of disinfecting compounds.

Vehicle of Disease Transmission: Any nonliving object or substance contaminated with pathogens.

Wastewater: The spent or used water from individual homes, a community, a farm, or an industry that contains dissolved or suspended matter.

Waterborne Disease: Water is a potential vehicle for disease transmission, and waterborne disease is one of the most preventable types of communicable illness. The application of basic sanitary principles and technology has virtually eliminated severe outbreaks of waterborne diseases in developed countries. The most prevalent waterborne diseases include *typhoid fever, dysentery, cholera, infectious hepatitis*, and *gastroenteritis*.

Note: Waterborne diseases are also referred to as *intestinal diseases*, as they impact the human intestinal tract. If pathogens in infected feces are accidentally ingested through contaminated water, the disease can spread further, potentially leading to an epidemic. Symptoms of intestinal disease include diarrhea, vomiting, nausea, and fever. Intestinal diseases may affect many individuals during an epidemic and can sometimes lead to fatalities among those infected. Untreated sewage in water is typically the primary source of this disease (Nathanson, 1997). In practice, "hazard identification in the case of pathogens is complicated because several outcomes—from asymptomatic infections to death—are possible, and their outcomes depend upon the complex integration between the agent and the host. This interaction, in turn, depends on the characteristics of the host as well as the nature of the pathogen. Host factors, for example, include preexisting immunity, age, nutrition, ability to mount an immune response, and other nonspecific host factors. Agent factors include types and strains of the organism as well as its capacity to elicit an immune response" (Gerba, 1996, pp. 357–358).

Water Softening: A chemical treatment method that uses either chemicals to precipitate or a zeolite to remove those metal ions (typically Ca^{2+}, Mg^{2+}, and Fe^{3+}) responsible for hard water.

Watershed: The land area that drains into a river, river system, or other body of water.

Wellhead Protection: The protection of the surface and subsurface areas surrounding a water well or well field supplying a public water system from contamination by human activity.

DID YOU KNOW?

Americans drink more than one billion glasses of tap water per day.

2.6 CLEAN, FRESH, AND PALATABLE: A HISTORICAL PERSPECTIVE

An early human, wandering alone from place to place, hunting and gathering to subsist, would have had little difficulty in obtaining drinking water, because such a person would—and could—only survive in an area where drinking water was available with little travail. The search for clean, fresh, and palatable water has been a human priority from the very beginning. We take no risk in stating that when humans first walked the Earth, many of the steps they took were in the direction of the water supply.

When early humans were alone or in small numbers, finding drinking water was a constant priority, but it is difficult for us to imagine today just how big a priority finding drinking water became as the number of humans proliferated. Eventually, communities formed, and with their formation came the increasing need to find clean, fresh, and palatable drinking water, as well as a means of delivering it from the source to the point of use.

Archaeological digs are replete with the remains of ancient water systems that reflect humans' early attempts to satisfy the never-ending need for drinking water. For well over 2,000 years, piped water supply systems have existed. Whether the pipes were fashioned from logs, clay, or carved from stone or other materials is not our point. The point is that they were fashioned to serve a vital purpose, one universal to the community needs of all humans: to deliver clean, fresh, and palatable water to where it was needed.

These early systems were not arcane. Their intended purpose is readily understood today. They were basic but effective, though they lacked two features we now consider standard. First, of course, they were not pressurized but instead relied on gravity flow, as the means to pressurize the mains was not known at the time. Even if such pressurized systems were known, they certainly would not have been used to pressurize water delivered via hollowed-out logs and clay pipes. They also lacked adequate sanitation, a basic expectation in today's industrialized world. Remember, to know the need for something exists (in this case, the ability to sanitize and disinfect water supplies), the nature of the problem must be defined. Not until the middle of the 1800s (after countless millions of deaths from waterborne disease over the centuries) did people realize that a direct connection between contaminated drinking water and disease existed. At that point, the sanitation of the water supply became an issue.

When the relationship between waterborne diseases and the consumption of drinking water was established, evolving scientific discoveries led the way toward the development of technology for processing and disinfection. Drinking water standards were developed by health authorities, scientists, and sanitary engineers.

With the current lofty state of effective technology that we in the United States and the rest of the developed world enjoy today, we could sit on our laurels, so to speak, and assume that because of the discoveries developed over time (and at the cost of countless people who died (and still die) from waterborne-diseases), all is well with us—that problems related to providing us with clean, fresh, palatable drinking water are problems of the past. Are they problems of the past? Have we solved all the problems related to ensuring that our drinking water supply provides us with clean, fresh, and palatable drinking water? Is the water delivered to our tap as clean, fresh, and palatable as we think it is . . . as we hope it is? Does anyone know?

What we do know is that we have made progress. We have come a long way from the days of gravity flow water delivered via mains of hollowed-out logs, clay, or stone. Many of us on this earth have come a long way from the days of cholera epidemics. We should ask those in Sydney who boiled water for weeks in 1998. Or better yet, we should speak with those who drank the "city water" in Milwaukee in 1993 or in Las Vegas, Nevada—those who suffered and survived the onslaught of *Cryptosporidium* from contaminated water out of their tap. Or if we could, we should ask these same questions of a little boy named Robbie, who died of acute lymphatic leukemia, the probable cause of which is far less understandable to us: toxic industrial chemicals, unknowingly delivered to him via his local water supply.

DID YOU KNOW?

Many bottled water drinkers drink water for its beauty benefits. That is, supposedly, drinking lots of water will make one's skin smoother (reduce wrinkles) and more youthful looking. However, there is no evidence that drinking water offers dermatological benefits or prevents wrinkling.

NOTE

1 From F.R. Spellman (2007). *The Science of Water*, 2nd ed. Boca Raton, FL: CRC Press.

REFERENCES

DeZuane, J. 1997. *Handbook of Drinking Water Quality*, 2nd ed. New York: John Wiley and Sons.

Foreman, J. 1998. No evidence of drinking. *Boston Globe*, May 11.

Gerba, C.P. 1996. Risk assessment. In *Pollution Science*, Pepper, Gerba, and Brusseau (Eds.). San Diego: Academic Press.

Hammer, M.J., and Hammer, M.J., Jr. 1996. *Water and Wastewater Technology*, 3rd ed. Englewood Cliffs, NJ: Prentice Hall.

Lewis, S.A. 1996. *Sierra Club Guide to Safe Drinking Water*. San Francisco (Ed.). CA: Sierra Club Books.

Meyer, W.B. 1996. *Human Impact on Earth*. New York: Cambridge University Press.

Nathanson, J.A. 1997. *Basic Environmental Technology: Water Supply, Waste Management, and Pollution Control*. Upper Saddle River, NJ: Prentice Hall.

Robbins, T. 1976. *Even Cowgirls Get the Blues*. Boston: Houghton Mifflin Company.

Spellman, F.R. 2007. *The Science of Water*, 2nd ed. Boca Raton, FL: CRC Press.

Tchobanoglous, G. 1991. *Wastewater Engineering: Treatment, Disposal, Reuse*, 3rd ed. New York: McGraw-Hill.

Varner, G.R. 2005. *Creatures in the Mist*. New York: Algora Publishing.

Part 3

Utility Management Basics

The most crucial arrow in any manager's quiver can be summed up in one word:

Credibility, Credibility, Credibility.

DOI: 10.1201/9781003655848-5

3 What Is Public Utility Management?

A leader [or manager] is best when people barely know he exists, when his work is done, his aim is fulfilled, they will say: We did it ourselves.

—Lao Tzu

3.1 INTRODUCTION

Public utility management, also known as public works management, oversees the operations, maintenance, safeguarding, and preservation of the systems responsible for water treatment, wastewater collection and treatment, and distribution of safe potable water to a localized community. Keep in mind, however, that public utilities also can include energy, telecommunications, and other local services; despite this, in this book, our focus is on water service by Water Managers. Concerning actual hands-on or eye-on operations, the public utilities manager supervises the staff, interacts (networks) with the public (ratepayers and stakeholders), is the principal guardian of ratepayers' and other contributors' funds, keeps an eye on plant and employee safety and health, stays in good standing with regulators—meets with liaisons from environmental and health agencies and with other government entities to discuss various issues affecting public utilities management—and coordinates with other municipal officials to ensure the efficient and professional function of facilities and the timely and reliable distribution of resources. The general and daily tasks a utility manager performs on a routine basis include assisting with long- and short-term project planning, design, maintenance, and operations in utility areas such as water, wastewater, collection systems, telecommunications, solid waste, and storm drain systems.

A public utility manager's attention typically centers on five key areas beyond routine tasks. These principal areas of primary focus, as shown in Figure 3.1, consist of staff, ratepayers, regulators, plant security, and the organization's governing body.

3.1.1 Key Terms and Definitions

To study any aspect of water management operations, you must master the language associated with technology. Each technology has its own terms with its own accompanying definitions. Many of the terms used in water management are unique; others combine words from many different technologies and professions. One thing is sure: Water managers without a clear understanding of the terms related to their profession are ill-equipped to perform their duties as required. Although this text includes a glossary of terms at the end of the work, a list of terms and definitions of many of the terms is presented right up front here—the ones relevant to this discussion.

DOI: 10.1201/9781003655848-6

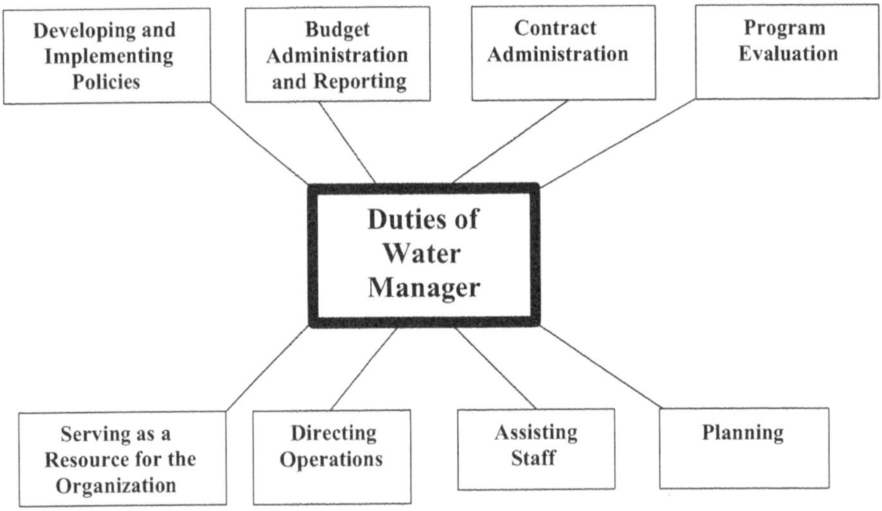

FIGURE 3.1 Public utility managers' principal areas of focus.

Experience has shown that an early introduction to keywords is a benefit to readers. Moreover, an up-front introduction to key terms facilitates a more orderly, logical, and systematic learning activity. Those terms not defined in this section are defined as they appear in the text.

The following terms are presented in this chapter or introduced here for a better understanding of the chapter's contents. These definitions provide a brief overview of their meaning (USEPA, 2008).

- **Attribute:** A characteristic or outcome of a utility that indicates effective performance.
- **Benchmarking:** Often used as a measurement tool in comparing similar operations and processes by measuring across organizations and/or sectors to identify best practices, set improvement targets, and simply measure progress.
- **Effective Utility Management:** Management that improves products and services, increases community support, and ensures a strong and viable utility into the future.
- **Gap analysis:** Defining the present state of an enterprise's operations, determining the desired or "target" state, and identifying the gap between them.
- **Internal trend analysis:** Comparison of outcomes or outputs relative to goals, objectives, baselines, targets, and standards.
- **Life-cycle cost:** The total of all internal and external costs associated with a product, process, or activity throughout its entire life cycle—from raw materials acquisition to manufacture/construction/installation, operation and maintenance, recycling, and final disposal.
- **Performance measurement:** Evaluation of status and trends; can also include comparison of outcomes or outputs relative to goals, objectives,

baselines, targets, standards, other organizations' performance, or processes (typically called benchmarking).

- **Operations and maintenance expenditure:** Expenses used for day-to-day operation and maintenance of a facility.
- **Operating revenue:** Revenue realized from the day-to-day operations of a utility.
- **Performance measure:** A particular value or characteristic designated to measure input, output, outcome, efficiency, or effectiveness.
- **Standard operating procedure:** A prescribed procedure to be followed routinely; a set of instructions having the force of a directive, covering those features of operations that lend themselves to a definite or standardized procedure without loss of effectiveness.
- **Strategic plan**: An organization's process of defining its goals and strategy for achieving those goals. Often it entails identifying an organization's vision, goals, objectives, and targets over a multi-year period, as well as setting priorities and making decisions on allocating resources, including capital and people, to pursue the identified strategy.
- **Stewardship:** The careful and responsible management of something entrusted to a designated person or entity's care; the responsibility to properly utilize its resources, including its people, property, and financial and natural assets.
- **Sustainability:** The use of natural, community, and utility resources in a manner that satisfies current needs without compromising future needs or options.
- **Watershed health:** The ability of ecosystems to provide the functions needed by plants, wildlife, and humans, including the quality and quantity of land and aquatic resources.

3.2 WATER MANAGER QUALIFICATIONS

What are the qualifications to be a water manager? A better question is what are the qualifications needed to be an effective water manager? Well, suppose you attend college to obtain a degree in management to find a management job in the water utility or public works field. In that case, you will be exposed to a wide assortment of management theories, concepts, models, systems, ideas, and philosophies. So, the question becomes: Which of all these concepts is the right one for achieving enlightenment and learning how to manage a water utility or other water enterprise?

Good question. The truth is, you will be exposed to a wide range of administrative theories, including the traditional standard bearer: Fayol's well-known Five Functions of Management. The Five Functions of Management are as follows:

1. Planning
2. Organizing
3. Commanding
4. Coordinating
5. Controlling

Although Fayol's Five Functions of Management have been around since 1916, many more current concepts and theories of management have been developed and employed since then, yet these functions remain relevant to organizations today. In a nutshell, these five functions focus on the relationship between personnel and their management. They offer reference points to solve problems efficiently and creatively.

In addition to exposure to and learning about Fayol's concepts in college, in many management courses on principles, functions, theories, and so forth, you may also be exposed to Group Dynamics, Max Weber's Bureaucratic Theory, McGregor's Theory X and Theory Y, Taylor's Scientific Management, Fayol's 14 Principles of Management, and/or French and Raven's Five Forms of Power.

The next question is which of these management theories is the best, is the most important, or is something the student must learn and must abide by? Students should be exposed to as many different management theories and practices as possible. Based on experience, to broaden a potential manager's spread of knowledge, it is best to generalize one's college education with a little of this and a little of that, along with the required courses needed to complete training and to graduate.

Again, based on experience, when a newly graduated college student with a degree in management or another area related to management completes their studies and enters the workforce, that person is not a manager. Being a college graduate is not a sure-fire element needed to manage or be a manager.

What does it take to be a manager? What does it take to be an effective manager? A college degree in management opens the door for a new employee to become a manager but does not guarantee success. Success is accomplished via education, on-the-job experience, and the natural ability to lead. Unless the new management graduate is a natural-born leader, a General Patton, Stonewall Jackson, Napoleon, Genghis Khan, or someone of that caliber and natural ability in the non-military sense, management must be learned; it must be groomed. Moreover, management must be generalized to the point where the prospective manager has an important mixture of managerial qualities, including physical, mental, moral, technical, and educational qualities, and experience as an effective manager.

In the education and on-the-job training model, what the new prospective manager is learning (hopefully) is how actual functions within the organization are integrated with one another. Learning these functions along with leadership ability and a leadership style that does not include the Gestapo mentality is the actual ticket punching that is required to become an effective manager.

Because they are important to managing and leading a public utility organization, it is important to take a closer look at each of the critical managerial functions within the public utility organization.

- **Strategic Planning**—The public utility manager who is not far-thinking, not cognizant of emerging trends, and has no vision of utility strategy is like a person in the middle of the ocean without sails, oars, food, and fresh water. In addition to being far-thinking, the manager must write a strategic plan by exploring internal resources and capabilities against an external backdrop of regulation, economic factors, customer preferences, modern technology, and competitive forces.

- **Establish Credibility**—When any manager directs his or her employees to perform this or that task and the manager has no idea what the workers' tasks are, there is a fundamental problem that develops with credibility. The quickest and surest way for any manager to lose respect and credibility is for that manager to not have a clue what he or she is talking about. So, what exactly is meant by a utility manager's lack of credibility? Well, an example would be if a football coach has no idea about what each player does on the field and their responsibilities. This would be like a utility manager not understanding plant nomenclature and a given employee's responsibility, with no knowledge of an employee's job description and requirements. Even more important regarding a utility manager's credibility is whether he or she understands the water operations. Can the manager, for example, identify and explain what water disinfection is all about? If not, employees will have little to no respect for such a manager.
- **Stakeholder Relations**—The Water Manager must have a strong relationship with stakeholders. First, the manager must find out who the stakeholders are. It is essential to find out what the stakeholders' work styles, priorities, and specific project responsibilities are. The Water Manager must meet the stakeholders; the manager cannot be a long-distance acquaintance. Instead, the manager should proactively engage with stakeholders and maintain open and transparent communication. Also, the Water Manager's interrelationship with the stakeholder(s) and everyone else can be summed up in one word: **Communicate, Communicate, Communicate!**
- **Make Decisions**—Have you ever worked for a manager, any manager at any level, who could not decide? Is there anything more frustrating? I think not. Many managers want to mull over decisions, and there is nothing wrong with that; that is wise in most circumstances. However, the manager who cannot or will not decide because he or she is afraid of failure is the failure itself. Being scared to fall on one's sword is cowardice to the extreme. Managers must make decisions and live with the consequences. If the decision is a bad one, bone up on it and move on. No one respects a coward, period.
- **Customer Issues**—Satisfying customers' needs is what the Water Manager must focus on. Water Managers sometimes overlook that their primary responsibility is to serve their customers. Customers have questions and needs. The Water Manager must address these questions and needs.
- **Finance, Rates, and Regulations**—Water Managers who neglect finance, rates, or regulations risk bankruptcy and legal issues due to noncompliance with local, state, and federal laws. With regard to rates, any Water Manager must be sensitive to the rates charged to ratepayers; if the utility charges too much for its service(s), it will find itself in constant conflict with the public it is designed to serve.
- **Resource Planning**—Water Managers must not only plan for whatever—chemicals, repair parts, equipment replacement, staffing—but also investments. Managers must be trained in financial analysis used for investment analysis. The manager must understand investing equity and/or the purchase of bonds to ensure the entity's economic success and stability. This financial acumen is critical to maintaining fair and equitable customer rates.

- **Public Speaking Skills**—If you have ever had to sit through a speaker's presentation that was presented in an evasive, vague, less than truthful, him/ha and roundabout way to avoid responding to a question or making a definite statement then you quickly understand the importance of being an effective speaker; one who can garner his or her audience's full attention and maintain it. The Water Manager who hems and haws around with his or her presentation will quickly lose his or her listeners. Instead of hemming and hawing, the speaker must adapt the presentation to the audience. The presentation must be of interest to the listeners. If the presentation is not spontaneous, it must be planned and practiced and full of energy and clarity. Presentations that include visual aids, humor, positives, quotes, or some remarkable facts are the most effective.
- **Leadership**—Leadership is simply another word for managing. Public Utility Managers who cannot lead by example, by doing, by presentation skills, and by force of personality will find leadership and/or management of people difficult. Public utilities responsible for water management must promote personnel based on merit, their demonstrated ability to lead, and **not** solely on seniority.
- **AI Savvy**—Leaders of today must be on board the digital train with computer operations. They must also be trained in artificial intelligence and its applicable subfields used in their organization. Leveraging AI's capabilities can improve water quality monitoring, leak detection, infrastructure maintenance, flood management, water resource planning, and water conservation.

SIDEBAR 3.1 ABOUT STAFFING WASTEWATER TREATMENT PLANTS

A specific connotation, implication, suggestion, label, or stereotype, and a false stigma exist about personnel (staff) who work in wastewater treatment plants and operations. Such a stigma is not rare. Think, for example, of exterminators, funeral and grave diggers, toilet cleaners, parking inspectors, and animal control officers. . . . "Oh, how can you do that dirty work?" the unenlightened average person might say.

Do you want a more expressive response from someone? Okay, simply say that you are a sewage rat, or that you treat raw sewage, or that you are a sewage operator. The immediate response:

Ugh!

Gross!

Oh, my god! Really?

Katie, bar the door . . . phew, let me out of here right now.

These responses are commonly known or attributed to what is known as the "Yuck Factor." Now, it is true that most people feel that someone must do it, but people can look down on that person who does the so-called dirty work. In contrast, if someone were to ask a water treatment plant operator what line of work he or she is in, the response could be quite different, with a pat on the back and a comment stating, "Keep up the good work."

3.2.1 A Regulator's View of Effective Water Management

After reviewing several explanations of what effective management of a public utility is and based on personal experience of working with several different public utility managers in water and wastewater operations, I found that the best explanation detailing the attributes of effectively managed water service utilities is given by the United States Environmental Protection Agency (USEPA) in its 2008 Effective Utility Management: A Primer for Water and Wastewater Utilities. USEPA developed its primer based on the collaboration of the Water Environment Federation, National Association of Water Companies, American Water Works Association, Association of Metropolitan Water Agencies, American Public Works Association, and National Association of Clean Water Agencies. For anyone who aspires to become a Public Utility Manager, this USEPA primer is highly recommended.

USEPA (2024) lists ten water management techniques: meter/measure/manage, optimize cooling towers, replace restroom fixtures, eliminate single-pass cooling, employ Water-Smart landscaping and irrigation, reduce steam sterilizer tempering water use, reuse laboratory culture water, control reverse osmosis system operation, recover rainwater, and recover air handler condensate.

- **Meter/Measure/Manage**: Metering and measuring facility water use helps to assure the equipment is run correctly and maintained properly to help prevent water waste from leaks or malfunctioning mechanical equipment.
- **Optimize Cooling Towers**: Cooling towers that provide air conditioning are large consumers of energy. Their operations can be optimized by carefully controlling the ratio of blowdown water to water evaporated. The ratio of evaporation to blowdown (discharge) is called the cycle of concentration. For maximum water efficiency, cooling towers should be operated at six or more cycles of concentration. Metering water filled into and discharged from the cooling tower ensures that the cooling water is operating correctly.
- **Replace Restroom Fixtures**: Many facilities have inefficient sanitary fixtures. For example, toilets used 3.5 gallons per flush (gpf). After the installation of water-efficient fixtures, new toilets have flow rates of 1.28 or 1.6 gpf. Urinals flush at 0.5 gpf or less, shower heads flow at 2.0 gallons per minute (gpm) or less, and faucet aerators flow at less than 2.2 gpm.
- **Eliminate Single-Pass Cooling**: This is the old once-and-done scenario; that is, the flow of water just once through the system for cooling purposes before it goes down the drain. Replacing this system with a recirculated chilled water loop can cut water use by up to 80%.
- **Use Water-Smart Landscaping and Irrigation**: Weather-based irrigation controllers or soil moisture sensors can be used to water only when plants need it. Moreover, planting native and drought-resistant plant species minimizes the need for supplemental irrigation.
- **Control Steam Sterilizer Water**: Steam sterilizers use cooling water to temper the steam condensate discharge from the sterilizer to the drain. Many of the older sterilizers discharge a continuous flow of tempering water to the drain, even when it is not needed. Wasteful sterilizers need

to be retrofitted with a tempering water control kit or replaced with older steam sterilizers that only apply tempering water when needed.

- **Reuse Culture Water:** Several laboratories need culture water for aquatic research, and in some cases, the water is pumped from local water bodies (lakes, rivers, or bays) and then dumped into treatment plants before emptying into the water body. Reusing the culture water is a better choice and conserves water.
- **Control RO System Operation**: Water consumption is related to the multi-step process of generating deionized purified water through reverse osmosis (RO). To save water, the idea is to carefully regulate purified water generation rates to meet demand and ensure the system is sized accordingly.
- **Recover Rainwater**: Recovery systems capture rainwater from roofs and redirect it to a storage tank. This water is used for flushing toilets, supplying cooling towers, irrigating the landscape, and watering gardens.
- **Recover Air Handler Condensate**: Air conditioning units produce condensate water from cooling coils. Many industries are capturing this water for use as cooling tower make-up water.

All of these management techniques are enhanced, in one way or another, by employing AI and its subfields where applicable in operational practices.

3.2.1.1 Attributes of Effectively Managed Water Sector Utilities

As stated earlier, USEPA (2008) has compiled ten attributes of what they judge as effectively managed water sector utilities. In this section, we briefly discuss each of the ten attributes.

3.2.1.1.1 Product Quality

Product quality refers to water operators tasked with producing water (the product) of suitable and safe quality as potable water, producing a treated effluent, and producing process residuals in full compliance with regulatory and reliability requirements, consistent with customer, public health, and ecological needs.

3.2.1.1.2 Customer Satisfaction

Customer satisfaction deals with whether the water service provides dependable, responsive, and affordable services in line with explicit, customer-accepted service levels. Additionally, does the service provider offer feedback in a timely manner to address customer needs and emergencies?

3.2.1.1.3 Financial Viability

The effective public utility manager understands the full life-cycle cost of the utility and establishes and maintains an effective balance between long-term debt, asset values, operations and maintenance expenditures, and operating revenues. The effective manager establishes predictable rates—consistent with community expectations and acceptability—adequate to recover costs, provides for reserves, maintains support from bond rating agencies, and plans and invests for future needs.

3.2.1.1.4 *Operational Optimization*

The effective utility manager ensures ongoing, timely, cost-effective, dependable, and sustainable performance improvements in all facets of operations. They minimize resource use, loss, and impacts from day-to-day operations. The manager also maintains awareness of information and operational technology developments to anticipate and support the timely adoption of improvements.

3.2.1.1.5 *Infrastructure Stability*

The effective utility manager understands the condition of and the costs associated with critical infrastructure assets. They maintain and enhance the condition of all assets over the long term at the lowest possible life-cycle cost and take on acceptable risks consistent with customer, community, and relation-supported service levels, as well as anticipated growth and system reliability goals. In addition, the effective manager ensures that asset repair, rehabilitation, and replacement efforts are coordinated within the community to minimize disruptions and other negative consequences.

3.2.1.1.6 *Operational Resiliency*

The effective utility manager ensures that utility leadership and staff work together to anticipate and avoid problems. The manager proactively identifies, assesses, establishes tolerance levels for, and effectively manages a full range of business risks (including legal, regulatory, financial, environmental, safety, security, and natural disaster-related) in an initiative consistent with industry trends and system reliability goals.

3.2.1.1.7 *Community Sustainability*

Effective utility managers are explicitly cognizant of and attentive to the impacts their decisions have on current and long-term community and watershed health and welfare. They manage operations, infrastructure, and investments to protect, restore, and enhance the natural environment; efficiently use water and energy resources; and promote economic vitality and engender overall community improvement. Explicitly considering a variety of pollution prevention, watershed, and source water protection approaches as part of an overall strategy to maintain and enhance ecological and community sustainability.

3.2.1.1.8 *Water Resource Adequacy*

Effective public utility managers ensure water availability consistent with current and future customer needs through long-term resource supply and demand analysis, conservation, and public education.

3.2.1.1.9 *Employee and Leadership Development*

Effective public utility managers recruit a workforce that is competent, motivated, adaptive, and safety-oriented. They build a collaborative organization focused on continuous learning and improvement, ensuring that employee knowledge is retained and enhanced over time. They also focus on and emphasize opportunities for professional and leadership development, striving to create an integrated and well-coordinated senior leadership team.

3.2.1.1.10 Stakeholder Understanding and Support

Effective utility managers cultivate understanding and support from oversight bodies, the community, watershed interests, and regulatory bodies for service levels, rate structures, operating budgets, capital improvement programs, and risk management decisions. The manager actively involves stakeholders in decisions that will affect them.

3.3 MEASURING ORGANIZATION EFFECTIVENESS

Measuring performance is one of the keys to water management success. This section provides ideas on how to approach measurement and then offers measures for each attribute to help understand an organization's status and progress.

You know, some public water managers sit back in their padded chairs, pat themselves on the back, and think to themselves, "Gee, I am doing a wonderful job here."

Really?

I've met some of them, but when asked about their plant, they avoid answering directly.

Have you ever witnessed such a response, such a manager?

Well, the public water manager's self-appreciation and pat on the back are warranted.

However, how does they know that? How did they compare their operation to others, to flagship operations, to any operation that is top-notch, the best in class?

Or did they?

That is the question—if management does not conduct measurements of its operations, how are they to know how they are doing? How are they to understand how they compare to other similar operations? If there are deficiencies, how are they to know that unless they measure and correct them?

Why should any public water manager worry about plant performance? Well, job security might be one reason; that is, their job security. And then there is the trend toward privatizing public water facilities and operations in the hope of saving money and increasing efficiency.

Privatizing a public water utility?

Yes, absolutely. It is a movement that is real and current, and an aspect that management had better be tuned into. Then there are those who might be too tuned in to the possibility of privatization. Consider the Waldrop Syndrome detailed in Case Study 3.1.

CASE STUDY 3.1 WALDROP SYNDROME

In 1995, G. (do not call me George) Daniel Waldrop, Director of Operations for one of the nation's largest sanitation districts, had a dilemma. The commission that governed his operation was bandying about the idea of privatizing the organization. The commissioners reasoned that public ratepayers would be better served by a private, efficient operation. In their view, private operation would not only save money but also bring in talented top managers. The commissioners felt that public service entities do not have a stellar history of attracting quality managers. For example, they felt their operation was a repository for a bunch of good old' boys, belonging to the ROAD (Retired On

Active Duty) Gang. Unfortunately for the commissioners (and anyone else for that matter) once entrenched, it is almost impossible, except for criminal behavior bordering on murder, to get rid of such no-loads.

Waldrop realized that a new privatizer would replace him and his team with experienced private-sector professionals. This was, of course, unacceptable to him. However, Waldrop had one ace in the hole: He knew that the commission, before making the decision to privatize or not, had decided to conduct a study (a pilot study—engineers love the words "pilot study") to determine if privatization made sense in their operation.

Waldrop understood that the results of the pilot study would help determine whether privatization would occur. Knowing that the metrics involved in the commissioners' pilot study would focus on staffing, value-added operations, and total costs, Waldrop decided to nip the situation in the bud. He knew that each of the 14 wastewater treatment plants he was responsible for was "over-manned." Each plant had a workforce of approximately 44 personnel. Of this number, each plant had 4–6 plant operator assistants. Many of these plant operator assistants were licensed operators at the lowest professional certification levels (i.e., Class 4 or 3 Wastewater Operator: 4 being the lowest certification level—although some states like Virginia have created Class 5 and Class 6 license levels, both of which are lower classifications). Annual salaries for each assistant plant operator average around $32,000, depending on longevity. In reviewing assistant operator pay and experience levels, Waldrop was surprised to find that their average continuous time served on the job was 12 years. And while Waldrop understood that these operator assistants performed essential functions at each of his treatment plants, they simply had to go. He believed that cutting the plant staff to under 40 would significantly reduce annual expenses for salaries, benefits, and equipment. Waldrop calculated that, including company-paid benefits like life insurance, retirement, and medical care at 1.4 times annual salary, each plant operator assistant costs his operation an average of $44,800 per year. He then took the total number of assistants, 56, and multiplied it by $44,800, which equals $2,508,800. Waldrop was surprised at this total cost figure; he could not fathom how a bunch of assistant operators could cost so much!

Before the upcoming budget submission deadline (July 1, 2016), Waldrop subtracted the cost of operator assistants, $2,508,800, generated and distributed a memo giving all assistant treatment plant operators notice that their services were no longer required after June 30, 2016 (last day of the fiscal year), and sat back in his chair chain-smoking his cigarettes, quite pleased with himself.

Waldrop's feelings about his actions and himself were short-term, however. As it turned out, after the new fiscal year began on July 1, 2016, the governor replaced the commission chairperson with a professional public administrator who had no intention (no personal motivation) for privatizing the district. So, the privatization effort evaporated before it ever started.

The results of Waldrop's downsizing of the treatment plant assistant operators did not sit well with the 14 plant managers who had lost their much-needed assistants.

Waldrop, being the dysfunctional public service manager that he was, had not bothered to discuss his downsizing plans with anyone, let alone with the plant managers, the ones most affected by the downsizing. If Waldrop had taken the time and had the common decency (in the view of the plant managers) to explain his downsizing plan to them, they would have adamantly and forcefully argued against such a shortsighted move. For example, the plant managers would have said to Waldrop that the assistant operators formed a pool of fully trained personnel to instantly draw from whenever it became necessary to replace full-time treatment plant operators (the average turnover rate of full-time operators was approximately three per year per plant).

In addition to losing their pool of trained assistant operators to fill vacant positions, the lack of fully qualified assistant operators caused another glaring problem. The district had an overly generous annual leave policy that allowed employees with more than 20 years of service to take six to eight paid weeks of annual leave, depending on pay grade and length of service. When the assistants were removed from the payroll, the natural fill-in personnel for operators wanting to use their annual leave each year were no longer available. Thus, it became increasingly harder for plant management personnel to allow key operators to take their annual leave each year.

Another unforeseen problem with doing away with the assistant operators was that when an operator became ill, injured, or encountered personal family problems, it was a significant burden on those plant personnel who had to fill the shoes of those who could no longer work. When the plants had a cadre of assistant plant operators to draw from to fill in for sick or otherwise unavailable employees, this was not a problem.

Waldrop's feelings of sitting back and riding the system until retirement did not work out as he had hoped.

Not being a far-thinker, he missed the digital train and its trend, its emergence in water management. He had no understanding or recognition that artificial intelligence (AI) and machine learning (ML) are revolutionizing water management by enabling more efficient and sustainable use of this critical resource. Waldrop was unaware of the potential AI technologies in water management. AI technologies, through predictive analysis, smart metering, and real-time monitoring, can optimize water resource management, improving access and distribution. When AI and ML training emerged as a water management tool, Waldrop missed the train and was eventually replaced by a digitized water manager who had a complete understanding of recent advancements in AI technologies that have resulted in a growing number of applications in support of integrated water resources management. The new digitized manager understood numerous applications ranging from surface water supply, groundwater modeling, hydropower generation, agriculture and irrigation, climate change risk and flood risk management, and water-energy-food nexus, among others. It also helped that the new manager was initiative-taking, far-thinking, and a natural-born leader.

Waldrop syndrome highlights Albert Einstein's idea that problems cannot be solved using the same mindset that created them.

3.4 BENCHMARKING

One of the best ways for a public utility (or any other agency or business) to measure its effectiveness is to benchmark. As illustrated in Case Study 3.1, it is primarily out of self-preservation (to retain their lucrative positions) that many utility directors work against the trend to privatize water, wastewater, and other public operations. Usually, the real work to prevent privatization is delegated to the individual managers in charge of each specific operation because they also have a stake in ensuring that their relatively secure careers are not affected by privatization. It can be easily seen that working against privatization by these "local" managers is in their self-interest and the interest of their workers because their jobs may be at stake.

The question is, of course, how does one go about preventing his or her water and wastewater operation from being privatized? Efficiency needs to increase while cutting operational costs. This is challenging in practice, but possible. For example, for those facilities under properly implemented and managed Total Quality Management (TQM), the process can be much easier.

The advantage TQM offers the plant manager is the variety of tools provided to help plan, develop, and implement water and wastewater efficiency measures. These tools include self-assessments, statistical process control, International Organization for Standardization (ISO) 9000 and 14000, process analysis, quality circle, and benchmarking (see Figure 3.2).

In this text, the focus is on the use of the benchmarking tool to improve water and wastewater operations' efficiency, based on personal experience characterized as a wonderful experience. *Benchmarking* is a process for rigorously measuring your performance against best-in-class operations and using the analysis to meet and exceed the best-in-class.

What Is Benchmarking?

1. Benchmarking versus best practices gives water and wastewater operations a way to evaluate their operations overall.
    a. how effective
    b. how cost-effective
2. Benchmarking shows plants how well their operations stack up and how well those operations are implemented.
3. Benchmarking is an objective-setting process.
4. Benchmarking is a new way of doing business.
5. Benchmarking forces an external view to ensure the correctness of objective setting.
6. Benchmarking forces internal alignment to achieve plant goals.
7. Benchmarking promotes teamwork by directing attention to those practices necessary to remain competitive.

Start---Plan---Research---Observe---Analysis---Adapt

FIGURE 3.2 Benchmarking process.

Potential Results of Benchmarking

1. Benchmarking may indicate the direction of required change rather than specific metrics.
    a. Costs must be reduced.
    b. Customer satisfaction must be increased.
    c. Return on assets must be increased.
    d. Improved maintenance
    e. Improved operational practices
2. Best practices could be translated into operational units of measure.

Targets

- Consideration of available resources converts benchmark findings into targets.
- A target is a realistic goal with a set deadline.
- Targets can show progress toward benchmark practices and metrics.
- Quantification of precise targets should be based on achieving a benchmark.

Note: Benchmarking can be performance-based, process-based, or strategy-based and can compare financial or operational performance measures, methods or practices, or strategic choices.

When forming a benchmarking team, the goal should be to provide a benchmark that evaluates and compares privatized and re-engineered water and wastewater treatment operations to make your operation more efficient, competitive, and continually improve. It is important to point out that benchmarking is more than simply setting a performance reference or comparison; it is a way to facilitate learning for continual improvements. The key to the learning process is looking outside one's plant to other plants that have discovered better ways of achieving improved performance.

Benchmarking Steps

As shown in Figure 3.2, the benchmarking process consists of five steps.

1. Planning—Managers must select a process (or processes) to be benchmarked. A benchmarking team should be formed. The process of benchmarking must be thoroughly understood and documented. The performance measure for the process should be established (i.e., cost, time, and quality).
2. Research—Information on the best-in-class performer must be determined through research. The information can be derived from the industry's network, industry experts, industry and trade associations, publications, public information, and other award-winning operations.
3. Observation—The observation step is a study of the benchmarking subject's performance level, processes, and practices that have achieved those levels, and other enabling factors.
4. Analysis—In this phase, comparisons in performance levels among facilities are determined. The root causes of the performance gaps are

studied. To make accurate and appropriate comparisons, the comparison data must be sorted, controlled for quality, and normalized.

5. Adaptation—This phase involves putting into action what has been learned throughout the benchmarking process. The findings of the benchmarking study must be communicated to gain acceptance, functional goals must be established, and a plan must be developed. Progress should be monitored, and as required, corrections in the process are made.

Note: Benchmarking should be interactive. It should also recalibrate performance measures and improve the process itself.

CASE STUDY 3.2 BENCHMARKING: AN EXAMPLE

To gain a better understanding of the benchmarking process, the following limited example (it is in outline and summary form only—discussion of a full-blown study is beyond the scope of this text) is provided.

RACHEL'S CREEK SANITATION DISTRICT

Introduction

In January 2007, Rachel's Creek Sanitation District formed a benchmarking team to provide a benchmark that evaluates and compares privatized and re-engineered wastewater treatment operations with Rachel's Creek operations to improve efficiency and competitiveness. After three months of evaluating wastewater facilities using the benchmarking tool, our benchmarking is complete. This report summarizes our findings and should serve as a benchmark for comparing and assessing Rachel's Creek Sanitation District operations.

Facilities

41 wastewater treatment plants throughout the United States.

The benchmarking team focused on the following target areas for comparison:

1. Reengineering
2. Organization
3. Operations and Maintenance
    a. Contractual services
    b. Materials and supplies
    c. Sampling and data collection
    d. Maintenance
4. Operational Directives
5. Utilities
6. Chemicals
7. Technology

8. Permits
   a. Water quality
   b. Solids quality
   c. Air quality
   d. Odor quality
9. Safety
10. Training and Development
11. Process
12. Communication
13. Public Relations
14. Reuse
15. Support Services
    a. Pretreatment
    b. Collection systems
    c. Procurement
    d. Finance and administration
    e. Laboratory
    f. Human resources

Summary of Findings:

Our overall evaluation of Rachel's Creek Sanitation District compared to our benchmarking targets is a good one; that is, we are in good standing compared to the 41 target facilities we benchmarked. In the area of safety, we compare quite favorably. Only plant 34, with its full-time safety manager, appeared to be better than we are. We were very competitive with the privatized plants in our usage of chemicals and far ahead of many public plants. We were also competitive in the use of power. Our survey of what other plants are doing to cut power costs showed that we clearly identified those areas of improvement, and our current effort to further reduce power costs is on track. We were far ahead in the optimization of our unit processes, and we were leaders in odor control.

There were also areas that we needed to improve. To Rachel's Creek employees, re-engineering applied only to the treatment department and had been limited to cutting staff, while plant practices and organizational practices were outdated and inefficient. Under the re-engineering section of this report, we have provided a summary of re-engineering efforts at the re-engineered plants visited. The experiences of these plants can be used to improve our re-engineering efforts. Next is our organization and staffing levels. A private company could reduce the entire treatment department staff by about 18% to 24%—based on the number of employees and not costs. In the organization section of this report, organizational models and their staffing levels are provided as guidelines for improving our organization and determining optimum staffing levels. The last wide area that we need to improve is the way we accomplish our work. Our people are not used efficiently because of outdated and inefficient policies and work practices. Methods to improve the way we work are found throughout this report. We noted that efficient work practices used by private companies allow plants to operate with small staffs.

Overall, Rachel's Creek Sanitation District's treatment plants are much better than other public service plants. Although some plants may have better equipment, better technology, and cleaner effluents, the costs of labor and materials are much higher than ours. Several of the public plants were in bad condition. Contrary to widespread belief, the privately operated plants had good to excellent operations. These plants met permit requirements, complied with safety regulations, maintained plant equipment, and kept the plant clean. Due to their efficiency and low staff, we felt that most of the privately operated plants were better than ours. We agreed that this needs to be changed. Using what we learned during our benchmarking effort, we can be just as efficient as a privately operated plant and still maintain our standards of quality.

CASE STUDY 3.3 WALDROP SYNDROME II

Approximately 30 years ago, Waldrop was sitting at his desk in his office, working through endless trail of paperwork that he had to read, agree to, or disagree with, and move on to the following written paperwork. A few days earlier, the new IT director had installed a desktop computer on his desk and the director had his team install similar computers at all 100+ workstations. Now, Waldrop is an engineer with a Master of Engineering degree specializing in water and wastewater treatment. But truth be told, he had no idea how to operate one of those new-fangled electronic devices. So, an IT person spent a few days training Waldrop in computer operations, and being a fast study, he caught on quickly.

As the weeks went by, he was able to use his office computer to communicate with all the work centers he was responsible for. He enjoyed his new computer and its ability to do so many things, which he was learning to use at warp speed.

One day, he received a computer message from one of his outside contractors who was exploring ways to make his water and wastewater operations more efficient and cost-effective. After a few days, the contractor sent Waldrop a message stating that if Waldrop agreed to install computerized operating systems within the plant to control the operation of valves, monitor parameters, and even take and analyze water samples automatically via automatic meters and valves connected to a central plant computer, it would solve all of Waldrop's needs.

At first, Waldrop brushed off the contractor's suggestions. He had little faith in a centralized computer plant operating system. "I have plant operators, plant maintenance operators, and assistants that do just fine in operating and maintaining plant operations . . . no need to shift to something else."

Well, after sleeping on it, Waldrop changed his mind. "Maybe it will be better to have computer-based operations of samplers and monitoring operations."

It took about six months for Waldrop to adopt computer-operated plant operations at all of his 14 water/wastewater treatment plants and to install supervisory control and data acquisition (SCADA) at his 103 pumping stations.

After a year of studying computerized plant operations, Waldrop determined the right-sized workforce he needed. He also reduced staffing by four to five employees through attrition from retirement or employee departure for whatever reason. Waldrop is now a lifelong advocate for computer, AI, IoT, and machine learning.

3.5 THE BOTTOM LINE

Measuring performance is one of the keys to public utility management success. Performance measurement can be conducted in-house by selected teams of employees who would be assigned to evaluate the current internal utility performance status and trends. Based on personal experience, I have found that benchmarking is the way to go when it comes to comparing your utility's performance with others. This comparison should only be made with utilities known for their effectiveness and best practices. Comparing your agency's practices with others can reveal valuable insights and guide improvements in operations and business. Artificial intelligence used in water management certainly provides a vivid vision of how technology from the fourth paradigm can manage mountains of data.

REFERENCES

USEPA. 2008. *Effective Utility Management: A Primer for Water and Wastewater Utilities*. Washington, DC: United States Environmental Protection Agency.

USEPA. 2024. *Water Management Plans and Best Practices at EPA*. Washington, DC: US Environmental Protection Agency.

Part 4

Drinking Water Supplies, Pollution, and Climate Change

4 Drinking Water Supplies

4.1 INTRODUCTION

Where do we get our drinking water? What is the source? Most public water comes from either groundwater or surface water—these are the primary sources for most supplies. From our earlier discussion of the hydrologic or water cycle, we know that from whichever of the two sources we obtain our drinking water, the source is constantly being replenished (we hope) with a supply of fresh water. This water cycle phenomenon was best summed up by Heraclitus of Ephesus, who said, "You could not step twice into the same rivers; for other waters are ever flowing on to you."

In this chapter, we discuss one of the Water Manager's primary duties—to find and secure a source of potable water for human use with the AI nexus.

DID YOU KNOW?

If water has a bad taste or odor, unless dying of thirst, people will not usually drink it.

If water is turbid (i.e., it is muddy or cloudy) and has a color, unless dying of thirst, people will not usually drink it. If water makes people sick, unless dying of thirst, they will not drink it.

4.2 WATER SOURCES[1]

In the real estate business, location is everything. We say the same when it comes to sources of water. The presence of water defines the "location" for communities. Although communities differ widely in character and size, all have the common concerns of finding water for industrial, commercial, and residential use. Freshwater sources are not always found where communities need them most. On land, a steady supply of drinkable water is the key factor determining which lifeforms are present. A map of the world immediately shows us that surface waters are not uniformly distributed over the Earth's surface. U.S. land holds rivers, lakes, and streams on only about 4% of its surface. The heaviest populations of any life forms, including humans, are found in regions of the United States (and the rest of the world) where potable water is readily available, because lands barren of water simply will not support large populations.

Note: The volume of freshwater sources depends on geographic, landscape, and temporal variations and on the impact of human activities.

DOI: 10.1201/9781003655848-8

TABLE 4.1
World Water Distribution

Location	Percent of Total
Land Areas	
Freshwater lakes	0.009
Saline lakes and inland seas	0.008
Rivers (average instantaneous volume)	0.0001
Soil moisture	0.005
Groundwater (above depth of 4,000 m)	0.61
Ice caps and glaciers	2.14
Total: Land areas	**2.8**
Atmosphere (water vapor)	0.001
Oceans	97.3
Total all locations (rounded)	**100**

Source: Adapted from Peavy et al. (1985), p. 12.

4.2.1 Just How Readily Available Is Potable Water?

Approximately 326 million cubic miles of water comprise Earth's entire water supply. Of this massive amount of water, though, while providing us indirectly with fresh water through evaporation from the oceans, only about 3% is fresh. Most of Earth's fresh water is found in polar ice caps and glaciers. The rest is held in lakes, flows through soil, and is found in river and stream systems. Only 0.027% of Earth's fresh water is available for human consumption (see Table 4.1 for the distribution percentages of Earth's water supply). We see from Table 4.1 that the significant sources of drinking water are from surface water, groundwater, and groundwater under the direct influence of surface water (i.e., springs or shallow wells).

DID YOU KNOW?

Of all the freshwater on Earth, only about 0.3% is contained in rivers and lakes—yet rivers and lakes are not only the water we are most familiar with but also where most of the water we use in our everyday lives exists.

4.3 WATER SUPPLIES

Most surface water originates directly from precipitation—rainfall or snow. To understand how runoff affects surface water in the United States, consider the national water balance. Over the U.S. mainland, rainfall averages about 4,250 billion gallons per day. Of this massive amount, about 66% returns to the atmosphere through evaporation directly from the surface of lakes and rivers and transpiration from plants. This leaves about 1,250 billion gallons per day to flow across or through the earth to return to the sea (see Table 4.2). Although municipal use accounts for a small proportion of the total water volume, per capita water consumption in the United States is approximately 150 gallons per person per day. Public water's low cost in the United States drives higher usage, while regions with scarce or expensive water see lower per capita consumption due to financial and conservation reasons.

The 1,250 billion gallons per day of surface runoff water, exposed and open to the atmosphere, results from the movement of water on and just beneath the Earth's surface (overland flow). Simply put, overland flow and surface runoff are the same—water flow that has not yet reached a definite stream channel. This occurs when the rate of precipitation exceeds either the rate of interception and evapotranspiration or the amount of rainfall readily absorbed by the Earth's surface. The total land area that contributes runoff to a stream or river is called a *watershed, drainage basin,* or *catchment area.*

Specific sources of surface water include the following:

- Rivers
- Streams
- Lakes
- Impoundments (human-caused lakes made by damming a river or stream
- Very shallow wells that receive input via precipitation
- Springs affected by precipitation (flow or quantity directly dependent upon precipitation)
- Rain catchments (drainage basins)
- Tundra ponds or muskegs (peat bogs)

Surface water has advantages as a source of potable water. Surface water sources are usually easy to locate; unlike groundwater, finding surface water does not require a geologist or hydrologist. Typically, surface water is not tainted with minerals precipitated from the earth's strata. Ease of discovery aside, surface water also presents

TABLE 4.2
Water Balance in the United States

Precipitation	4,250
Evaporation and transpiration	3,000
Runoff	1,250
Withdrawal	310
Irrigation	142
Industry (principally utility cooling)	142
Municipal	26
Consumed (principally irrigation loss)	90
Returned to streams	220

Source: National Academy of Sciences—National Research Council Publ. 1000-B, 1962.

some disadvantages: Surface water sources are easily contaminated (polluted) with microorganisms that can cause waterborne diseases and with chemicals that enter from surrounding runoff and upstream discharges. Water rights can also present problems.

As mentioned, most surface water is the result of surface runoff. The amount and flow rate of this surface water are highly variable, which comes into play for two main reasons: (1) human interferences (influences) and (2) natural conditions. In some cases, surface water runs quickly off land surfaces. From a water resources standpoint, this is undesirable because quick runoff does not provide enough time for the water to infiltrate the ground and recharge groundwater aquifers. Surface water that quickly runs off land also causes erosion and flooding problems. The main advantage of rapidly running surface water is its minimal mineral content due to limited contact time. Slow surface water off land has all the opposite effects.

Drainage basins collect surface water and direct it on its gravitationally influenced path to the ocean. The drainage basin is generally characterized as an area measured in square miles, acres, or sections. Obviously, if a community is drawing water from a surface water source, the size of its drainage basin is an important consideration. Surface water runoff typically moves along the path of least resistance. Surface water within the drainage basin normally flows toward one primary watercourse (river, stream, brook, creek, etc.), unless a canal or pipeline redirects the flow. Surface water runoff from land surfaces depends on several factors (see Figure 4.1):

- *Rainfall duration*: Prolonged light rain can eventually saturate soil, leading to runoff.
- *Rainfall intensity*: With increases in intensity, the surface of the soil quickly becomes saturated. This saturated soil can hold no more water; as more rain falls and water builds up on the surface, it creates surface runoff.
- *Soil moisture*: The amount of existing moisture in the soil has a definite impact on surface runoff. Soil already wet or saturated from a previous rain causes surface runoff to occur sooner than if the soil were dry. Because frozen ground is impermeable, all snowmelt or rain becomes surface runoff.

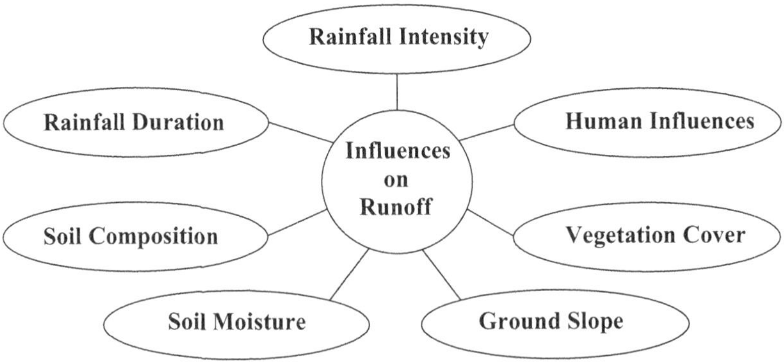

FIGURE 4.1 Influence on runoff.

- *Soil composition*: Surface soil makeup directly impacts runoff levels. For example, hard rock surfaces, obviously, result in 100% runoff. Clay soils have very small void spaces that swell when wet; the void spaces close and do not allow infiltration. Coarse sand contains substantial void spaces that facilitate efficient water permeability, resulting in adequate drainage even during heavy rainfall.
- *Vegetation cover*: Groundcover limits runoff. Vegetation roots, pine needles, pinecones, leaves, and branches create a layer of decomposing organic material that is permeable and lies above the soil. This porous "organic" sheet readily allows water into the soil. Vegetation and organic waste also function as cover to protect the soil from heavy rainfall, which can compact bare soils, close off void spaces, and increase runoff. Vegetation and groundcover help retain soil moisture by improving infiltration and reducing evaporation.
- *Ground slope*: On steep slopes, over 80% of rainfall can turn into surface runoff. Gravity moves the water down the surface more quickly than it can infiltrate the soil. On flat land, water runs off more slowly, letting more rain soak into the soil.
- *Human influences*: Various human activities have a definite impact on surface water runoff. Most human activities tend to increase the rate of water flow. Canals and ditches are built for steady flow, while agriculture removes groundcover that would otherwise slow runoff. In contrast, humans usually construct dams to slow runoff.

DID YOU KNOW?

Paved streets, tarmac, paved parking lots, and buildings are impervious to water infiltration, greatly increasing the amount of stormwater runoff from precipitation events. These human-made surfaces (which work to hasten the flow of surface water) often cause flooding to occur, sometimes with devastating consequences. In poorly planned areas, even relatively light precipitation can cause local flooding. Impervious surfaces not only present flooding problems, but they also do not allow water to percolate into the soil to recharge groundwater supplies, which can be another devastating blow to a location's water supply.

Paved streets, tarmac, paved parking lots, and buildings are impervious to water infiltration, greatly increasing the amount of stormwater runoff from precipitation events. These man-made surfaces (which work to hasten the flow of surface water) often cause flooding to occur, sometimes with devastating consequences. In badly planned areas, even relatively light precipitation can cause local flooding. Impervious surfaces not only present flooding problems, they also do not allow water to percolate into the soil to recharge groundwater supplies—often delivering another devastating blow to a location's water supply.

4.4 GROUNDWATER SUPPLY

Unbeknownst to most of us, our Earth possesses an unseen ocean. This ocean, unlike the surface oceans that cover most of the globe, is freshwater: the groundwater that lies contained in aquifers beneath Earth's crust. This gigantic water source forms a reservoir that feeds all the natural fountains and springs of the Earth. But how does water travel into the aquifers that lie under the Earth?

Groundwater sources are replenished from a percentage of the average of approximately three feet of water that falls to earth each year on every square foot of land. Water falling to Earth as precipitation follows three courses. About six inches of water every three feet flows into rivers and streams before reaching the sea. Evaporation and transpiration through vegetation take up about two feet. The remaining six inches seep into the ground, entering and filling every interstice, hollow, and cavity. Although groundwater comprises only 1/6 of the total (1,680,000 miles of water), if we could spread this water over the land, it would blanket it to a depth of 1,000 feet.

Most groundwater continuously moves through aquifer pores and crevices. The water table usually mirrors the ground's contours rather than staying flat. Groundwater flows downhill along the sloping water table. The water table sometimes intersects low points of the ground, where it seeps out into springs, lakes, or streams.

Usual groundwater sources include wells and springs that are not influenced by surface water or local hydrological events.

As a potable water source, groundwater has several advantages over surface water. Unlike surface water, groundwater is not easily contaminated. Groundwater sources are usually lower in bacteriological contamination than surface waters. Groundwater quality and quantity typically stay consistent year-round. In the United States, groundwater is available in most locations.

As a potable water source, groundwater does present some disadvantages compared to surface water sources. Operating costs are usually higher because groundwater supplies must be pumped to the surface. Any contamination is often hidden from view. Removing any contaminants is very difficult. Groundwater typically exhibits elevated mineral content and, consequently, greater hardness due to prolonged contact with geological minerals. Groundwater near coasts can be affected by saltwater intrusion.

Note: The quality of its source influences groundwater quality. Changes in source waters or degraded quality of source supplies may seriously impair the quality of the groundwater supply.

4.5 WATER SUPPLY AND THE AI CONNECTION

AI in water supply and treatment allows real-time detection of potential failures before they occur. AI and its subfields provide much faster alerts than traditional water supply troubleshooting and quality measurements. Traditional lab testing for contaminants is continuously analyzed. Water distribution networks can fail at any time. For example, AI enhances water distribution networks by providing the Internet of Things sensors to detect distribution system leaks, again, in real time. AI also elevates water quality monitoring and, in turn, is revolutionizing environmental protection. This is important, of course, because drinking water is of little use and potentially dangerous to life if its quality is badly degraded. Clean water is essential for life, but judging a water's particular quality is challenging. AI quickly analyzes water data and predicts failures, improving monitoring.

AI provides more accurate mitigation and prediction of flooding. This makes one wonder if, what if, AI sensors had been installed along the Guadalupe River, Texas, before July 4, 2025, when flooding levels at more than 26 feet above normal caused more than 100 people to be killed, and still counting, many of whom were children and campers—at present (July 2025), almost 200 more people are missing?

AI can provide energy efficiency in wastewater treatment. Wastewater treatment plants using the activated sludge unit process break down organic contaminants and ensure the safety of the water for reuse or discharge into the environment by using aeration. In turn, aeration in the activated sludge process uses up to 60% of the energy used in the treatment train.[2] Although aeration is a critical process in wastewater treatment, it comes with significant operational costs and energy consumption.

AI and other advanced computational technologies enable more accurate drought prediction and water allocation. This allows policymakers, planners, and farmers to make better decisions. Accurate forecasting and the implementation of adaptive strategies are critical for mitigating disasters or shortages and enhancing survival initiatives. What this does is allow for the outperformance of long-used forecasting methods of the past with new indices in accuracy.

4.6 THE BOTTOM LINE

Our freshwater supplies are constantly renewed through the hydrologic cycle, but the balance between the normal ratio of freshwater and saltwater is not subject to our ability to change. As our population grows and we move into lands without ready freshwater supplies, we place ecological strain upon those areas and on their ability to support life. Communities established in regions lacking sufficient local water sources may face challenges during emergencies. Proper attention to our surface and groundwater sources, including remediation, pollution control, and water reclamation and reuse, can help to ease the strain. Still, technology cannot fully replace adequate local freshwater supplies, whether from surface or groundwater sources. AI is quickly being used in public utilities to help manage water supplies, and the real bottom line is that water managers need to board the AI train.

NOTES

1 Adapted from F.R. Spellman's (2008). *The Science of Water*, 2nd ed. Boca Raton, FL: CRC Press.

2 From F.R. Spellman (2020). *Handbook of Operations for Water and Wastewater Operations*, 4th ed. Boca Raton, FL: CRC Press.

REFERENCE

Peavy, H.S., et al. 1985. *Environmental Engineering*. New York: McGraw-Hill, Inc.

5 Water Pollution

This country's waterways have been transformed by **omission**. In the absence of beavers, water flows rapidly toward the sea; without prairie dogs, water remains on the surface instead of permeating into the aquifer. Similarly, the loss of bison results in fewer groundwater recharged ponds within grasslands, causing increased disturbance in riparian zones. Additionally, without alligators, the transition between aquatic and terrestrial habitats becomes less complex. Without forests, the water runs unfiltered to the waterways, and there is less deadwood in the channel, reducing stream productivity. Without floodplains and meanders, the water moves more swiftly, and silt carried in the water is more likely to be swept to sea.

The beaver, the prairie dog, the bison, and the alligator have been scarce for so long that we have forgotten how plentiful they once were. Beaver populations are controlled because they flood fields and forests, while the acreage of wetlands decreases annually. Prairie dogs are poisoned because they compete with cattle for grass, while the grasslands grow more barren year by year. Buffalo are often viewed as picturesque relics, while alligators' reptilian nature makes them less appealing. But all these animals once shaped the land in ways that improve water quality.

—Alice Outwater, Water: A Natural History, pp. 175–76, 1996

5.1 INTRODUCTION

Is water contamination severe? In answer to the first part of the question, we can say that it depends on where your water comes from. For the second part of the question, we refer you to a book (or the film based on the book) that concerns a case of toxic contamination—one you might be familiar with—*A Civil Action*, written by Jonathan Harr (1995). The book and film portray the legal repercussions connected with polluted water supplies in Woburn, Massachusetts. Two wells became polluted with industrial solvents, in all apparent likelihood causing 24 of the town's children, who lived in neighborhoods supplied by those wells, to contract leukemia and die.

Many who have read the book or seen the movie may mistakenly get the notion that Woburn, a toxic "hot spot," is a rare occurrence. Nothing could be further from the truth. Toxic "hot spots" abound. Certain areas with cancer clusters include the following:

- Woburn, where about two dozen children were stricken with leukemia over 12 years, a rate several times the national average for a community of its size.
- Storrs, Connecticut, where wells polluted by a landfill are suspected of sickening and killing residents in nearby homes.
- Bellingham, Washington, where pesticide-contaminated drinking water is thought to be linked to a sixfold increase in childhood cancers.

DOI: 10.1201/9781003655848-9

As Schlichtmann[1] points out, these are only a few examples of an underlying pathology that threatens many other communities. Meanwhile, cancer is now the primary cause of childhood death from disease.

Drinking water contamination, along with a shortage of clean, safe drinking water, is a Water Manager's worst nightmare problem—a very serious problem. In this chapter, we discuss a wide range of water contaminants, the contaminant sources, and their impact on drinking water supplies from both surface water and groundwater sources.

5.2 SOURCES OF CONTAMINANTS

Listing all contaminant sources and types, with brief descriptions, would require an entire book. To illustrate the scope of the problem, we compiled a "short list" of key sources and contaminants:

Note: Keep in mind that when we specify "water pollutants," we are, in most cases, speaking about pollutants that somehow get into the water (by whatever means) from the interactions of the other two environmental media: air and soil. A clear example is acid rain: pollutants released into the atmosphere return to earth, impacting soil and water. Consider that 69% of the anthropogenic lead and 73% of the mercury in Lake Superior reach it by atmospheric deposition (Hill, 1997).

Subsurface Percolation—including hydrocarbons, metals, nitrates, phosphates, microbes, and cleaning agents (trichloroethylene, TCE). The following lists the sources of subsurface and surface water pollution.

1. *Injection Wells*—hydrocarbons, metals, non-metals, organics, organic and inorganic acids, microorganisms, and radionuclides.
3. *Land Application*—Nitrogen, phosphorus, heavy metals, hydrocarbons, microorganisms, and radionuclides.
4. *Landfills*—organics, inorganics, microorganisms, and radionuclides.
5. *Open Dumps*—organics, inorganics, and microorganisms.
6. *Residential (Local) Disposal*—organic chemicals, metals, non-metal inorganics, inorganic acids, and microorganisms.
7. *Surface Impoundments*—organic chemicals, metals, non-metal inorganics, inorganic acids, microorganisms, and radionuclides.
8. *Waste Tailings*—arsenic, sulfuric acid, copper, selenium, molybdenum, uranium, thorium, radium, lead, manganese, vanadium.
9. *Waste Piles*—arsenic, sulfuric acid, copper, selenium, molybdenum, uranium, thorium, radium, lead, manganese, vanadium.
10. *Materials Stockpiles*—coal pile: aluminum, iron, calcium, manganese, sulfur, and traces of arsenic, cadmium, mercury, lead, zinc, uranium, and copper. Other material piles: metals/non-metals and microorganisms.
11. *Graveyards*—metals, non-metals, and microorganisms.
12. *Animal Burial*—contamination is site-specific—depending on disposal practices, surface and subsurface hydrology, proximity of the site to water sources, type and amount of disposed material, and cause of death.

13. *Above-Ground Storage Tanks*—organics, metal/non-metal inorganics, inorganic acids, microorganisms, and radionuclides.
14. *Underground Storage Tanks*—organics, metal, inorganic acids, microorganisms, and radionuclides.
15. *Containers*—organics, metal/non-metal inorganics, inorganic acids, microorganisms, and radionuclides.
16. *Open Burning and Detonating Sites*—inorganics, including heavy metals; organics, including TNT.
17. *Radioactive Disposal Sites*—radioactive cesium, plutonium, strontium, cobalt, radium, thorium, and uranium.
18. *Pipelines*—organics, metals, inorganic acids, and microorganisms.
19. *Material Transport and Transfer Operations*—organics, metals, inorganic acids, microorganisms, and radionuclides.
20. *Irrigation Practices*—fertilizers, pesticides, naturally occurring contamination, and sediments.
21. *Pesticide Applications*—1,200–1,400 active ingredients. Contamination has already been detected: alachlor, aldicarb, atrazine, bromacil, carbofuran, cyanazine, DBCP, DCPA, 1,2-dichloropropane, dyfonate, EDB, metolachlor, metribyzen, oxalyl, siazine, and 1,2,3-trichloropropane. The extent of groundwater contamination cannot be determined with current data.
22. *Animal Feeding Operations*—Nitrogen, bacteria, viruses, and phosphates.
23. *De-Icing Salt Applications*—chromate, phosphate, ferric ferrocyanide, Na-ferrocyan, chlorine.
24. *Urban Runoff*—suspended solids and toxic substances, especially heavy metals and hydrocarbons, bacteria, nutrients, and petroleum residues.
25. *Percolation of Atmospheric Pollutants*—sulfur and nitrogen compounds, asbestos, and heavy metals.
26. *Mining and Mine Drainage*—
    - Coal: acids, toxic inorganics (heavy metals), and nutrients. Phosphate: radium, uranium, and fluorides.
    - Metallic Ores: sulfuric acid, lead, cadmium, arsenic, sulfur, cyan.
27. *Production Wells*—
    - Oil—1.2 million abandoned production wells
    - Irrigation—farms
    - All—potential to contaminate installation, operation, and plugging techniques.
28. *Construction Excavation*—pesticides, diesel fuel, oil, salt, and a variety of others.
29. *PPCPs*—any product used by an individual for personal health or cosmetic reasons or used by agribusiness to enhance the growth or health of livestock.

Note: Before discussing specific water pollutants, we must examine several terms important to the understanding of water pollution. One of these is a *point source*. A *point source* is identified as any single identifiable source of pollution from which pollutants are discharged, for example, a pipe, ditch, ship, or factory smokestack. For example, the outlet pipes of an industrial facility or a municipal wastewater treatment plant are point sources. In contrast, *nonpoint sources* are widely dispersed sources and are a significant cause of stream pollution. An example of a nonpoint source of pollution is rainwater carrying topsoil and chemical contaminants into a river or stream. Some of the significant sources of nonpoint pollution include water runoff from farming, urban areas, forestry, and construction activities. The word *runoff* signals a nonpoint source that originated on land. Runoff may carry a variety of toxic substances and nutrients, as well as bacteria and viruses. Nonpoint sources now comprise the largest source of water pollution, contributing approximately 65% of the contamination in quality-impaired streams and lakes.

SIDEBAR 5.1 SICK WATER

The United Nations introduced the term *sick water* in 2010 to highlight the urgent need to address global water pollution. The gist of the UN's report pointed out that transforming waste from a significant health and environmental hazard into a clean, safe, and economically attractive resource is emerging as a key challenge in the twenty-first century. As practitioners of environmental health, we certainly support the UN's view on this critical topic.

When we discuss sick water, in the context of this text and many others we have authored on the topic, we go a few steps further than the UN in describing the real essence and tragic implications of potable water that makes people or animals sick or worse. "Sick" water refers to polluted wastewater, containing fertilizer runoff, sewage, and various animal, industrial, and agricultural wastes. In addition to the listed wastes of concern, other wastes that are beginning to garner widespread attention are certainly worth our consideration in research on the potential problems related to these so-called other wastes.

What are these other wastes? Any waste or product that we dispose of in our waters, that we flush down the toilet, pour down the sink or bathtub drain, or pour down the drain of a worksite deep sink. For example, many substances we send to wastewater treatment plants or septic tanks are pollutants, even though we may not view them as waste.

Each morning, a family of four, two adults and two teenagers, wakes up and prepares for the day that lies ahead. Fortunately, this family has three upstairs bathrooms to accommodate everyone's needs, and each day the family's natural wastes, soap suds, cosmetics, hair treatments, vitamins, sunscreen, fragrances, and prescribed medications end up down the various drains. Cat and dog waste left overnight is regularly collected and flushed. Here is a brief list of what this family of four used or discarded during their morning routine:

- Toilet-flushed animal wastes
- Prescription and over-the-counter therapeutic drugs
- Veterinary drugs
- Fragrances

- Soap
- Shampoo, conditioner, and other hair treatment products
- Body lotion, deodorant, and body powder
- Cosmetics
- Sunscreen products
- Diagnostic agents
- Nutraceuticals (e.g., vitamins, medical foods, and functional foods)

Even though these bioactive substances have been around for decades, today we group all of them (with the exception being animal wastes) under the title of *pharmaceuticals and personal care products*, or PPCPs.

Let us go back to a family of four. After having applied, used, or ingested the various substances mentioned earlier, they also add at least traces of these products (PPCPs) to the environment through excretion (the elimination of waste material from the body) and bathing, as well as through disposal of any unwanted medications to sewers and trash, but other sources of PPCPs should also be recognized. For example, residues from pharmaceutical manufacturing; residues from hospitals, clinics, doctors' or veterinary offices, or urgent care facilities; illicit drug disposal [i.e., police knock on the door and the frightened user flushes the illicit drugs down the toilet (along with $100 bills, weapons, dealers' phone numbers, etc.) and into the wastewater stream]. Veterinary drug use, especially antibiotics and steroids, and agribusiness are all contributors of PPCPs in the environment.

In our examination of the personal deposit of PPCPs for the environment and the local wastewater supply, let us return to that family of four. After having applied or ingested the various substances mentioned earlier, the four individuals involved unwittingly add traces (or more than traces) of these products, PPCPs, to the environment through excretion (the elimination of waste material from the body) and when bathing, and then possibly through disposal of any unwanted medications to sewers and trash. How many of us have found old medical prescriptions in the family medicine cabinet and decided they were no longer needed? Many people have disposed of unused medications by flushing them down the toilet. Many of these medications (e.g., antibiotics) are not typically found in the environment. Earlier, we noted that wastewater consists of fertilizer runoff, sewage, and various animal, industrial, agricultural, and other wastes. When we factor in PPCPs to this cocktail, we can state analogously that we are simply adding mix to the mix.

This mixed-waste cocktail raises many questions: Does the disposal of antibiotics or other medications into the local wastewater treatment system pose risks to anyone or anything else downstream? When we drink locally treated tap water, are we also ingesting flushed-down-the-toilet or rinsed-down-the-drain antibiotics, other medications, illicit drugs, animal excretions, cosmetics, vitamins, personal or household cleaning products, sunscreen products, diagnostic agents, crankcase oil, grease, fats, and veterinary drugs and hormones?

How about the family of four? Suppose we were to show them the list of pharmaceuticals and personal care products they used and disposed of during their morning routine. In that case, they might be impressed, surprised, or give the list and the whole account of their morning routine the old ho-hum, followed by that tell-all yawn of total boredom. However, if after showing them the list of products they used and disposed of in the waste stream that left their home, we asked them if they would be willing to drink from that same waste stream, it is likely that they would no longer be bored—one thing is sure: "ho hum" would not be replaced by "yum yum"—but instead their attention would be piqued and they might be experiencing that queasy feeling in their stomachs—commonly called the Yuck Factor influence—and we could expect their response to drinking the waste they produced to be along the lines of "No way, Paco, Jose and Maria . . . no way."

Ok, first, the so-called yuck factor of drinking human-caused waste streams is overrated. The fact is, we have been drinking from these waste streams from time immemorial; that is, ancient beyond time or memory. Let us clarify this statement. For instance, Hercules—often seen as the first environmental engineer—was tasked by Eurystheus with his fifth labor: cleaning King Augeas' stables. Hercules, faced with a mountain of horse and cattle waste piled high in the stable area, had to devise some method to dispose of the waste; he did. He diverted a couple of river streams to the inside of the stable area so that all the animal waste could be deposited into the river streams: Out of sight, out of mind. The waste flowed downstream. Hercules understood the principal point in pollution control technology that is pertinent to this very day and this discussion; that is, *dilution is the solution to pollution*.

In addition to Hercules' sensible waste disposal methods for his era, another important aspect is the practice of de facto water recycling. When our four family members and other people say they would never drink toilet water, they have no idea what they are saying. As pointed out in my textbook, *The Science of Water*, 4th edition, the fact is we drink recycled wastewater every day via the de facto water cycle, which turns the yuck factor into the "aww" factor. Wastewater treatment plants throughout the industrialized world treat wastewater, or used water, and then outfall (discharge) the treated water to the major rivers (or other water bodies) in their respective regions. Many of those regions' rivers are sources of local drinking water supplies. Even local groundwater supplies are routinely infiltrated with surface water inputs, which, again, are commonly supplied by treated wastewater (and sometimes infiltrated by raw sewage that is accidentally spilled). The water cycle helps address water pollution, but the fate of PPCPs remains uncertain.

The effects of PPCPs in drinking water remain uncertain. We do not know what we do not know about the fate of PPCPs or their impact on the environment once they enter our wastewater treatment systems, the water cycle, and eventually our drinking water supply systems. Some PPCPs degrade quickly in the body or environment, but particular wastes remain problematic for extended periods.

There is another issue related to PPCPs and other contaminants in our water systems. It is an issue many of us never think about. All living organisms, including wildlife, birds, seabirds, and aquatic life, rely on water. The fact is that water pollution causes immense damage to marine, avian, and land animals. In addition to PPCPs, other contaminants of our water bodies include:

- Raw sewage running into lakes, rivers, or streams
- Industrial waste spills contaminating groundwater
- Radiation spills or nuclear accidents
- Illegal dumping of substances or items within bodies of water
- Biological contamination, such as bacterial growth
- Farm runoff into nearby bodies of water

It is also important to point out that drinking water sources and animal usage are a double-edged sword. First, animals have a right to drink from unpolluted and safe drinking water sources. However, secondly, wildlife, farm animals, and other animals are also contributors to drinking water pollution. This is why, when in the woods or at campsites or other remote locations, it is always wise to filter river, stream, or lake water with a 1-μ filter before drinking.

Returning to PPCPs, these substances represent a contemporary challenge in the field of pollution management. Although Hercules was able to dispose of animal waste into a running water system, where eventually the water's self-purification process cleaned the stream, he did not have to deal with today's pharmaceuticals and the hormones that are given to many types of livestock to enhance health and growth.

Pharmaceuticals have been detected in national water bodies. Further research suggests that certain drugs may cause ecological harm. The EPA and other research agencies are committed to investigating this topic and developing strategies to help protect the health of both the environment and the public. To date, scientists have found no hard evidence of adverse human health effects from PPCPs in the environment. Others might argue that even if PPCPs were present today or in ancient (and mythical) times, the amount in local water systems and other supply areas would represent only a small fraction (ppt—parts per trillion, 10^{-12}) of the total volume of water. Critics would be quick to point out that when we are speaking of parts per trillion (ppt), we are speaking of a proportion equivalent to 1/20 of a drop of water diluted into an Olympic-size swimming pool. One student in our environmental health classes stated that he did not think the water should be termed "sick water," as it was evident to him that if the water contained so many medications, how could it be sick? Instead, it might be termed "well water," with the potential to make anyone who drinks it well, cured, no longer sick, etc.

We point out that the term *sick water* can be applied not only to PPCP-contaminated water but also to any filthy, dirty, contaminated, polluted, pathogen-filled drinking water sources. The fact is that dirty or sick water means that, worldwide, more people now die from contaminated and polluted water than from all forms of violence, including wars (UN, 2010[2]). The United Nations also points out that dirty or sick water is also a key factor in the rise of deoxygenated dead zones that have been emerging in seas and oceans across the globe.

DID YOU KNOW?

PPCPs are the primary ingredients in the sick conglomeration of products we use each day. Why the concern, you might ask. There is not just one concern, but instead there are many:

- Large quantities of PPCPs can enter the environment after use by individuals or domestic animals.
- Wastewater systems are not equipped for PPCP removal. Currently, there are no municipal wastewater treatment plants that are engineered specifically for PPCP removal or for other unregulated contaminants. Effective removal of PPCPS from treatment plants varies based on the type of chemical and on the individual wastewater treatment facilities.
- The risks are uncertain. The risks posed to aquatic organisms and to humans are unknown, largely because the concentrations are so low.
- The number of PPCPs is growing. In addition to antibiotics and steroids, over 100 individual PPCPs have been identified (as of 2007) in environmental samples and drinking water (EPA, 2011).

5.3 RADIONUCLIDES

When radioactive elements decay, they emit alpha, beta, or gamma radiation caused by the transformation of the nuclei to lower energy states. Radioactivity in drinking water comes from either natural or artificial radionuclides. These radioactive substances in water are of two types: radioactive minerals and radioactive gases. The USEPA reports that some 50 million Americans face increased cancer risk because of radioactive contamination of their drinking water.

Because of their occurrence in drinking water and their effects on human health, the natural radionuclides of chief concern are radium-226, radium-228, radon-222, and uranium. These naturally occurring radioactive minerals often come from regions with current or former mining activity. Mining activities expose rock strata, most of which contain some amount of radioactive ore. Uranium mining, for example, produces runoff. Another source of natural radioactive contamination occurs when underground streams flow through various rock beds and geologic formations containing radioactive materials. Other naturally occurring sources where radioactive minerals may enter water supplies are smelters and coal-fired electrical generating plants. Sources of human-caused radioactive minerals in water are nuclear power plants, nuclear weapons facilities, radioactive materials disposal sites, and mooring sites for nuclear-powered ships. Hospitals also contribute to radioactive pollution when they dump low-level radioactive wastes into sewers. Some of these radioactive wastes eventually find their way into the water supply systems.

Though uranium and radium in water can pose health risks, radon is an even greater danger. *Radon* is a colorless, odorless gas created by (a byproduct of) the natural decay of minerals in the soil. Usually present in all water in minute amounts, radon is especially concentrated in water that has passed through rock strata of granite, uranium, or shale.

Radon enters homes from the soil beneath the house, through cracks in the foundation, crawl spaces, and unfinished basements, as well as in tainted water. It is considered the second leading cause of lung cancer in the United States, with about 20,000 cases each year, following cigarette smoking. Contrary to widespread belief, radon is not a threat from surface water (lake, river, or above-ground reservoir) because radon dissipates rapidly when water is exposed to air. Even if the water source is groundwater, radon is still not a threat if the water is exposed to air (aerated) or if it is processed through an open tank during treatment. Though studies show that where high concentrations of radon occur within the air in a house, most of the radon comes through the foundation and from the water. Radon present in tap water, showers, baths, and during cooking with hot water can result in increased concentrations of radon in the air.

Note, however, that radon is a threat from groundwater taken directly from an underground source—either a private well or a public water supply—and whose treatment of the water does not include exposure to air. Because radon in water evaporates quickly into the air, the primary danger is from inhaling it in a house, not from drinking it.

5.4 THE CHEMICAL COCKTAIL

In previous pages, we referred to a glass of water filled to the brim from the household tap. If we were to examine a full glass of water, several possibilities regarding its contents might arise. Cloudy or discolored contents may indicate the water is unsafe to drink. The contents appear normal but smell strongly of chlorine. Most often, when we take the time to look at water drawn from the tap, it simply looks like water, and we drink it or use it to cook with, or whatever.

The fact is, typically, a glass of treated water is a chemical cocktail (Kay, 1996). Water utilities treat raw water with chemicals to ensure public health, resulting in drinking water that contains both these chemicals and their by-products. For example, the water treatment facility typically adds chlorine for disinfection, but chlorine can produce contaminants. Another concoction is formed when ammonia is added to disinfect. Alum and polymers are added to the water to settle out various pollutants. To safeguard the water distribution system and its appurtenances from pipe corrosion and to manage water hardness, the water treatment facility introduces caustic soda, ferric chloride, and lime. These agents subsequently contribute to elevated levels of aluminum, sulfates, and salts in the treated water. When we look at a glass of clear water, we see a mixture of chemicals dissolved in it—a chemical cocktail.

The most common chemical additives used in water treatment are chlorine, fluorides, and flocculants. Because we have already discussed fluorides, we focus our discussion in the following sections on the by-products of chlorine and flocculant additives.

5.4.1 By-Products of Chlorine

To lessen the potential impact of the water cocktail, the biggest challenge today is to ensure that the old standby—chlorine—does not create as many new contaminants as it destroys. Currently, weighing the balance of the argument, arguing against chlorine and the chlorination process is difficult. Chlorine has been used in the United States since 1908 to eliminate microorganisms that cause cholera, typhoid fever, and other waterborne diseases. In the 1970s, scientists found that chlorine itself does not cause cancer in lab animals but can produce carcinogenic by-products during water treatment. The by-products of chlorine—organic hydrocarbons called *trihalomethanes* (usually discussed as total trihalomethanes or THMs)—present the most significant health concern.

The USEPA classifies three of these THM by-products—chloroform, bromoform, and bromodichloromethane—as probable human carcinogens. The fourth, dibromochloromethane, is classified as a possible human carcinogen.

The USEPA set the first THM limits in 1979. Most water companies met these standards initially, but the standards were tightened after the 1996 SDWA Amendments. The USEPA is continuously studying the need to regulate other cancer-causing contaminants, including haloacetic acids (HAAs), also produced by chlorination.

Most people concerned with protecting public health applaud the USEPA's efforts in regulating water additives and disinfection by-products. However, some of those in the water treatment and supply business express concern. A common problem often heard from water utilities having a tough time balancing the use of chlorine without going over the regulated limits revolves around the necessity of meeting regulatory requirements by lowering chlorine amounts to meet by-product standards and, at the same time, ensuring that all the pathogenic microorganisms are killed off. Some argue that although there is no confirmed evidence linking disinfection by-products to cancer in humans, historical cases indicate that the absence of water chlorination can lead to an increase in waterborne illnesses and fatalities.

Because chlorine and chlorination are now prompting regulatory pressure and compliance with new, demanding regulations, many water treatment facilities are looking for options. Choosing an alternative disinfection chemical process is feeding a growing enterprise. One alternative that is currently being given widespread consideration in the United States is ozonation, which uses ozone gas to kill microorganisms. Ozonation is widely used in Europe and avoids THMs. However, the USEPA does not currently recommend fully replacing chlorine or chlorination systems with ozone. The USEPA points out that ozone also has problems: It does not produce a residual disinfectant in the water distribution system, it is much more expensive, and in salty water, it can make another carcinogen, bromate. Later, we will discuss disinfection alternatives in greater detail.

Currently, drinking water professionals are working to optimize water treatment processes in practical settings. What it all boils down to is a delicate balancing act. The drinking water professional does not want to cut back disinfection—if anything, they would prefer to strengthen it.

The compound question is, "How do we bring into parity the microbial risks versus the chemical risks? How do you reduce them both to an acceptable level?" The answer? No one is quite sure how to do this. The problem revolves around the enigma associated with a "we don't know what we don't know" scenario.

The disinfection by-products problem stems from the fact that most U.S. water systems produce unwanted by-products when chlorine reacts with decayed organics, such as vegetation and other carbon-containing materials in water. Communities that take drinking water from lakes and rivers have a tougher time keeping the chlorine by-products out of the tap than those that use clean groundwater.

In some communities, when there is a lot of debris in the reservoir, the water utility switches to alternate sources—wells, for example. In other facilities, chlorine is combined with ammonia in a disinfection method called *chloramination*. This method is not as potent as pure chlorination, but it stops the production of unwanted THMs.

In regions where precipitation carries leaves, trees, and grasses into local water sources such as lakes or rivers, elevated summer temperatures can promote algal blooms. This increase in organic matter may subsequently lead to higher production of THMs. Spring runoff in many communities exacerbates the problem. With increased runoff comes agricultural waste, pesticides, and excess nutrients entering the water, which must be addressed.

Nature's conditions in summer diminish some precursors for THMs—the bromides in salty water.

The irony is that under such conditions, nothing unusual is visible in the drinking water. However, cloudy water from silt (dissolved organics from decayed plants) is enough to create THMs.

With the advent of the new century, most cities will strain out the organics from their water supplies before chlorinating to prevent the formation of THMs and HAAs.

In other communities, the move is already on to switch from chlorine to ozone and other disinfectant methods. The National Resources Defense Council predicts that most U.S. systems will adopt ozone to kill microbes like cryptosporidium within 15–20 years, aligning with European practices. When this method is employed, the finishing touch is usually accomplished by filtering through granulated activated carbon, which increases the cost slightly (estimated at $100 or more per year per hookup) that customers must pay.

5.4.2 Trihalomethane Regulations[3]

In ensuring safe drinking water supplies in the United States, in December 1998, President Clinton announced these rules.

- **Stage 1 Disinfectants and Disinfection By-product Rule: Federal Register Notice (HTML)**
- **Interim Enhanced Surface Water Treatment Rule: Federal Register Notice (HTML)**

A major challenge for drinking water practitioners is how to balance the risks from microbial pathogens and disinfection by-products. Protecting from these microbial pathogens while simultaneously decreasing health risks to the population from disinfection by-products (DBPs) is essential. The Safe Drinking Water Act (SDWA) Amendments, signed by the President in August 1996, required the USEPA to develop rules to achieve these goals. The new Stage 1 Disinfectant and Disinfection By-product Rule and Interim Enhanced Surface Water Treatment Rule are the first in a set of rules under the Amendments.

These new rules are a product of six years of collaboration between the water industry, environmental and public health groups, and local, state, and federal governments. The schedule for the Microbial-Disinfectant and Disinfection By-products Rules is as follows:

Schedule of M-DBP Rules

November 1998—Final Rule

Interim Enhanced Surface Water Treatment Rule and Stage 1 Disinfection By-product Rule

August 2000—Final Rule Filter Backwash Recycling Rule

November 2000—Final Rule Long Term 1 Enhanced Surface Water Treatment Rule and Groundwater Rule

May 2002—Final Rule—Stage 2 Disinfection By-product Rule and Long Term 2 Enhanced Surface Water Treatment Rule

Stage 2: Community Water System serving at least 100,000

October 1, 2006—Submit monitoring plans, system-specific study plans for 40/30 certification

Stage 2: Community Water System serving 50,000–99,000

March 31, 2009—Complete an initial distribution evaluation

Note: Other pertinent Stage 2 DBPR rules are contained in applicable EPA regulations www.epa.gov/safetwater/lt2/pdfs/fact_lt2.pdf.

Most Americans drink tap water that meets all existing health standards all the time. These new rules will further strengthen existing drinking water standards and thus increase protection for many water systems.

The USEPA's Science Advisory Board concluded in 1990 that exposure to microbial contaminants such as bacteria, viruses, and protozoa (e.g., *Giardia lamblia* and *Cryptosporidium*) remained a significant health challenge for drinking water suppliers. Acute health effects from exposure to microbial pathogens are documented, and associated illnesses can range from mild to moderate cases lasting only a few days to more severe infections that can last several weeks and may result in death for those with weakened immune systems.

While disinfectants are effective in controlling many microorganisms, they react with natural organic and inorganic matter in source water and distribution systems to form potential DBPs. Many of these DBPs have been shown to cause cancer and reproductive and developmental effects in laboratory animals. More than 200 million people consume water that has been disinfected. Due to the large population exposed, health risks associated with DBPs, even if small, need to be taken seriously. Regulations include the following:

- **Microbial Contaminants**: The Surface Water Treatment Rule, promulgated in 1989, applies to all public water systems using surface water

sources or groundwater sources under the direct influence of surface water. It establishes maximum contaminant level goals (MCLGs) for viruses, bacteria, and *Giardia lamblia*. It also includes treatment technique requirements for filtered and unfiltered systems specifically designed to protect against the adverse health effects of exposure to these microbial pathogens. The Total Coliform Rule, revised in 1989, applies to all PWSs and establishes a maximum contaminant level (MCL) for total coliforms.
- **Disinfection By-products**: In 1979, the USEPA set an interim MCL for total THMs of 0.10 mg/l as an annual average. This applies to any community water system serving at least 10,000 people that adds a disinfectant to the drinking water during any part of the treatment process. In December 1998, the EPA established the Stage 1 Disinfectants/Disinfection By-products Rule, which required public water systems to use treatment measures to reduce the formation of disinfection by-products and to meet the specific standards such as the following:

Disinfection By-products	Treatment Measures
Total trihalomethanes	80 ppb
Haloacetic acids	60 ppb
Bromate	10 ppb
Chlorite	1.0 ppm

5.4.3 Information Collection Rule

To support the M-DBP rulemaking process, the **Information Collection Rule** establishes monitoring and data reporting requirements for large public water systems (PWSs) serving at least 100,000 people. This rule is intended to provide the USEPA with information on the occurrence in drinking water of microbial pathogens and DBPs. The USEPA is collecting engineering data on how PWSs currently control such contaminants as part of the Information Collection Rule.

5.4.4 Interim Enhanced Surface Water Treatment Rule

The USEPA finalized the Interim Enhanced Surface Water Treatment Rule and Stage 1 Disinfectants and Disinfection By-products Rule in November 1998, as required by the 1996 Amendments to the Safe Drinking Water Act, Section 1412(b)(2)(C). The final rules resulted from formal regulatory negotiations with a wide range of stakeholders that took place in 1992–1993 and 1997.

The Interim Enhanced Surface Water Treatment Rule applies to systems using surface water or groundwater under the direct influence of surface water that serve 10,000 or more persons. The rule also includes provisions for states to conduct sanitary surveys for surface water systems regardless of system size. The rule was built upon the treatment technique requirements of the Surface Water Treatment Rule, with the following key additions and modifications:

- Maximum contaminant level goal (MCLG) of zero for *Cryptosporidium*
- 2-log *Cryptosporidium* removal requirements for systems that filter
- Strengthened combined filter effluent turbidity performance standards
- Individual filter turbidity monitoring provisions
- Disinfection profiling and benchmarking provisions
- Systems using groundwater under the direct influence of surface water are now subject to the new rules dealing with *Cryptosporidium*
- Inclusion of *Cryptosporidium* in the watershed control requirements for unfiltered public water systems
- Requirements for covers on new finished water reservoirs
- Sanitary surveys, conducted by states, for all surface water systems regardless of size

The Interim Enhanced Surface Water Treatment Rule, with tightened turbidity performance criteria and required individual filter monitoring, was designed to optimize treatment reliability and to enhance physical removal efficiencies to minimize the *Cryptosporidium* levels in finished water. The rule also includes disinfection benchmark provisions to assure continued levels of microbial protection while facilities take the necessary steps to comply with new DBP standards.

5.4.5 Stage 1 Disinfectants and Disinfection By-Products Rule

The final Stage 1 Disinfectants and Disinfection By-products Rule applies to community water systems and non-transient non-community systems (including those serving fewer than 10,000 people) that add a disinfectant to the drinking water during any part of the treatment process.

The final Stage 1 Disinfectants and Disinfection By-products Rule includes the following key provisions:

- Maximum residual disinfectant level goals for chlorine (4 mg/l), chloramines (4 mg/l), and chlorine dioxide (0.8 mg/l)
- Maximum contaminant level goals (MCLGs) for four THMs [chloroform (zero), bromodichloromethane (zero), dibromochloromethane (0.06 mg/l), bromoform (zero), trichloroacetic acid (0.3 mg/l), bromate (zero), and chlorite (0.8 mg/l)]
- MRLDs for three disinfectants [chlorine (4.0 mg/l), chloramines (4.0 mg/l), and chlorine dioxide (0.8 mg/l)]
- MCLs for total THMs—a sum of the four listed above (0.080 mg/l), haloacetic acids (HAA5) (0.060 mg/l)—a sum of the two listed above plus monochloroacetic acid and mono- and dibromoacetic acids, and two inorganic disinfection by-products [chlorite (1.0 mg/l)] and bromate (0.010 mg/l)
- A treatment technique for the removal of DBP precursor material

The terms MRDLG and MRDL (not included in the SDWA) were created during the negotiations to distinguish disinfectants (because of their beneficial use) from contaminants. The final rule provides monitoring, reporting, and public notification requirements for these compounds. This final rule also describes the best available technology upon which the MRDLs and MCLs are based.

5.4.6 Long Term 1 Enhanced Surface Water Treatment Rule

While the Stage 1 Disinfectants and Disinfection By-products Rule applied to systems of all sizes, the Interim Enhanced Surface Water Treatment Rule only applied to systems serving 10,000 or more people. The Long Term 1 Enhanced Surface Water Treatment Rule strengthens microbial controls for small systems (i.e., those systems that serve fewer than 10,000 people). The rule also prevented a significant increase in microbial risk when small systems implement the Stage 1 Disinfectants and Disinfection By-products Rule.

The USEPA believes that the rule will track the approaches in the Interim Enhanced Surface Treatment Rule for improved turbidity control, including individual filter monitoring and reporting. The rule also addresses disinfection profiling and benchmarking. The Agency is considering what modifications of some large system requirements may be appropriate for small systems.

5.4.7 Long Term 2 Enhanced Surface Water Treatment Rule

The SWDA (as amended in 1996) requires the USEPA to finalize a Stage 2 Disinfectants and Disinfection By-products Rule by May 2002. Although the 1996 Amendments do not require the USEPA to finalize a Long Term 2 Enhanced Surface Water Treatment Rule along with the Stage 2 Disinfectants and Disinfection By-products Rule, the USEPA believes that finalizing these rules together to ensure a proper balance between microbial and DBP risks is important.

The rules intend to provide additional public health protection (if needed) from DBPs and microbial pathogens.

5.4.8 Groundwater Rule

The USEPA published the Ground Water Rule in the Federal Register on November 8, 2006. The purpose of the rule is to provide for increased protection against microbial pathogens in public water systems that use groundwater sources. EPA is particularly concerned about groundwater systems that are susceptible to fecal contamination, since disease-causing pathogens may be found in fecal contamination. The Groundwater Rule specifies the appropriate use of disinfection and, equally importantly, addresses other components of groundwater systems to ensure public health protection. More than 158,000 public or community systems serve almost 89 million people through groundwater systems. Ninety-nine percent (157,000) of groundwater systems serve fewer than 10,000 people. However, systems serving more than 10,000 people serve 55%—more than 60 million—of all people who get their drinking water from public groundwater systems.

5.4.9 Filter Backwash Recycling Rule (FBRR)

The 1996 SDWA Amendments required that the USEPA set a standard on recycling filter backwash within the treatment process of public water systems by August 2000. The actual Filter Backwash Recycling Rule (FBRR) 66 FR 31086, Vol. 66, No. 111, was implemented June 8, 2001. The objective of the Rule is to enhance public health protection by evaluating and, where necessary, modifying recycling practices to achieve more effective control of contaminants, with particular emphasis on microbial contaminants. The FBRR typically requires recycling systems to return designated recycled flows through all processes of the existing filtration system or another state-approved location. The Rule applies to public water systems that use surface water or groundwater under the direct influence of surface water, practice conventional or direct filtration, and recycle spent filter backwash, thickener supernatant, or liquids from dewatering processes.

Note: This section provides a summary of federal drinking water requirements; to ensure full compliance, consult the federal regulations at 40 CFR 141 and any approved state requirements.

The USEPA encourages public input into regulation development. Public meetings and opportunities for public comment on M-DBP rules are announced in the Federal Register. EPA's Office of Groundwater and Drinking Water also provides this information for the M-DBP rule and other programs in its online Calendar of Events (www.epa.gov).

In addition to chlorine and sometimes fluoride, water treatment plants often add several other chemicals, including flocculants, to improve the efficiency of the treatment process—and they all add to the cocktail mix. *Flocculants* are chemical substances added to water to make particles clump together, which improves the effectiveness of filtration. Some of the most common flocculants are polyelectrolytes (polymers), chemicals with constituents that cause cancer and birth defects, and are banned for use by several countries. Although the USEPA classifies them as "probable human carcinogens," it still allows their continued use. Acrylamide and epichlorohydrin are two flocculants used in the United States that are known to be associated with probable cancer risk (Lewis, 1996).

5.5 GROUNDWATER CONTAMINATION

Note that groundwater under the direct influence of surface water comes under the same monitoring regulations as does surface water (i.e., any water exposed to the air and affected by surface runoff). The legal definition of *groundwater under the direct influence of surface water* is any water beneath the surface of the ground with (1) significant occurrence of insects or microorganisms, algae, or large-diameter pathogens such as *Giardia lamblia*, or (2) significant and relatively rapid shifts in water characteristics such as turbidity, temperature, conductivity, or pH, which closely correlate to climatological or surface water conditions. Individual sources must determine direct influence by criteria established by the state. The state determines individual sources based on criteria established by the state. This determination may be based on site-specific measurements of water quality and/or documentation of well construction characteristics and geology, along with field evaluation.

Most U.S. groundwater is of high quality and provides essential amounts. The full magnitude of groundwater contamination in the United States is, however, not fully documented, and federal, state, and local efforts continue to assess and address the problems (Rail, 1985).

Groundwater supplies about 25% of the fresh water used for all purposes in the United States, including irrigation, industrial uses, and drinking water (about 50% of the U.S. population relies on groundwater for drinking water). John Chilton (1998b) points out that the groundwater aquifers beneath or close to Mexico City provide the area with more than 3.2 billion liters per day. But, as groundwater pumping increases to meet water demand, it can exceed the aquifers' rates of replenishment, and in many urban aquifers, water levels show long-term decline. With excessive extraction comes a variety of other undesirable effects, including:

- increased pumping costs
- changes in hydraulic pressure and underground flow directions (in coastal areas, this results in seawater intrusion)
- saline water drawn up from deeper geological formations
- poor-quality water from polluted shallow aquifers leaking downwards.

A serious deterioration in their quality often compounds severe depletion of groundwater resources. Groundwater contamination should be a key concern for water professionals providing potable supplies to communities.

Despite our firm reliance on groundwater, it has for many years been one of the most neglected natural resources. Why? Good question. Groundwater has been ignored because it is less visible than other environmental resources—rivers or lakes, for example. People do not worry about or consider what they cannot see. However, recent publicity about events concerning groundwater contamination is making the public more aware of the problem, and the regulators have also taken notice.

Are natural contaminants a threat to human health—harbingers of serious groundwater pollution events? No, not really. The main problem concerning severe groundwater pollution has been human activities. When we (all of us) improperly dispose of waste or spill hazardous substances onto/into the ground, we threaten groundwater, and in turn, the threat passes on to public health.

Let us take a closer look at a few sources of groundwater contamination.

5.5.1 Sources of Groundwater Contamination

Several sources of groundwater contamination present cause for concern for the drinking water practitioner. Why? Consider the importance of groundwater. People depend upon groundwater in every state, and its usage accounts for approximately one-fourth of all water used. This consumption includes about 35% of the water withdrawn for municipal water supplies.

5.5.1.1 Underground Storage Tanks (USTs)

If we were to look at a map of the United States marked with the exact location of every underground storage tank (UST) (we wish such a map existed!), except for isolated areas, most of us would be surprised by the substantial number of tanks buried underground. Given the substantial number of buried tanks, structural failures from various causes over the years are unsurprising. Subsequent leaking has become a massive source of contamination that affects the quality of local groundwaters.

Note: A UST is any tank, including any underground piping connected to the tank, that has at least 10% of its volume below ground (USEPA, 1987).

Petroleum and its products often leak from USTs more frequently than commonly perceived. This widespread problem has been a significant concern and priority in the United States for well over a decade. In 1987, the USEPA promulgated regulations for many of the nation's USTs, and much progress has been made in mitigating this problem to date.

When a UST leak or past leak is discovered, the contaminants released to the soil and thus to groundwater, for the average person, would seem relatively straightforward to identify as fuel oil, diesel, and gasoline. However, even though it is true that these are the most common contaminants released from leaking USTs, others also present problems. For example, in the following section, we discuss one such contamination, a by-product of gasoline—one that is not well known—to help illustrate the magnitude of leaking USTs.

5.5.1.2 MtBE

In December 1997, the USEPA issued a drinking water advisory titled *Consumer Acceptability Advice and Health Effects Analysis on Methyl Tertiary-Butyl Ether (MtBE)*. The purpose of the advisory was to guide communities exposed to drinking water contaminated with MtBE.

Note: A USEPA advisory is usually initiated to provide information and guidance to individuals or agencies concerned with potential risk from drinking water contaminants for which no national regulations currently exist. Advisories serve as guidance, not enforceable standards. They are not legally enforceable and are subject to revision as additional information becomes available. The USEPA's Health Advisory program is recognized in the Safe Drinking Water Act Amendments of 1996, which state in section 102(b)(1)(F):

> The Administrator may publish health advisories (which are not regulations) or take other appropriate actions for contaminants not subject to any national primary drinking water regulation.

As its title indicates, this advisory includes consumer acceptability advice as "appropriate" under this statutory provision, as well as a health effects analysis.

MtBE is a volatile organic chemical. Since the late 1970s, MtBE has been used as an octane enhancer in gasoline. Because it promotes more complete burning of gasoline (thereby reducing carbon monoxide and ozone levels), it is commonly used as a gasoline additive in localities that do not meet the National Ambient Air Quality Standards.

In the Clean Air Act of 1990, Congress mandated the use of reformulated gasoline in areas of the country with the worst ozone or smog problems. RFG is required to comply with technical specifications established by the Act, such as maintaining a designated oxygen content. Ethanol and MtBE are the main oxygenates used to fulfill oxygen content requirements. MtBE is used in about 84% of RFG supplies. Currently, 32 areas in a total of 18 states are participating in the RFG program, and RFG accounts for about 30% of gasoline nationwide.

Studies identify significant air quality and public health benefits that directly result from the use of fuels oxygenated with MtBE, ethanol, or other chemicals. The refiners' 1995/1996 fuel data submitted to the USEPA indicate that the national emissions benefits exceeded those required. According to the 1996 Air Quality Trends Report, levels of toxic air pollutants experienced a substantial decrease from 1994 to 1995. Early analysis indicates that this progress may be attributable to the use of RFG. From 2000 onward, emission reductions required were approximately 27% for volatile organic compounds, 22% for toxic air pollutants, and 7% for nitrogen oxides.

Note: MtBE in oxygenated gasoline dissolves readily when it contacts water. At 25°C, the water solubility of MtBE is about 5,000 milligrams per liter for a gas that is 10% MtBE by weight. In contrast, for a non-oxygenated gasoline, the total hydrocarbon solubility in water is typically about 120 milligrams per liter. MtBE sorbs only weakly to soil and aquifer material; therefore, sorption will not significantly retard MtBE's transport by groundwater. In addition, the compound resists degradation in groundwater (Squillace et al., 1998).

A limited number of instances of significant contamination of drinking water with MtBE have occurred because of leaks from underground and aboveground petroleum storage tank systems and pipelines. Because of its small molecular size and water solubility, MtBE migrates into groundwater more quickly than other gasoline components. Public and private wells have been contaminated in this manner. Nonpoint sources, including recreational watercraft, are considered the primary contributors to minor contamination observed across numerous shallow aquifers and surface water bodies. Air deposition through precipitation of industrial or vehicular emissions may also contribute to surface water contamination. The extent of any potential for buildup in the environment from such deposition is uncertain.

Based on the limited sampling data currently available, most concentrations at which MtBE has been found in drinking water sources are unlikely to cause adverse health effects. However, the USEPA is continuing to evaluate the available information and is doing additional research to seek more definitive estimates of potential risks to humans from drinking water.

There is no data on how MtBE-contaminated water affects humans. In laboratory tests on animals, cancer and noncancer effects occur at elevated levels of exposure. These tests were conducted by inhalation exposure or by introducing the chemical in oil directly into the stomach. The tests support a concern for potential human hazard. Because the animals were not exposed to drinking water, significant uncertainties exist concerning the degree of risk associated with human exposure to low concentrations typically found in drinking water.

MtBE has a very unpleasant taste and odor, and these properties make contaminated drinking water unacceptable to the public. The Advisory recommends control levels for taste and odor acceptability that will also protect against potential health effects.

Studies conducted on the concentrations of MtBE in drinking water determined the level at which individuals can detect the odor or taste of the chemical. Humans vary widely in the concentrations they can detect. Some who are sensitive can detect very low concentrations. Others do not taste or smell the chemical, even at much higher concentrations. The presence or absence of other natural or water treatment chemicals sometimes masks or reveals the taste or odor effects.

Studies to date have not been extensive enough to describe the extent of this variability completely or to establish a population response threshold. Nevertheless, we conclude from the available studies that keeping concentrations in the range of 20 to 40 micrograms per liter (μg/l) of water or below will avert unpleasant taste and odor effects, recognizing that some people may detect the chemical below this threshold.

Concentrations in the range of 20 to 40 μg/l are about 20,000 to 100,000 (or more) times lower than the range of exposure levels in which cancer or noncancer effects were observed in rodent tests. This margin of exposure lies within the range of margins of exposure typically provided to protect against cancer effects by the National Primary Drinking Water Standards under the Federal Safe Drinking Water Act—a margin greater than such standards normally offer to protect against noncancer effects. Protection of the water source from unpleasant taste and odor, as recommended, also protects consumers from potential health effects.

The USEPA also notes that occurrences of groundwater contamination observed at or above this 20–40 μg/l taste and odor threshold—that is, contamination at levels that may create consumer acceptability problems for water supplies—have, to date, resulted from leaks in petroleum storage tanks or pipelines, not from other sources.

Public water systems that conduct routine monitoring for volatile organic chemicals can evaluate MtBE at little additional cost, and some states are already moving in this direction.

Public water systems detecting MtBE in their source water at problematic concentrations can remove MtBE from water using the same conventional treatment techniques that are used to clean up other contaminants originating from gasoline releases—air stripping and granular activated carbon, for example. However, because MtBE is more soluble in water and more resistant to biodegradation than other chemical constituents in gasoline, air stripping and GAC treatment require additional optimization. They must often be used together to remove MtBE from water effectively. The costs of removing MtBE are higher than when treating gasoline releases that do not contain MtBE. The oxidation of MtBE through UV/peroxide/ozone treatment is a method, although it involves greater capital and operating expenses compared to air stripping and GAC.

Note: Of the 60 volatile organic compounds (VOCs) analyzed in samples of shallow ambient groundwater collected from eight urban areas during 1993–1994 as part of the U.S. Geological Survey's National Water Quality Assessment program, MtBE was the second most frequently detected compound (after trichloromethane (chloroform)) (Squillace et al., 1998).

5.5.1.3 Industrial Wastes

Since industrial waste represents a significant source of groundwater contamination, drinking water practitioners and others spend an increasing amount of time abating or mitigating pollution events that damage groundwater supplies.

Groundwater contamination from industrial wastes usually begins with the practice of disposing of industrial chemical wastes in surface impoundments—unlined landfills or lagoons, for example. Fortunately, these practices are behind us. Today, we know better. For example, we now know that what is most expedient or least expensive does not work for industrial waste disposal practices. We have found through experience that the long run has proven just the opposite—for society (concerning health hazards and the costs of cleanup activities), to ensure clean or unpolluted groundwater supplies is very expensive—and utterly necessary.

5.5.1.4 Septic Tanks

Septage from septic tanks is a biodegradable waste capable of affecting the environment through water and air pollution. The potential environmental problems associated with the use of septic tanks are magnified when you consider that subsurface sewage disposal systems (septic tanks) are used by almost one-third of the U.S. population.

Briefly, a septic tank and leaching field system traps and stores solids while the liquid effluent from the tank flows into a leaching or absorption field, where it slowly seeps into the soil and degrades naturally.

The problem with subsurface sewage disposal systems, such as septic tanks, is that most of the billions of gallons of sewage that enter the ground each year are not properly treated. Some systems malfunction due to poor construction or maintenance.

Experience has shown that septic disposal systems are frequently sources of fecal bacteria and virus contamination of water supplies taken from private wells. Many septic tank owners dispose of detergents, nitrates, chlorides, and solvents in their septic systems or use solvents to treat their sewage waste. A commonly used septic tank cleaner contains TCE, an organic solvent and potential human carcinogen that can contaminate groundwater near septic systems.

5.5.1.5 Landfills

Humans have been disposing of waste by burying it in the ground since time immemorial. In the past, this practice was uncontrolled, and the disposal sites (i.e., garbage dumps) were places where municipal solid wastes were dumped on and into the ground without much thought or concern. Even in this modern age, landfills have been used to dispose of trash and waste products at controlled locations that are then sealed and buried under the earth. Nonetheless, such practices are increasingly seen as a less-than-satisfactory disposal method because of the long-term environmental impact of waste materials in the ground and groundwater.

Unfortunately, many of the older (and even some of the newer) sites were in low-lying areas with high groundwater tables. *Leachate* (seepage of liquid through the waste), high in BOD, chloride, organics, heavy metals, nitrate, and other contaminants, has little difficulty reaching the groundwater in such disposal sites. In the United States, thousands of inactive or abandoned dumps like this exist.

5.5.1.6 Agriculture

Fertilizers and pesticides are the two most significant groundwater contaminants that result from agricultural activities. The impact of farming practices that typically use fertilizers and pesticides is dependent upon local soil conditions. If, for example, the soil is sandy, nitrates from fertilizers are easily carried through the porous soil into the groundwater, contaminating private wells.

Pesticide contamination of groundwater is a matter of national importance because groundwater is used for drinking water by approximately 50% of the nation's population. This especially concerns people living in the agricultural areas where pesticides are most often used, as about 95% of that population relies upon groundwater for drinking water. Before the mid-1970s, the common thought was that soil functioned as a protective filter, one that stopped pesticides from reaching groundwater. Studies have now shown that this is not the case. Pesticides can reach water-bearing aquifers below ground from applications onto crop fields, seepage of contaminated surface water, accidental spills and leaks, improper disposal, and even through the injection of waste material into wells.

Pesticides are mostly modern chemicals. Many hundreds of these compounds are used, and extensive tests and studies of their effect on humans have not been completed. That leads us to ask, "Just how concerned should we be about their presence in our drinking water?" Indeed, treating pesticides as potentially dangerous and thus managing them with care would be wise. We can say they pose a potential danger if they are consumed in massive quantities, but as any experienced scientist knows, you cannot draw factual conclusions unless scientific tests have been done. Some pesticides have had a designated Maximum Contaminant Limit (MCL) in drinking water set by USEPA, but many have not. Another serious point to consider is the potential effect of combining more than one pesticide in drinking water, which might be different than the effects of each pesticide alone. We lack enough scientific data to make reliable conclusions—we do not know.

5.5.1.7 Saltwater Intrusion

In many coastal cities and towns, as well as in island locations, the intrusion of salty seawater presents a serious water-quality problem. Because fresh water is lighter than saltwater (the specific gravity of seawater is about 1.025), it will usually float above a layer of saltwater. When an aquifer in a coastal area is pumped, the original equilibrium is disturbed and saltwater replaces the freshwater (Viessman and Hammer, 1998). Growing population, urbanization, and industrialization further strain groundwater resources. In such areas, while groundwater is heavily drawn upon, the quantity of natural groundwater recharge is decreased because of the construction of roads, tarmac, and parking lots, which prevent rainwater from infiltrating, lowering the groundwater table elevation.

In coastal areas, the natural interface between the fresh groundwater flowing from upland areas and the saline water from the sea is constantly under attack by human activities. Since seawater is approximately 2.5 times denser than freshwater, a high-pressure head of seawater occurs (compared to freshwater), which results in a significant rise in the seawater boundary. Potable water wells near this area of rising sea level may have to be abandoned because of saltwater intrusion.

DID YOU KNOW?

According to USGS (2002), one-third of the water supply for coastal areas of Greater Los Angeles comes from local groundwater sources. Saltwater has penetrated a part of the supply, and a significant part of the remaining supply is at risk. USGS scientists, working in cooperation with local water agencies, are studying the connection between coastal aquifers and the offshore geology to better understand the processes and pathways of saltwater intrusion.

5.5.1.8 Other Sources of Groundwater Contamination

To this point, we have discussed only a few of the many sources of groundwater contamination. For example, we have not discussed mining and petroleum activities that lead to groundwater contamination, or contamination caused by activities in urban areas. Both are essential sources. Urban activities, such as the application of road salt for winter ice management, contribute to the contamination of groundwater resources. Underground injection wells used to dispose of hazardous materials can lead to groundwater contamination. As we have discussed, underground storage tanks (USTs) are also significant contributors to groundwater pollution. Other groundwater contamination sources are listed as follows:

- Waste tailings
- Residential disposal
- Urban runoff
- Hog wastes
- Biosolids
- Land applied wastewater
- Graveyards
- Deicing salts

- Surface impoundments
- Waste piles
- Animal feeding operations
- Natural leaching
- Animal burial
- Mine drainage
- Pipelines
- Open dumps
- Open burning
- Atmospheric pollutants

We did not include raw sewage, as it is no longer dumped into wells or soil in the country. Sewage treatment plants effectively treat wastewater so that it can be safely discharged to local water bodies. In fact, the amount of pollution being discharged from these plants has been cut by over one-third during the past 20 years, even as the number of people served has doubled.

In some areas, raw sewage spills persist due to blocked, broken, or undersized sewer lines or heavy rainfall that overwhelms system capacity, causing overflows into streets and streams. Some of this sewage finds its way to groundwater supplies.

5.6 THE BOTTOM LINE

Today, Water Managers are not limited, hamstrung, or restrained in their duty to ensure drinking water served to the public is safe—the use of computers, IT networks, and AI are tools he or she can use to their benefit as well as for the rest of us—us being the public. I have always felt that the best way to prevent groundwater pollution is to stop it from occurring in the first place. Unfortunately, a perception held by many is that natural purification of chemically contaminated ground can take place on its own—without the aid of human intervention. To a degree, this is true; however, natural purification functions on its own time, not on human time. Natural purification could take decades and centuries. The alternative? Remediation. However, remediation and mitigation efforts require significant financial investment. When groundwater is contaminated, the cleanup efforts are sometimes much too expensive to be practical.

The USEPA has established the Groundwater Guardian Program. The program is a voluntary way to improve drinking water safety. Established and managed by a nonprofit organization in the Midwest and strongly promoted by the USEPA, this program focuses on communities that rely on groundwater for their drinking water. It provides special recognition and technical assistance to help communities protect their groundwater from contamination. Groundwater Guardian programs have operated in almost 100 communities across 31 states since 1994 (USEPA, 1996).

The AI connection to preventing water pollution issues includes the manager's ability to conduct water monitoring in real time, a capability previously unattainable before the digitalization of work environments. The digitalization used in computer-operated water treatment unit processes has enabled managers to identify and prevent leaks in the water distribution system. AI supports the optimization of water treatment unit processes, minimizes waste, and facilitates initiative-taking maintenance by identifying and addressing potential issues before they occur. This function assists managers in performing their responsibilities more efficiently and effectively. Moreover, AI in water management functions works to predict and manage flooding. Another benefit of AI in water management is that it integrates data points for sustainable water management.

NOTES

1 This account, written by Jan Schlichtmann (1999). Appeared. *USA Today*, February 4.
2 UNEP/GRID-Avendal and UN-HABITAT. E. Corcoran, et al. (ed.) (2010). *Sick Water*.
3 The information contained in this section is from USEPA's (1998). *Drinking Water Priority Rulemaking: Microbial and Disinfection by-Products Rules*. United States: Environmental Protection Agency, EPA 815-F-95-0014.

REFERENCES

EPA. 2011. *Pharmaceutical and Personal Care Products (PPCPs)*. Accessed 10/16/2011 @ www.Epa.gov/ppcp/faq.html.

Hill, M.K. 1997. *Understanding Environmental Pollution*. Cambridge, UK: Cambridge University Press.

Kay, J. 1996. Chemicals used to clean water can also cause problems. *San Francisco Examiner*, October 3.

Lewis, S.A. 1996. *The Sierra Club Guide to Safe Drinking Water*. San Francisco: The Sierra Book Club.

Rail, C.D. 1985. Groundwater Monitoring Within An Aquifer—A Protocol. *Journal of Environmental Health*, 48(2): 125–132.

Squillace, P.J., et al. 1998. Environmental behavior and fate of methyl tertiary-butyl ether. *Water Online*, November 4 @ www.wateronline.com.

UN. 2010. *Sick Water*. Washington, DC: United Nations.

USEPA. 1987. *Proposed Regulations for Underground Storage Tanks: What's in the Pipeline?* Washington, DC: Office of Underground Storage Tanks.

USEPA. 1996. *Targeting High Priority Problems*. Washington, DC: epa.gov.

USGS. 2002. *Saltwater Intrusion in Los Angeles Area Coastal Aquifers—the Marine Connection—B. Edwards and K Evans—Fact Sheet 030-02* @ http://walrus.wr.usgs.gov/reports/facts030-02.html.

Viessman, W., Jr., and Hammer, M.J. 1998. *Water Supply and Pollution Control*, 6th ed. Menlo Park, CA: Addison-Wesley.

6 Climate Change Impact on Water Resources

6.1 IS IT REAL?[1]

Is global warming a hoax? Is Earth's climate changing? Are warmer times on the way? Are colder times on the way? Is the greenhouse effect going to affect our climate, and if so, do we need to worry about it? Will the tides rise and flood New York City? Does the ozone hole portend disaster right around the corner? Do we need to eliminate all cows and other ruminants because of their methane production?

The bottom line: Are humans responsible for climate change?

These and many other questions related to climate change have come to our attention. We are inundated by a constant barrage of newspaper headlines, magazine articles, and television news reports on these topics. And recently, since the latest election, we have the uninformed and uneducated coming out and stating that what we need is the so-called "New Green Deal."

Are you kidding me?

Apparently not.

Anyway, getting back to the facts regarding the real world . . . recently, we've seen report after report on El Niño and its devastation of the West Coast of the United States (and Peru and Ecuador)—along with its reduction of the number, magnitude, and devastation of hurricanes that annually blast the East Coast of the United States.

Scientists have also been warning us of the catastrophic harm that can be done to the world by atmospheric warming. One view states that the effect could bring record droughts, record heat waves, record smog levels, and an increasing number of forest fires.

Another caution put forward warns that the increasing atmospheric heat could melt the world's icecaps and glaciers, causing ocean levels to rise to the point where some low-lying island countries would disappear. At the same time, the coastlines of other nations would be drastically altered for ages—or for all time.

What is going on? We hear plenty of theories put forward by doomsayers, but are they correct?

If they are correct, what does it all mean?

Does anyone really know the answers?

Should we be concerned?

Should we invest in waterfront property in Antarctica?

Should we panic?

No. No one truly knows the answers—it is a classic case of "we don't know what we don't know"—and though concern is warranted, there is no need to panic.

Should we consider taking specific action by developing prompt solutions and creating a plan to address these issues?

DOI: 10.1201/9781003655848-10

What needs to be done?

What can we do?

Is there anything we can do?

The single most important question to answer here is "What needs to be done?"

Studying the facts, issues, and consequences is essential, but effectively addressing these problems requires a careful and honest evaluation. We need to maintain scientific fact, sound judgment, and composure. Shooting from the hip is not called for and makes little sense—and could have Titanic consequences for us all.

The other question that has merit here is, "Will we take the correct actions before it is too late?" The keywords here are: "correct actions." Eventually, we may have to take some action (beyond hiding in a cave somewhere). But we do not yet know what those actions could be or should be.

Through time and experience, we have learned (yes, teachers learn, too) that whether we call it global warming, global climate change (humankind-induced global warming, under a broader label), or an inconvenient truth, the topic is a conundrum (riddle, the answer of which is a pun). As such, before diving into the many emotionally charged, heated class discussions about this "hot" topic (pun intended), we are reminded of two celebrated statements of just how complicated a conundrum can be. These celebrated conundrums are:

> What is black and white and read all over? A newspaper.
>
> "Why is a man in jail like a man out of jail?" There is no available solution to this matter.
>
> (Charles Dickens, 1843, *Martin Chuzzlewit* London, GB: Everyman's Library)

Consider this: Any damage we do to our atmosphere affects the other three environmental media: water, soil, and biota (us—all living things). Thus, the endangered atmosphere (if it is endangered) is a major concern (a life-and-death concern) to all of us.

This book, as the title states, is about artificial intelligence (AI) and its impact on water management. The government and NGOs have substantial amounts of data about water. AI steps in to help here. AI in water management helps conserve supplies, cut contaminants, ensure fair distribution, improve quality, and protect against floods and droughts.

AI is transforming water management amid climate change by optimizing usage and strengthening infrastructure resilience. Moreover, the key is to adjust to any change, including climate change. Water Management personnel must be far-thinkers and ask a whole bunch of what-if questions—often, daily.

Key Point: Climate change is likely to increase people's demands on water while also decreasing water supplies in some parts of the country.

6.2 METHODS FOR IDENTIFYING WATER SUPPLY ISSUES

Water Managers typically assume that specific components in their water-management equipment and infrastructure must be replaced or repaired within set periods of time.

There is a problem with this assumption. Specifically, following this maintenance approach can cause a component to be upgraded or replaced before required or push it past its service life and cause a catastrophic failure.

The first and most crucial water management concern is whether a clean, safe, and accessible source of water is available. If not available, then this lack is the Water Manager's worst nightmare. He or she must be able to find out: Where can I procure the water we need? The Water Manager's next nightmare is: Where to find water that meets water quality standards? The evidence shows that a change in climate affects water quality. For example, increased precipitation can lead to excessive runoff of sediments, nutrients, pathogens, and other substances into water bodies. A second problem with this second nightmare is the increase in nutrient runoff, along with warming temperatures, which leads to harmful algal blooms, which can kill fish, shellfish, and other animals. They can also lead to drinking water sources that are unsafe for people and pets.

Another element in this second nightmare is climate change, threatening to increase the salinity of water bodies and groundwater through saltwater intrusion. In many coastal cities and towns, as well as in island locations, the intrusion of salty seawater presents a serious water-quality problem. Because fresh water is lighter than saltwater (the specific gravity of seawater is about 1.025), it will usually float above a layer of saltwater. When an aquifer in a coastal area is pumped, the original equilibrium is disturbed, and saltwater replaces the freshwater (Viessman and Hammer, 1998). Population growth, urbanization, and industrialization are putting greater pressure on groundwater resources. In such areas, while groundwater is heavily drawn upon, the quantity of natural groundwater recharge is decreased because of the construction of roads, tarmac, and parking lots, which prevent rainwater from infiltrating, lowering the groundwater table elevation. Population growth, urbanization, and industrialization are putting greater pressure on groundwater resources.

In coastal areas, the natural interface between the fresh groundwater flowing from upland areas and the saline water from the sea is constantly under attack by human activities. Because seawater has about 2.5 times the density of freshwater, a higher-pressure head is generated for seawater compared to freshwater. This difference leads to an elevation in the boundary where seawater meets freshwater. Potable water wells near this area of rising sea level may have to be abandoned because of saltwater intrusion.

DID YOU KNOW?

According to USGS (2002), one-third of the water supply for coastal areas of Greater Los Angeles comes from local groundwater sources. Saltwater has penetrated a part of the supply, and a significant part of the remaining supply is at risk. USGS scientists, working in cooperation with local water agencies, are studying the connection between coastal aquifers and the offshore geology better to understand the processes and pathways of saltwater intrusion.

Climate change is the third greatest concern for water managers regarding infrastructure. In my ongoing research into water infrastructure throughout the United States, I have, along with many of these Water Managers, become concerned about the age and continuous aging of the nation's water supply infrastructure (including water treatment plants). Much of the U.S. water infrastructure is nearing the end of its planned life, and climate change impacts, such as more extreme weather events, will further strain its ability to operate well (Lall et al., 2018). Climate change induced by heavy rainfall can cause levees and dams to breach (Lall et al. (2018). On the basis of experience, we know that climate impacts on other segments of society can indirectly influence water utilities. For example, damaging storms that knock down or put out of operation water and wastewater plants are a real possibility; storms can disrupt power systems and shut down water supply or treatment plants. Note that there are other needs provided by surface reservoirs, which include free demands as a drinking water supply as well as a source for generating electricity.

6.3 THE WHITE KNIGHT VERSUS CLIMATE CHANGE

To address climate change impacts on water management, water managers have resources available that help them adapt to most climate anomalies.

The White Knight?

Yes. And in combating the effects of climate change on water supplies and infrastructure, artificial intelligence (AI) is used.

How?

Good question, which is answered in the following text.

Water management must incorporate and combine AI, the Internet of Things (IoT), edge computing, and other technologies to enable predictive maintenance. These technologies allow Water Management to use these virtual models of actual water systems and let managers compare data gathered now, in real time, to historical data from thousands of similar systems. With this information at hand, utilities can predict when, rather than how, specific systems will behave.

Water Management must not only focus on its responsibilities but also stay in tune with its surroundings and the workings of Mother Nature. Water Management cannot ignore climate change and push it aside without any further thought. Climate change can directly impact both water supply and demand by altering the water cycle, potentially increasing demand while limiting supply.

Although climate change may concern the Water Manager, their main priority is ensuring that the plant operates smoothly without interruptions. Predictive methods help water managers optimize uptime and extend component lifespan. Applying AI algorithms to data allows them to repair or replace components before they fail, but not so early as to waste time or resources unnecessarily. Water Managers can also plan for and schedule work for times when demand is low, reducing outages and the impact on daily users.

Water managers must keep water supplies and equipment in optimal condition to address ongoing uncertainties, regardless of and including climate change.

AI in Water Management can identify trends in large datasets. The data can reveal early signs of climate anomalies that impact water supply, helping Water Managers predict demand and optimize distribution throughout the day.

Machine learning algorithms can process data in real time to regulate water flow and pressure within a network. This approach helps maintain a steady water supply during repairs or climate-related adjustments to reduce losses and improve overall network performance.

AI enables water managers to monitor quality, track usage, and detect potential infrastructure problems. With timely insights, they can address issues before they escalate into crises.

Water Management typically uses IoT and similar technologies to gather data required for this type of analysis. What this requires are trained teams of data scientists and others who can use technology to extract the value that lies within the data. Due to the challenge of sourcing the required talent for these roles, Water Managers often choose to provide in-house training and upskill their existing workforce. It is essential to include edge computing, cloud computing, and other technologies to equip employees to support water management's goals.

After receiving appropriate training, employees will be equipped to monitor pH levels, turbidity, contaminants, and additional parameters in real time. Water Management can also use AI to analyze sensor data to detect anomalies, trigger alerts, and guide corrective actions to maintain safe, clean, and reliable water supplies.

One of the most worrisome, catastrophic, and life-threatening events is flooding. It is impossible to prevent all flooding simply because floods are a normal part of the water cycle. Not all flooding is bad or unexpected. For example, the floods of the Nile and Mekong River deltas occur on a predictable schedule. However, in less accommodating areas, Water Managers must juggle data on weather, river levels, climate change, and historical flood patterns to predict and mitigate the effects of floods.

It was not that long ago when we had to do this work on a calculator and write down calculations and all the other information using paper and pencil. No longer. Now we leverage IoT, generative AI, and different combinations of AI-enabled water management technologies to predict and mitigate the risks associated with floods. The bottom line is that during critical situations, these technologies can provide the Water Manager with the necessary information.

These technologies can also help to reduce electricity costs. Water and wastewater treatment plants, for example, use up 40% of their total electrical demand. Along with saving costs by optimizing electricity usage, AI-driven optimization algorithms can help Water Managers optimize treatment processes, adjust aeration rates, and minimize energy-intensive operations to reduce energy consumption.

AI water management addresses the challenges of supplying clean drinking water to a growing population amid climate change. By analyzing previously unrelated data, Water Managers can now better assess climate risks and create adaptive strategies for water supply management.

IoT applications analyze data from sensors embedded across the landscape, while AI helps to create virtual replicas that mirror the response to various stimuli in real time. Predicting droughts is difficult due to the need to juggle substantial amounts of disparate forms of data on weather, water tables, soil moisture, vegetation health, drought patterns, and other historical climate information.

This extensive data analysis is made-to-order for AI in general and Gen AI in particular. Water managers can use IoT and AI-enabled platforms to help address drought in their regions. These technologies enable the identification and implementation of more efficient water usage practices, supporting efforts to manage limited supplies effectively.

It would be tough to name and list operations that need water more than agriculture. All crops need water, some more than others, but not every inch of the field needs to be watered in the same way, at the same time, or with the same amount of water.

In the past, farmers used experience laced with intuition to determine how much water to apply to their crops. However, large-scale agriculture prevents such judgments from being possible. Today, AI-powered IoT irrigation systems integrate data on crops, weather, soil conditions, water tables, vegetation health, and sunlight to optimize water use on industrial-scale farms. The beauty of this is that these systems then deliver the precise amount of water when and where the plants need it, reducing waste and maximizing crop yields. And with exploding growth in populations worldwide and climate anomalies, this is a good thing.

Achieving the level of precision required to adjust to increasing populations and climate anomalies. Water Managers must train their employees to enable them to know how to combine the right tech stack with the right sensors, applications, platforms, and data. Based on personal experience, it is difficult to find properly "digitized" personnel involved in water management at any level. Thus, on-the-job training in computer operations and other digital operations is essential and is soon becoming the point best classified as "critical." We must adjust to the times, driven by technological advancements that are just at the budding stage. By training in-house, these employees not only possess "local knowledge" but also will have the opportunity to assemble the tools they need to grow with the advances in AI and its subfields, enabling the water management operation to keep pace with the advancements in computer operations.

6.4 THE BOTTOM LINE

Note that climate change affects where, when, and how much water is available. These effects vary by region and can harm the health of people and ecosystems. To proactively react to climate change and to mitigate negative aspects, we need to ensure that we create resilient water utilities, build sustainable water infrastructure, use water wisely, reduce runoff, learn about local water quality, work with nature, and finally incorporate AI in all aspects of water management.

NOTE

1 Based on F.R. Spellman's (2020). *Understanding Climate Change*. Lanham, MD: Bernan Press.

REFERENCES

Lall, U., et al. 2018. Ch 3: Water. In *Impacts, Risks, Adaptation in the United States: Fourth National Climate Assessment*, vol. 2. Washington, DC: U.S. Global Change Research Program, p. 154.

USGS. 2002. *Saltwater Intrusion in Los Angeles Area Coastal Aquifers—the Marine Connection—B. Edwards and K Evans—Fact Sheet 030-02* @ http://walrus.wr.usgs.gov/reports/facts030-02.html.

Viessman, W., Jr., and Hammer, M.J. 1998. *Water Supply and Pollution Control*, 6th ed. Menlo Park, CA: Addison-Wesley.

Part 5

Water Management Applications

7 Dams, Levees, and Reservoirs

7.1 WHAT IS CRITICAL INFRASTRUCTURE?

For the United States of America, 9/11 was a slap in the face, a punch in the gut (actually, the ultimate sucker punch), and the most serious wake-up call. The 9/11 wake-up call generated several reactions on our part—obviously, protecting ourselves from further attack became (and hopefully still is) priority number one. Considering this critical need (i.e., the survival of our way of life), the Department of Homeland Security was created. According to Governor Tom Ridge, "You may say Homeland Security is a Y2K problem that doesn't end Jan. 1 of any given year" (Henry, 2002). And, according to Barack Obama (2007), the Department of Homeland Security does "the work that ensures no other family members have to lose a loved one to a terrorist who turns a plane into a missile, a terrorist who straps a bomb around her waist and climbs aboard a bus, a terrorist who figures out how to set off a dirty bomb in one of our cities."

Among other things, the new emphasis on homeland security pointed to the need to protect and enhance the security of the nation's critical infrastructure. Critical infrastructure can be defined or listed in many ways. Governments typically use the term to refer to assets, systems, and services crucial to an economy's operation and public confidence. Destruction or compromise of any of these systems or services would have a debilitating impact on the area either directly, through interdependence, or from cascading effects. For this text, critical infrastructure refers to physical assets, key resources, and computer or service systems essential to basic government and economic operations. Critical infrastructures (in the author's opinion) include the following:

- Agriculture
- Banking and finance
- Chemical and hazardous materials
- Defense industrial base
- Emergency services
- Energy
- Organizations
- Postal and shipping
- Public health
- Strategies and assessments
- Telecommunications
- Transportation
- Water and water treatment systems
- Dams

DOI: 10.1201/9781003655848-12

- Information technology
- National monuments and icons
- Commercial assets
- Nuclear power plants

DID YOU KNOW?

Over 85% of the critical infrastructure within the United States is owned and operated by the private sector.

Although we did not list cyberspace and all ancillaries (excluding, of course, the listing of information technology (IT)) involved in or with digital operations (e-technology), in this current era, we can state without equivocation, doubt, ambiguity, or vagary that AI and digital connection today are the glue that holds all critical infrastructure together. This is the case, of course, because all separate infrastructures are interconnected in one way or another. It might come as a surprise, but we're simply discussing current e-technology, not rocket science. It would be hard to imagine that any of the above-listed infrastructure sectors could operate today without e-technology.

For example, consider Grand Coulee Dam, which is obviously a stellar member of the U.S. Dam Sector and is part of a sector where reliability is critical; it must never go offline unexpectedly (Stocks, 2014). Like many other Dam Sector infrastructures, Grand Coulee used SCADA for electronic control and security for many years. The SCADA system is still used extensively in many different applications. Still, it is considered to be an aging system. Therefore, for better, more accurate, and modern control and security applications, Grand Coulee Dam was recently retrofitted with a new power control system and is now operational using RTI, the real-time communications platform for the Internet of Things company. "The Grand Coulee Dam control system delivers extreme availability, fault tolerance, performance, security, and wide-area communications. When fully deployed, it must control 24 power turbines and 12 pumps. It comprises four major power plants, three high voltage transmission switchyards, and a fully-redundant control room" (Stocks, 2014).

We could go down the list of critical infrastructures and easily point out where e-technology and the specific industry interface. However, since this is a discussion focusing on the dam sector, we will keep the focus on e-technology as it relates to or interfaces with the dam sector. If you are not familiar with the dam sector—dam projects, hydropower plants, navigation locks, levees, dikes, hurricane barriers, mine tailing and other industrial waste impoundments, and other similar water retention and water control facilities in the Water Manager's quiver—it might surprise you to know that the entire modern hydropower and distribution grid is operated by e-technology from computer operation stations (manned by people, of course) with various digital proximity switches and other devices strategically positioned throughout the process to operate generators, pumps, valves, switches, monitor critical parameters, and provide automatic emergency shutdown procedures.

7.2 THE DAM, RESERVOIR, AND LEVEE AI CONNECTION[1]

Security is just one of the needs for AI and its subfields to help make dams, reservoirs, and levees safe. Before going further herein, let us explain what dams are.

Dams—Facilities that include some or all of the following components to perform their intended purposes safely: a conventional dam or retaining structure (the structural sections that hold back the water), a reservoir (the body of water impounded by the dam), spillways (the structures that pass normal and/or flood flows in a manner that protects the structural integrity of the dam), outlet works (the combination of structures and equipment required for the safe operation and control of water released from a reservoir), a powerhouse (the structure that houses turbines, generators, and associated control equipment for the production of hydroelectricity), and a canal/aqueduct (a constructed channel, usually open, that conveys water by gravity or is used for navigation).

Dam infrastructure is critical to the economy of the twenty-first century. The sector is a vital and beneficial part of the nation's infrastructure and continuously provides a wide range of economic, environmental, and social benefits, including hydroelectric power, river navigation, water supply, wildlife habitat, waste management, flood control, and recreation, to name a few. Some examples of the benefits derived from dam sector assets are discussed as follows:

- **Water Storage and Irrigation**: Dams create reservoirs throughout the United States that supply water for a multitude of industrial, municipal, agricultural, and recreational uses. Ten percent of American cropland is irrigated using water stored behind dams, and thousands of jobs are tied to producing crops grown with irrigation water.
- **Electricity Generation**: The United States is one of the largest producers of hydropower in the world, second only to Canada. Dams in the United States produce more than 270,000 gigawatt-hours, contributing 7% of the Nation's electricity and representing 70% of the nation's renewable energy generation.
- **"Black Start" Capabilities**: There are 4,316 megawatts of "incremental" hydropower available at sites with existing hydroelectric facilities. Incremental hydropower is defined as capacity additions or improved efficiency at existing hydro projects. During the August 2003 blackout in the Northeast, hydropower projects in New York and several other states were able to quickly start generating electricity, leading the way to restoring power to millions of Americans.
- **Recreation**: Dams and other sector assets provide prime recreational facilities throughout the United States. In 2002, a total of 105.7 million recreation user-days and nights were provided at hydropower projects licensed by the Federal Energy Regulatory Commission. In addition, about 400 million people annually visit a project of the U.S. Army Corps of Engineers (USACE), and about 90 million visit a project of the Bureau of Reclamation in a year.
- **Navigation**: Navigational projects constitute an essential component of the U.S. water system, which includes 236 lock chambers at 192 lock sites owned and/or operated by USACE. The principal value of the inland and intracoastal navigation system is the ability to convey large

volumes of bulk commodities moving long distances efficiently. If the cargo transported on the inland waterways each year had to be moved by another mode, it would take an additional 6.3 million rail cars or 25.2 million trucks to carry the load. The ability to move more cargo per shipment makes barge transport both fuel-efficient and environmentally advantageous.
- **Flood Risk Reduction**: Many dams and levees function as flood control projects, thereby reducing the potential human health and economic impacts of flooding. Reservoirs and levees built by USACE prevented more than $19 billion in potential damage during the 1993 Midwest Flood. USACE levee systems currently provide a 6:1 return on flood damages prevented compared to initial costs, and robust levee systems provide a 24:1 return on investments. Levees and hurricane barriers reduce flood damage to rural communities, as well as major metropolitan areas.
- **Sediment Control**: Some dams enhance environmental protection by controlling detrimental sedimentation.
- **Impoundment of Mine Tailings and Industrial Waste Materials**: More than 1,500 mine tailings and industrial waste impoundments controlled by dams in the Nation facilitate mining and processing of coal and other vital minerals and manufacturing while protecting the environment.

The potential risks in the event of asset failures, regardless of causation, within the Dams Sector are considerable and could result in significant destruction, including loss of life, massive property damage, and severe long-term consequences. Simply put, without a stable dam infrastructure sector, health and welfare are threatened, and the U.S. economy cannot function effectively.

The dam infrastructure sector includes subsections or elements such as dam projects, hydropower plants, navigation locks, levees, dikes, hurricane barriers, mine tailings and other industrial waste impoundments, as well as other similar water retention and water control facilities. It is important to note that these subsections or elements of the dam sector encompass physical, human, cyber, and, as noted, security elements.

7.3 DAMS SECTOR ELEMENTS

As mentioned, the Dams Sector is comprised of physical, human, cyber, and security elements. The multiplicity and complexity of sector asset functions dictate the importance of these elements as described later.

7.3.1 Physical Elements

Dams in the United States range in height from a few feet to more than 700 feet, and in length from a few tens of feet to more than one mile. Normal reservoir storage (i.e., the volume of water stored in a reservoir for most of the time) behind U.S. dams can range from a few acre-feet to 30 million acre-feet. Dams in the U.S. territories range from 30 feet to 274 feet high, with reservoir sizes ranging from 100 acre-feet to 55,000 acre-feet. Dams are also located near or along the borders with Canada and Mexico.

In addition to the main impounding structure, one or several appurtenant structures may be associated with the multiple purposes of the facility. Examples of the types of structures that could be appurtenant to a dam include free overflow and gated spillway structures, outlet works, water supply features, penstocks, power generation features, navigation locks, embankments associated with the main structure (also known as saddle dams, dikes, and freeboard dams), and different types of control structures. Each dam type and each appurtenant structure has its purpose and varying degrees of vulnerability. In general, each functional component may be associated with different protective configurations.

Levees (and flood risk reduction systems in general) may feature multiple components, including embankment sections, floodwall sections, pumps and pumping stations, interior drainage works, flood damage reduction channels, closure structures, penetrations (e.g., utilities and pipes), and transitions (i.e., sections where the materials constituting the levee system change, e.g., from a conventional floodwall section to an earthen section).

7.3.2 Human Element

The safe operation and security of the Nation's critical infrastructure depend heavily on the human element. Dams Sector facilities can be either staffed or unstaffed. Many smaller or remote assets have minimal requirements for operation and are operated without personnel on-site for lengthy periods of time (e.g., a week). Some facilities are operated remotely from central operations centers and may or may not require on-site personnel daily. Some facilities are staffed on a "business day schedule" (i.e., eight- or ten-hour schedules for five to seven days a week). Some are larger or more complex facilities that require staff to be present on a 24/7 basis.

Due to the specialized nature of the work performed in the Dams Sector, the workforce for sector assets such as hydropower facilities or navigation locks has considerable stability compared to other industries, and each facility has a small workforce.

DID YOU KNOW?

Because many of the Dams Sector operations require small workforces, owners and managers are usually well acquainted with their workers and schedules. As a consequence, the potential for criminal infiltration is very limited in impostors, infiltrators, or inside threats, which would be easier to recognize than is the case for other types of infrastructure. In addition, due to this smaller and more easily recognizable workforce, the work environment is significantly easier to monitor, and unusual activity, such as surveillance, terrorist attacks, and other criminal activities, is much easier to distinguish and respond to in the early stages of a potential situation. However, this may not necessarily be applicable to other types of facilities, such as levees, which are spatially distributed and require easy and open access for maintenance purposes.

Regarding hydropower dams, the number of personnel required to operate and maintain the asset can vary significantly. The need for staff depends on the complexity of the dam and the benefits derived from it. Larger hydropower dams may require adjustments in operations several times per hour to respond to generation needs and economics. They may require several operators for each shift; a crew of maintenance and monitoring personnel; specialized staff that respond to hourly, daily, or weekly adjustments to power demands; marketing specialists; transmission line crews; and security personnel and guards. Simple, less complex facilities may require only one person to operate the structure and perform monitoring and maintenance duties. For example, recreational dams and stock ponds have minimal staffing needs. Staffing at levees and flood control projects is similar to that of dams but is more focused on operations and maintenance activities (O&M) and will vary depending on water levels and weather conditions.

Because many of these positions are highly technical and specialized in nature, they require personnel with three or more years of apprenticeship work or advanced training. Operations at some dams may also require the full-time or corporate sharing of management and IT personnel. Personnel succession plans and continuity-of-operations plans help ensure ongoing operation in the event of work stoppages and extended periods of illness, such as during pandemics. In addition, the schedule for noncritical operations can be modified to respond to situations that may be understaffed (e.g., establishing a deferred maintenance schedule).

7.3.3 Cyber Element

The efficiency and safety of dam sector O&M are constantly improving as technology in industrial control systems continues to advance. Control systems vary widely based on the age and generation of the system; thus, they also differ with respect to complexity and sophistication. Some systems are closed and use isolated networks, as well as proprietary communication protocols, while others use open architectures and standard communication protocols and rely on the Internet. In addition, the control systems at dams may also be connected to the electric power grid. Most dam owners use computer security measures such as authentication, encryption, firewalls, antivirus software, and anti-spyware for their primary terminal units, data servers, and historians.

Note that many of the industrial control systems used today were designed for operability and reliability during an era when security received low priority; that is, before 9/11 and the rise in worldwide terrorism. These systems operated in isolated environments and typically relied on proprietary software, hardware, and/or communications technology. Infiltrating and compromising these systems often required specific knowledge of individual system's architecture and physical access to system components.

In contrast, newer control systems are highly network-based and use common and open standards for communication protocols, such as Internet Protocol addressable. Many controllers are Internet Protocol addressable. Asset owners and operators have gained immediate benefits by extending the connectivity of their systems and have increasingly adopted commercial off-the-shelf technologies that provide the higher levels of interoperability required among today's modern infrastructure. Standard operating systems such as Windows or UNIX are increasingly used in central supervisory stations, which are now typically connected to remote controls via public-private networks provided by telecommunications companies. Common telecommunications technologies, such as the Internet, public-switched telephone networks, or cable or wireless networks, are often used.

DID YOU KNOW?

As a general proposition, levee operations are conducted by smaller entities and are considerably less dependent on information technology than dams. They rely on basic telecommunications and information infrastructures that can be supplemented with manual operations.

7.3.4 Security Element

The unique characteristics of the Dams Sector influence the implementation of security measures. First, the spatial dimensions of dam projects are significantly larger than those facilities within other sectors. Second, there are often multiple components critical for project purposes that are not next door (non-adjacently located). Each of these components may potentially be associated with different security configurations. Third, many dams are in remote areas; thus, a response force belonging to local law enforcement authorities may take longer to arrive once an alarm is received. For some attack types, the appropriate response may not arrive in time to thwart the attempt.

The security configuration of sector assets may include a combination of features, such as fences and gates, vehicle-resistant perimeters, boat barriers, intrusion detection systems, electronic entry control systems, closed-circuit television (CCTVs), appropriate lighting, patrols, guards, and hardening of select components.

In addition to these electronic and physical security features, Dams Sector partners also rely on robust security plans that define operational security procedures, such as information protection, threat and suspicious incident analysis centers, planned responses to varying threat levels, and coordination with law enforcement agencies. Owners and operators evaluate Emergency Action Plans to help protect populations at risk and maintain response and recovery plans to bring project benefits back online rapidly.

7.4 THE NUTS AND BOLTS ABOUT DAMS

As mentioned earlier, the Dams Sector is composed of assets that include dam products, hydropower generation facilities, navigation locks, levees, dikes, hurricane barriers, mine tailings, other industries' waste impoundments, and other similar water retention and control facilities. The Dams Sector is a vital part of the nation's infrastructure. It provides a wide range of economic, environmental, and social benefits, including hydroelectric power, river navigation, water supply, wildlife habitat, waste management, flood control, and recreation.

The Dams Sector has interdependencies with a wide range of other factors, including:

- **The Emergency Services Sector**—Emergency response may depend on Dams Sector assets for firefighting, emergency water supply, and waterway access when land routes are inaccessible during disasters.
- **The Energy Sector**—Hydropower dams generate 8%–12% of national power.
- **The Food and Agriculture Sector**—Dams sector assets are a source of water for our nation's food and agriculture production.
- **The Transportation Systems Sector**—Dams and locks manage navigable waters throughout inland waterways for shipping and recreation.
- **The Water and Wastewater Systems Sector**—Water and wastewater management systems rely on Dams Sector assets to provide water to large, populated areas and commercial facilities.

7.4.1 Dams Sector Assets

As previously described, the dams sector assets encompass a broad spectrum of facilities. These facilities are found throughout the Nation.

7.4.1.1 Dams

Dams are facilities that include some or all of the following components to perform their intended purposes: a conventional dam or impounding structure (structural sections that hold back the water), a reservoir (a body of water impounded by the dam), spillways (structures that facilitate the discharge of normal and/or flood flows in a manner that protects the structural integrity of the project), outlet works (a combination of structures and equipment required for safe operation and control of water released from a reservoir), a powerhouse (a structure that houses turbines, generators, and associated control equipment for the production of hydroelectricity), penstocks (pipelines or conduits used to convey water under pressure to turbines or pump units), and canals/aqueducts (constructed channels, usually open, that convey water by gravity and may also be used for navigational purposes).

In general, canals convey water through an open channel from a source to a destination and serve a variety of functions. Some canals, constructed as part of water supply or transportation projects, have an embankment section forming one of the sides of the water conveyance structure and acting as a water retention structure. Water can be conveyed for a variety of uses (e.g., for water supply—domestic, municipal, industrial, or agricultural—or for transporting boats, barges, and other vessels. In some cases, dams also include navigation locks.

As pointed out earlier, one of the significant components of a hydroelectric dam is the area behind the dam, known as its reservoir. The water temporarily stored there is called gravitational potential energy. The water is stored above the rest of the dam facility to allow gravity to carry the water down to the turbines. Because this higher altitude is different from where the water would naturally be, the water is at an altered equilibrium. The result is stored energy of position, that is, gravitational potential energy. The water has the potential to do work because of its position. As shown in Figure 7.1, gravity will force the water to fall to a lower position through the intake and the control gate. Installed within the dam, when the control gate is opened, the water from the reservoir goes through the intake and becomes translational kinetic energy as it falls through the next central part of the system, the penstock. Translational kinetic energy is the energy due to motion from one location to another. The water is moving (falling) from the reservoir toward the turbines through the penstock; the water is carried through the penstock's long shaft toward the turbines, where the kinetic energy becomes mechanical energy. The force of the water is used to turn the turbines, which turn the generator shaft. The generators convert the energy of water into electricity, and then step-up transformers increase the voltage produced to higher voltage levels.

Dams may be classified according to the type of construction materials used, the methods of construction, the slope or cross section, the way the structure resists the forces of the water pressure behind it, the means used for controlling seepage, the purpose, and the hazard potential. Materials used for construction include earth, rock, tailings from mining or milling, concrete, masonry, steel, timber, miscellaneous material such as plastic or rubber, or any combination of these materials. The two most common types of dams in use today are embankment dams and concrete dams (Association of State Dam Safety Officials (ASDSO), 2015).

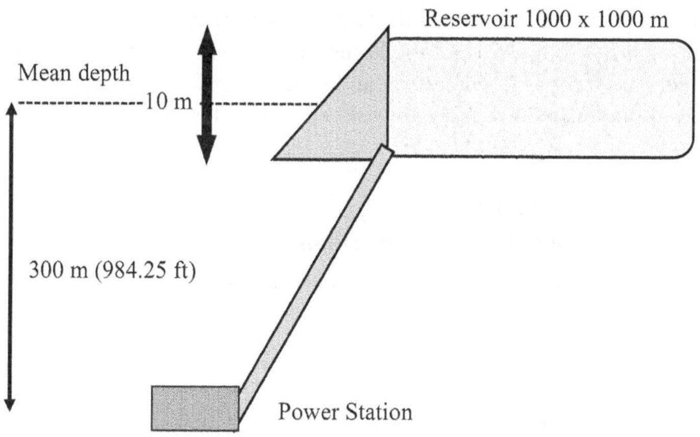

FIGURE 7.1 Schematic representation of a Hydro scheme.

Source: Adapted from F.R. Spellman (2010). *Dams and Homeland Security*. Lanham, MD: Bernan Press.

Embankment dams are the most common type of dam in use today. An embankment dam is termed an "earth fill" or "rockfill" dam, depending on whether it is composed of compacted earth or mostly compacted or dumped rock. The ability of an embankment dam to resist the reservoir water pressure is primarily a result of the mass weight and the type and strength of the materials from which the dam is made.

Concrete dams may be categorized as gravity or arch dams. Concrete gravity dams use their width and the friction at the foundation to withstand water pressure from the reservoir. Gravity dams are constructed of vertical blocks of concrete with flexible seals in the joints between the blocks. A buttress dam is a specific type of gravity dam in which the large mass of concrete is reduced, and the forces are diverted to the dam foundation through vertical or sloping buttresses. Concrete arch dams are characterized by thinner cross sections in which the reservoir water pressure is carried laterally into the abutments. The shape of the arch may resemble a segment of a circle or an ellipse, and the arch may be curved in the vertical plane as well. Such dams are usually constructed of a series of thin vertical blocks that are keyed together; barriers to stop water from flowing are provided between the blocks. Variations of arch dams include multi-arch dams in which more than one curved section is used and arch-gravity dams that combine some features of the two types of dams.

The National Inventory of Dams currently lists 82,642 dams in the United States; more than 70,000 of them are State-regulated, 4,639 are regulated by federal agencies, and the remainder are not regulated by any government agency. The ownership of dams in the United States makes them a unique part of the national infrastructure. Although most infrastructure facilities, such as roads, bridges, and sewer (wastewater) systems, are owned by public entities, most dams (almost 65%) in the United States are owned by private entities. Table 7.1 illustrates the ownership structure, and Table 7.2 shows the primary purpose of dams in the United States.

Although Table 7.2 shows that only about 2% of the dams within the United States are used for hydroelectric production purposes, it is essential to point out that with regard to the importance of critical infrastructure applications the production of electrical power and hydropower (i.e., the harnessing of water to perform work) in general is just as important as flood control and water supply. The power of falling water has been used in industry for thousands of years over the ages as listed in Table 7.3.

TABLE 7.1
Ownership of U.S. Dams

Owner	Percentage of Owners
Private	65%
Local government	20%
State government	5%
Unknown	4%
Federal government	4%
Public utility	2%

TABLE 7.2
Dams by Primary Purpose

Primary Purpose	Percentage of Purpose
Recreation	34%
Flood control and storm water Management	17%
Fire protection, stock, or small farm Pond	15%
Irrigation	10%
Water supply	7%
Other	7%
Unknown	4%
Hydroelectric	~2%
Fish and wildlife pond	>2%
Tailings	>1.5%
Debris control	>1.0%
Navigation	<0.5%

TABLE 7.3
History of Hydropower

Date	Hydropower Event
B.C.E.	Hydropower was used by the Greeks to turn water wheels for grinding wheat into flour more than 2,000 years ago.
Mid-1770s	French hydraulic and military engineer Bernard Forest de Bélidor wrote a four-volume work describing vertical- and horizontal-axis machines.
1775	The U.S. Army Corps of Engineers was founded with the establishment of the Chief Engineer for the Continental Army.
1880	Michigan's Grand Rapids Electric Light and Power Company, generated electricity by a dynamo belted to a water turbine at the Wolverine Chair Factory, lighting up 16 brush-arc lamps.
1881	Niagara Falls powered city street amps with hydropower.
1882	The world's first hydroelectric power plant began operation on the Fox River in Appleton, Wisconsin.
1886	There are about 45 water-powered electric plants in the United States and Canada.
1887	San Bernardino, CA, opened the first hydroelectric plant in the West.
1889	Two hundred electric plants in the United States use water power for some or all of their generation.
1901	The first Federal Waterpower Act.
1902	Bureau of Reclamation was established.
1907	Hydropower provided 15% of U.S. electrical generation.
1920	Hydropower provided 25% of U.S. electrical generation. The Federal Power Act establishes the Federal Power Commission's authority to issue licenses for hydro development on public lands.
1933	Tennessee Valley Authority was established.

TABLE 7.3
History of Hydropower

Date	Hydropower Event
1935	The authority of the Federal Power Commission was extended to all hydroelectric projects built by utilities engaged in interstate commerce.
1937	The Bonneville Dam, the first federal dam, began operation on the Columbia River, and the Bonneville Power Administration was established.
1940	Hydropower provides 40% of electrical generation. Conventional capacity tripled in the United States since 1920.
1980	Conventional capacity has nearly tripled in the United States since 1900.
2003	About 10% of U.S. electricity comes from hydropower. Today, there is about 80,000MW of conventional capacity and 18,000 MW of pumped storage.

Source: EERE (2008).

DID YOU KNOW?

The potential energy in a specific slug or parcel of water is expressed in newton-meters (N·m). The newton is the standard unit of force, equivalent to a meter per second squared (1 m/s^2). This is the product of the mass of the slug in kilograms, the acceleration of gravity in meters per second squared (about 9.8 m/s^2), and the elevation of the parcel in meters (the vertical distance it falls as its energy is harnessed). The equivalent kinetic-energy unit is the joule (J), which is in effect equal to a watt-second (W·s).

DID YOU KNOW?

One of the problems encountered in a hydraulic system is storing the liquid. Unlike air, which is readily compressible and can be stored in large quantities in relatively small containers, a liquid such as water cannot be compressed. Therefore, it is not possible to store a large amount of water in a small tank—62.4 lbs of water occupies a volume of one cubic foot, regardless of the pressure applied to it.

$$\mathbf{1\ ft^3\ H_2O = 62.4\ lb}$$

Both cubic feet and pounds are used to describe a volume of water. There is a defined relationship between these two methods of measurement. The specific weight of water is defined relative to a cubic foot. One cubic foot of water weighs 62.4 pounds. This relationship is true only at a temperature of 4°C and at a pressure of one atmosphere (known as standard temperature and pressure; STP)—14.7 lbs. per square inch at sea level containing 7.48 gallons). The weight varies so little that, for practical purposes, this weight is used from a temperature of 0°C to 100°C. One cubic inch of water weighs 0.0362 pounds. Water one foot deep will exert a pressure of 0.43 pounds per square inch on the bottom area (12 in × 0.0362 lb/in^3). A column of water two feet high exerts 0.86 psi, one 10 feet high exerts 4.3 psi, and one 55 feet high exerts

$$55 \text{ ft} \times 0.43 \text{ psi / ft} = 23.65 \text{ psi.}$$

A column of water 2.31 feet high will exert 1.0 psi. To produce a pressure of 50 psi requires a water column of

$$50 \text{ psi} \times 2.31 \text{ ft / psi} = 115.5 \text{ ft.}$$

Remember: The important points being made here are as follows:

1. 1 ft^3 H_2O = 62.4 lb
2. A column of water 2.31 ft high will exert 1.0 psi.

Another important relationship is

$$1 \text{ gallon } H_2O = 8.34 \text{ pounds.}$$

Table 7.4 lists the largest U.S. hydroelectric power stations.

7.4.1.2 Flow/Discharge Rate: Water in Motion

Fluid flow is more complex than fluids at rest, but understanding it is essential since hydropower relies on moving water, such as that which drives turbine blades.

Discharge (or flow) is the quantity of water passing at a given point in a pipe or channel during a given period. Stated another way for open channels: The flow rate through an open channel is directly related to the velocity of the liquid and the cross-sectional area of the liquid in the channel.

TABLE 7.4
Largest U.S. Hydroelectric Power Stations

Name	Year of Completion	Total Capacity (Megawatts, MW)
Grand Coulee	1942/1980	6,809
Bath County PSP	1945	3,003
Chief Joseph Dam	1958/1973/1979	2,620
Robert Moses Niagara PP	1961	2,515
John Day Dam	1949	2,160
Hoover Dam	1936/1961	2,080
The Dalles Dam	1981	2,038

Source: Bureau of Reclamation. (http://www.usbr.gov/pn/news/02new/gcrecrd.html) Retrieved 7/11/25.

$$Q = A \times V \tag{7.1}$$

where

Q = Flow (discharge in cubic feet per second, cfs)
A = Cross-sectional area of the pipe or channel (ft^2)
V = Water velocity in feet per second (fps or ft/sec)

EXAMPLE 7.1

Problem:
The channel is six feet wide, and the water depth is three feet. The velocity in the channel is four feet per second. What is the discharge or flow rate in cubic feet per second?

Solution:

$$\text{Flow, cfs} = 6 \text{ ft} \times 3 \text{ ft} \times 4 \text{ ft/second} = 72 \text{ cfs}$$

Discharge or flow can be recorded as gallons per day (gpd), gallons per minute (gpm), or cubic feet per second (cfs). Many hydropower systems manage large flows, often measured in million gallons per day (MGD). The discharge or flow rate can be converted from cfs to other units such as gpm or million MGD by using appropriate conversion factors.

7.4.1.3 Reservoir Stored Energy

One of the significant components of a hydroelectric dam is the area behind the dam, its reservoir. The water temporarily stored there is called gravitational potential energy. Water is held above the dam to let gravity move it down to the turbines. Because this higher altitude is different from where the water would naturally be, the water is at an altered equilibrium. The result is stored energy of position; that is, gravitational potential energy. The water has the potential to do work because of its position. Gravity will force the water to fall to a lower position through the intake and the control gate. Installed within the dam, when the control gate is opened, the water from the reservoir goes through the intake and becomes translational kinetic energy as it falls through the next main part of the system, the penstock. Translational kinetic energy is the energy due to motion from one location to another. The water is moving (falling) from the reservoir toward the turbines through the penstock. The water is carried through the penstock's long shaft toward the turbines, where the kinetic energy becomes mechanical energy. The force of the water is used to turn the turbines, which in turn rotate the generator shaft. The generators convert the energy of water into electricity, and then step-up transformers increase the voltage produced to higher voltage levels.

7.4.1.4 Hydro Turbines

According to EERE (2008), there are two main types of hydro turbines: impulse and reaction. The type of hydropower turbine selected for a project is based on the height of standing water—as mentioned, this is referred to as "head"—and the flow, or volume of water, at the site. Cost, desired efficiency, and the depth at which the turbine must be set are other deciding factors.

Impulse Turbine—Impulse turbines use the velocity of the water to move the runner and discharge to atmospheric pressure. The water stream hits each bucket on the runner. There is no suction on the downside of the turbine, and water flows out the bottom of the turbine housing after hitting the runner. An impulse turbine is suitable for high-head, low-flow applications.

Reaction Turbine—Reaction turbines develop power form the combined action of pressure and moving water. The runner is placed directly in the water stream flowing over the blades rather than striking each individually. Reaction turbines are used for sites with lower head and higher flows compared to impulse turbines.

7.4.2 Navigation Locks

A "navigation lock" refers to projects that include one or more lock chambers, the related dam (unless in a canal), and supporting structures like approach walls, control rooms, and administration or maintenance buildings. The U.S. waterway system includes 236 lock chambers at 192 lock sites owned and/or operated by the USACE. Many navigation projects may serve multiple purposes beyond their primary navigation function. For example, 46 lock-associated dams currently produce hydropower.

These facilities make many inland waterways viable, year-round transportation corridors. Natural riverbeds are often long, uneven downhill slopes with shallow areas and deep pools. Water depths can vary with seasonal rainfall and snow melting; dry periods can significantly affect navigation. Locks on navigable rivers enable commercial and recreational river traffic to move safely from one pool level to the next. The lift of the locks—the difference between raising and lowering from one pool to the next—can range from just a few feet to 110 feet (e.g., John Day Lock and Dam on the Columbia River in Oregon and Washington).

JOHN DAY DAM

The John Day Dam is a concrete gravity run-of-the-river dam spanning the Columbia River on the Washington/Oregon border (BPA, 2010). The dam features a navigation lock and fish ladders on both sides and has the highest lift (110 feet) of any U.S. lock (USGS, 2015). Lake Umatilla is the reservoir impounded by the dam, and it runs 76.4 miles (123 kilometers) up the river channel to the foot of McNary Dam (USGS, 2015).

Multiple locks may be part of one navigable river system. The Tennessee River system is managed through a series of dams owned by the United States and operated by Tennessee Valley Authority and the U.S. Army Corps of Engineers. The system's nine main and five auxiliary locks allow smooth movement on the river as it drops 513 feet in elevation from its beginning to the point at which it joins the Ohio River. The Mississippi River is made navigable through a total of 30 main and four auxiliary locks.

A principal value of the inland and intracoastal navigation systems is the ability to convey large volumes of bulk commodities moving long distances efficiently. Towboats push barges lashed together to form a "tow." A 15-barge tow is standard on larger rivers with locks, such as the Ohio, upper Mississippi, Illinois, and Tennessee rivers. Such tows are a highly efficient mode of transportation, moving about 22,500 tons of cargo as a single unit. A single 15-barge tow is the equivalent of about 225 rail cars or 870 tractor-trailer trucks. If the cargo transported on the inland waterways each year had to be moved by another mode, it would take an additional 6.3 million rail cars or 25.2 million trucks to carry the load. The ability to move more cargo per shipment makes barge transport both fuel-efficient and environmentally advantageous. On average, a gallon of fuel allows one ton of freight to be shipped 155 miles by truck, 413 miles by railway, and 576 miles by barge. As a result, cargo transported by barge results in an average transportation savings of about $11 per ton over the cost of shipping by alternative modes. Environmental advantage can also be measured by comparing the greenhouse gas (GHG) emissions produced during fuel consumption. On average, the tons of GHG emissions produced per million-ton miles are 71.61 by truck, 26.88 by railway, and 19.27 by barge.

Coal is the largest commodity by volume moving on the inland waterways. The Nation's electric industry depends on inland waterways for more than 20% of the coal it consumes to produce electricity. Petroleum is the next largest group, including crude oil, gasoline, diesel fuel, heavy fuel oils, and asphalt. Another large group includes grain and other farm products, most of which move by waterway to ports on the Lower Mississippi or Columbia Rivers for export overseas. Sixty percent of the country's farm exports travel through inland waterways. Other major commodities include aggregates, such as stone, sand, and gravel used in construction; chemicals, including fertilizers; metal ores, minerals, and products, such as steel; and many other manufactured products.

7.5 LEVEES

The term "levee" is used to denote manmade flood risk reduction systems (e.g., embankment sections, floodwalls, closure structures, pumps, interior drainage works, and flood damage reduction channels) whose primary purpose is to furnish flood protection from seasonal high water and which are therefore subject to water loading for periods of only a few days or weeks a year. Embankments that are subject to water loading for prolonged periods (longer than normal flood protection requirements) or are permanently water loaded should be designed following earth dam criteria rather than the levee criteria given in Table 7.5.

Even though levees are similar to small earth dams, they differ from earth dams in the following important respects: (a) A levee embankment may become saturated for only a short period beyond the limit of capillary saturation, (b) levee alignment is dictated primarily by flood protection requirements, which often results in

TABLE 7.5
Levee Major and Minimum Requirements

Step	Procedure
1.	Conduct a geological study based on a thorough review of available data, including analysis of aerial photographs. Initiate preliminary subsurface explorations.
2.	Analyze preliminary exploration data and, from the analysis, establish preliminary soil profiles, borrow locations, and embankment sections.
3.	Initiate final exploration to provide a. Additional information on soil profiles. b. Undisturbed strengths of foundation materials. c. More detailed information on borrow areas and other required excavations.
4.	Using the information obtained in Step 3: a. Determine both embankment and foundation soil parameters and refine preliminary sections where needed, noting all problem areas. b. Compute rough quantities of suitable material and refine borrow area location
5.	Divide the entire levee into reaches of similar foundation conditions, embankment height, and fill material and assign a typical trial section to each reach.
6.	Analyze each trial section as needed for: a. Underseepage and through seepage. b. Slope stability. c. Settlement. d. Trafficability of the levee surface.
7.	Design special treatments to preclude any problems as identified from Step 6. Determine surfacing requirements for the levee based on its anticipated future use.
8.	Based on the findings of Step 7, establish final sections for each reach.
9.	Compute final quantities needed, determine final borrow area locations.
10.	Design embankment slope protection.

construction on poor foundations, and (c) borrow is generally obtained from shallow pits or from channels excavated adjacent to the levee, which produce fill material that is often heterogeneous and far from ideal. The selection of the levee section is often based on the properties of the poorest material that must be used.

Regarding levee design, numerous factors must be considered. These factors may vary from project to project, and no specific step-by-step procedure covering details of a particular project can be established. However, it is possible to present general, logical steps based on successful past projects that can be followed in levee design and can be used as a base for developing more specific procedures for any project. Such a procedure is given in Table 7.6.

CAPILLARY ACTION

Suppose we were to mention the term "capillary action" to the man or woman on the street. In that case, they might instantly nod their heads and respond that their bodies are full of them—that capillaries are the tiny blood vessels that connect the smallest arteries and the smallest of the veins. This would be true, of course. But in the context of water science, a capillary is something different than capillary action in the human body.

Even if you've never heard of capillary action, it is still essential in your life. Capillary action is necessary for moving water (and all the things that are dissolved in it) around. It is defined as the movement of water within the spaces of a porous material due to the forces of adhesion, cohesion, and surface tension.

Surface tension is a measure of the strength of the water's surface film. The attraction between the water molecules creates a strong film, which, among other common liquids, is only surpassed by that of mercury. This surface tension permits water to hold up substances heavier and denser than itself. A steel needle carefully placed on the surface of a glass of water will float. Some aquatic insects, such as the water strider, rely on surface tension to walk on water.

Capillary action occurs because water is sticky; thanks to the forces of cohesion (water molecules like to stay closely together) and adhesion (water molecules are attracted and stick to other substances). So, water tends to stick together, as in a drop, and it sticks to glass, cloth, organic tissues, and soil. Dip a paper towel into a glass of water, and the water will "climb" onto the paper towel. In fact, it will keep going up the towel until the pull of gravity is too much for it to overcome.

DID YOU KNOW?

Levees, as opposed to dams, are geospatially extended systems that can run for multiple miles; however, the levee system may consist of several segments under different governance units. This creates opportunities for multiple failure modes and points throughout the system. Proper levee maintenance is often overlooked because it appears to be pretty basic—repairing erosion damage, protecting against unwanted vegetation growth or animal burrowing, and so forth—so it may be given low priority or ignored completely.

The method of construction must also be considered. In the past, levees were built using methods of compaction that varied from none to carefully controlled compaction. The local economic situation also affects the selection of a levee section. Traditionally, in areas with high property values, high land use, and good foundation conditions, levees have been built with steep slopes using controlled compaction. In areas with lower property values, poor foundations, or high rainfall during the construction season, uncompacted or semi-compacted levees with flatter slopes are more typical. This is evident by comparing the steep slopes of levees along the industrialized Ohio River Valley with those along the Lower Mississippi River, which have much broader sections with gentler slopes. Levees built with smaller sections and steeper slopes require more comprehensive investigation and analysis than do levees with broad sections and flatter slopes whose design is more empirical. Where rainfall and foundation conditions permit, the trend in the design of levees is toward sections with steeper slopes. Levee maintenance is another factor that often has a considerable influence on the selection of a levee section.

7.5.1 Levee Types According to Location

Levees are broadly classified according to the area they protect as either urban or agricultural levees because of the different requirements for each. In this book, urban and agricultural levees are defined as follows:

(1) Urban levees. Levees that provide protection from flooding in communities, including industrial, commercial, and residential facilities.
(2) Agricultural levees. Levees that provide protection from flooding in lands used for agricultural purposes.

7.5.2 Levee Types According to Use

Some of the more common terms used for levees serving a specific purpose in connection with their overall purpose of flood protection are listed in Table 7.6.

TABLE 7.6
Classification of Levees According to Use

Type	Definition
Mainline and tributary levees	Levees that lie along a mainstream and its tributaries, respectively.
Ring levees	Levees that completely encircle an area subject to inundation from all directions.
Setback levees	Levees that are built landward of existing levees, usually because the existing levees have suffered distress or are in some way being endangered, as by river migration.
Sublevees	Levees that are built for the purpose of underseepage control. Sublevees encircle areas behind the main levee that are subject, during high-water stages, to high uplift pressures and the development of sand boils. They normally tie into the main levee, thus providing a basin that can be flooded during high-water stages, thereby counterbalancing excess head beneath the top stratum with the basin. Sublevees are rarely employed as the use relief wells, or seepage berms make them unnecessary except in emergencies.
Spur levees	Levees that project from the main levee and serve to protect the main levee from the erosive action of stream currents. Spur levees are not true levees but training dikes.

HEAD

Head is defined as the vertical distance water must be lifted from the supply tank to the discharge, or as the height a column of water would rise due to the pressure at its base. A perfect vacuum plus atmospheric pressure of 14.7 psi would lift the water 34 ft. If the top of the sealed tube is opened to the atmosphere and the reservoir is enclosed, the pressure in the reservoir is increased, and the water will rise in the tube. Because atmospheric pressure is essentially universal, we usually ignore the first 14.7 psi of actual pressure measurements and measure only the difference between the water pressure and the atmospheric pressure; we call this *gauge pressure*. For example, water in an open reservoir is subjected to the 14.7 psi of atmospheric pressure, but subtracting this 14.7 psi leaves a gauge pressure of 0 psi. This shows that the water would rise 0 feet above the reservoir surface. If the gauge pressure in a water main were 120 psi, the water would rise in a tube connected to the main:

$$120 \text{ psi} \times 2.31 \text{ ft/psi} = 277 \text{ ft} \quad \text{(rounded)}$$

The *total head* includes the vertical distance the liquid must be lifted (static head), the loss due to friction (friction head), and the energy required to maintain the desired velocity (velocity head).

$$\text{Total Head} = \text{Static Head} + \text{Friction Head} + \text{Velocity Head} \quad (3.2)$$

DID YOU KNOW?

Levee ownership and governance patterns can sometimes be politically, socially, and economically more complex than those for dams. Levee activities may be cost-shared among agencies at several levels of government, which can contribute to delays in repairs, response, and mitigation actions. Actual ownership, especially for older levees, may be unclear and may require lengthy litigation before the necessary repairs can be undertaken. Unclear ownership of levees and the levee segments that contribute to a levee system complicates the coordination and communication required for levee safety and resilience.

7.6 HURRICANE BARRIERS

A *hurricane barrier*, *flood barrier*, *surge barrier*, or *storm surge barrier* is a specific type of floodgate designed to restrict the inflow of storm surges and extratropical storms into bays, estuaries, rivers, and streams, and/or onto the adjacent lowlands. These barriers consist of gates that are closed during storms, either fully or partially open, depending on the degree of routine use that they also serve.

Hurricane barriers in the Northeast that protect against flooding in large bays are normally left fully open and are closed only during a storm. Barriers in southern Florida serve a routine water control purpose and are completely closed during a hurricane event.

Hurricane barriers are typically large steel structures. Each of the movable gates that span the Providence River in Providence, Rhode Island, is 40 feet wide and weighs 53 tons. Hurricane barriers are usually composed of at least two gates—one on each bank of the channel. However, more complex configurations may include multiple gates that can be independently raised or closed depending on the surge volume.

7.7 TAILINGS DAMS

To obtain the metals and other minerals needed for industrial processes, fertilizers, homes, cars, and other consumer products, enormous quantities of rock are mined, crushed, pulverized, and processed to recover metal and other mineral values. A fine grind is often necessary to release metals and minerals, so the mining industry produces enormous quantities of fine rock particles, in sizes ranging from sand-sized down to as low as a few microns. These fine-grained wastes are known as "tailings."

In recent decades, many mines were small underground operations with correspondingly modest requirements for tailings disposal. Since that time, due to increasing demand, it has become economical to mine large lower-grade deposits by utilizing advances made by mining equipment manufacturers and developments in mining and mining technology. This has led to more tailings and waste being produced by both individual mines and the entire industry (USEPA, 1994).

There are approximately 1,000 active metal mines in the United States (USGS, 2003). Many of these have at least one tailings impoundment and often several impoundments grouped together in cells. The USEPA estimates that there may be several thousand tailings impoundments associated with active non-coal mining, and tens of thousands of inactive or abandoned impoundments.

By far the most significant proportion of ore mined in most industry sectors becomes tailings that must be disposed of. In the gold industry, for example, only a few hundredths of an ounce of gold may be produced for every ton of dry tailings generated. Similarly, the copper industry and others typically mine low-grade ores that contain less than a few percent of metal values; the residue becomes tailings. Thus, tailings disposal is a significant part of the overall mining and milling operation at most hard rock mining projects. There are several methods used for tailings disposal. These include disposal of dry or thickened tailings in impoundments or free-standing piles, backfilling underground mine workings and open pits, subaqueous disposal, and the most common method, the disposal of tailings slurry in impoundments. Modern tailings impoundments are engineered structures for permanently disposing of the fine-grained waste from mining and milling operations. At some projects, tailings embankments reach several hundred feet in height, and the impoundments cover several square miles.

Historically, tailings were often dumped in nearby water bodies or drainages for convenience and cost savings. As local concerns arose about sedimentation in downstream watercourses, water use, and other issues, mining operations began impounding tailings behind earthen dams, which were often constructed of tailings and other waste materials. The impoundments served the dual purpose of containing the tailings and, particularly in the arid West, allowing the re-use of scarce water.

More recently, concerns have been raised about the stability and environmental performance of tailings dams and impoundments. Stability concerns are raised in part by the use of tailings material in tailings dams/embankments. To mitigate these concerns, such embankments often rely on a certain amount of controlled seepage to enhance stability, which in turn affects environmental performance. Perhaps triggered by the initial attention to uranium impoundments, the increasing concern for ecological performance has led to better engineering design of tailings dams in other mining industry sectors, focusing on both stability and environmental performance. For instance, experience gained with leach pad liners is being transferred to linings for tailings ponds, and the use of synthetic lining materials is growing (although the use of liners is still far from being the industry norm). In addition, the use of cyanide and other toxic reagents in mill processes has raised special concerns for some tailings, leading to increased treatment before disposal and increased attention to containment. Finally, continuing concerns over acid mine drainage are resulting in a growing body of research and emerging concepts of long-term control or mitigation.

Inactive tailings impoundments are also receiving more attention due to the long-term effects of windblown dispersal, groundwater contamination, and acid drainage. In many cases, the costs of remediation can be considerable, exceeding the costs of the original design and operation of the tailings impoundment.

Design depends on the quantity and the individual characteristics of the tailings produced by the mining and milling operation, as well as the climatic, topographic, geologic, hydrogeologic, and geotechnical characteristics of the disposal site, and on regulatory requirements related to dam safety and environmental performance. What may work for one type of tailings may not work for another kind and may not work for the same tailings at different sites. Hence, each situation requires its own design process. The estimated quantity of tailings to be disposed of is particularly important given the evolving nature of most mining projects. Tailings quantity estimates are based on estimated reserves that change continuously as mine development progresses. Accordingly, the final size and design of tailings impoundments can differ from initial projections. This poses significant challenges for federal land managers, state permit writers overseeing tailings impoundments, and the public, who may bear the costs of environmental damage caused by errors.

Under normal circumstances, the Water Manager's calculation of the safety factor for a tailings embankment required an analysis of the potential failure surfaces of the embankment. Embankments are susceptible to several standard failure modes. These include slope failure from rotational slide, overtopping, foundation failure, erosion, piping, and liquefaction. Each failure mode may result in partial or complete embankment failure. Moreover, the commission of an act (or acts) of deliberate sabotage can not only lead to these detrimental results but, at the very least, worsen or hasten damage effects. Simply put, acts of terrorism have changed the calculus related to failure modes of tailings embankments and other Dams Sector infrastructure.

7.8 RESERVOIRS

In addition to flood control, communities are reliant on reservoir storage for drinking water, agricultural use, and industrial use. Reservoirs are used to store raw or untreated water. They can be located underground (buried), at ground level, or on an elevated surface. Reservoirs can vary significantly in size; small reservoirs can hold as little as 1,000 gallons, while larger reservoirs may hold many millions of gallons.

Reservoirs can be either natural or human-made. Natural reservoirs can include lakes or other contained water bodies, while human-made reservoirs usually consist of some engineered structure, such as a tank or other impoundment structure. Besides the water containment structure, reservoir systems may include equipment for water treatment and distribution, such as intakes, pumps, piping, and chemical storage areas.

Reservoirs are vulnerable to contamination of the stored water, either through direct contamination of the storage area or through infiltration of the equipment, piping, or chemicals associated with the reservoir. For example, because many drinking water reservoirs are designed as aboveground, open-air structures, they are potentially vulnerable to airborne deposition, bird and animal wastes, human activities, and dissipation of chlorine or other treatment chemicals. However, one of the most serious potential threats to the system is the direct contamination of the stored water through dumping contaminants into the reservoir. Utilities have taken various measures to mitigate this type of threat, including fencing off the reservoir, installing cameras to monitor intruders, and monitoring for changes in water quality. Another option for enhancing security is covering the reservoir using a manufactured cover to prevent intruders from gaining physical access to the stored water. Implementing a reservoir cover may or may not be practical, depending on the size of the reservoir (e.g., covers are not typically used on natural reservoirs because they are too large for the cover to be technically feasible and cost-effective). Keep in mind that while covers can enhance the reservoir's security, covering a reservoir typically changes the reservoir's operational requirements. For example, vents must be installed in the cover to ensure gas exchange between the stored water and the atmosphere.

Water managers thoroughly research before installing a reservoir cover. They learn that a reservoir cover is a structure installed on or over the surface of the reservoir to minimize water quality degradation. The three basic design types for reservoir covers are as follows:

- Floating
- Fixed
- Air-supported

A variety of materials are used in manufacturing a cover, including reinforced concrete, steel, aluminum, polypropylene, chlorosulfonated polyethylene, or ethylene interpolymer alloys. Several factors affect a reservoir cover's effectiveness and its ability to protect the stored water. These factors include the following:

- The location, size, and shape of the reservoir
- The ability to lay/support a foundation (e.g., footing, soil, and geotechnical support conditions)
- The length of time the reservoir can be removed from service for cover installation or maintenance
- Aesthetic considerations
- Economic factors, such as capital and maintenance costs

For example, it may not be practical to install a fixed cover over a reservoir if the reservoir is too large or if the local soil conditions cannot support a foundation. A floating or air-supported cover may be more appropriate for these types of applications.

In addition to the practical considerations for the installation of these types of covers, there are several O&M concerns that affect the utility of a cover for specific applications. These include how different cover materials will withstand local climatic conditions, what types of cleaning and maintenance will be required for each particular type of cover, and how these factors will affect the cover's lifespan and its ability to be repaired when it is damaged.

The primary feature affecting the security of a reservoir cover is its ability to maintain its integrity. Any cover, no matter what its construction material, will provide good protection from contamination by rainwater or atmospheric deposition, as well as from intruders attempting to access the stored water with the intent of causing intentional contamination. The covers are large and heavy, and it isn't easy to circumvent them to get into the reservoir. At the very least, it would take a determined intruder, as opposed to a vandal, to defeat the cover.

7.9 THE BOTTOM LINE

Again, it is important to point out that the Dams Sector provides services that are essential to the U.S. economy and is a major contributor to the so-called good life (standard of living) that we presently enjoy. Dam managers recognize their vulnerabilities and are making significant efforts to improve planning and preparedness. Cooperation through industry groups has resulted in substantial information sharing of effective and best practices across the sector. Many sector owners and operators have experience with infrastructure protection and have more recently focused their attention on AI to optimize resource allocation, conduct predictive modeling, reduce costs, increase optimization, and improve cyber and site security. In addition to the economic consequences of a successful homegrown or foreign terrorist attack against Dams Sector facilities, there is also the potential threat to public health and safety and the environment.

NOTE

1 This section is adapted from F.R. Spellman (2020). *Infrastructure Protection Plan: Dam Sector*. Lanham, MD: Bernan Press.

REFERENCES

Association of State Dam Safety Officials (ASDSO). 2010. *Introduction to Dams*. Lexington, KY.

BPA. 2010. *The Columbia River System Inside Story*. Accessed 05/11/2025 @ http://www.bpa.gov/power/pg/columbia_river_inside_sotry.pdf.

EERE. 2008. *History of Hydropower*. Accessed 06/15/2025 @ www.eere.energy.gov/.

Henry, K. 2002. New face of security. *Government Security*, April, 30–37.

Obama, B. 2007. *Homeland Security*. Washington, DC: White House Government.

Stocks, C. 2014. Grand Coulee Dam. *Waterpowermagazine.com*. Accessed 08/18/2025 @ www.andrytz.com.

USEPA. 1994. *Design and Evaluation of Tailings Dams*. Washington, DC: U.S. Environmental Protection Agency.

USGS. 2003. *Active Mines and Mineral Processing Plant in the United States*. Washington, DC. United States Geological Survey.

USGS. 2015. *Lake Umatilla*. Accessed 05/11/2525 @ http://geonames.usgs.gov/apex/f?p=gnispq:3:::NO::{#_P3_FID:1513298.

8 Flooding, Drought, Water Overuse

An oyster settled on the seabed millions of years ago and later emerged a mile high in the Grand Canyon.

—Ciardi, J., p. 270, 1997

8.1 FLASH FLOOD ALLEY

A wall of water swamped teens and camp counselors at the summer camp as they slept or tried to sleep. On July 17, 1987, a flash flood on Texas' Guadalupe River swept away campers in tents and recreational vehicles, resulting in ten fatalities. This event is distinct from the occurrence on July 4, 2025, when a large volume of water (more than a trillion gallons) moved down the Guadalupe River in Texas Hill Country, leading to at least 130 deaths and over 150 people reported missing as of July 13, 2025, and still counting. What we know for sure at this point is that between July 4 and July 7, 2025, a large volume of water—over 1.8 trillion gallons—moved through the Kerr County region, resulting in significant loss of life and widespread damage. The actual death toll from the July 4 flood may remain unknown for some time. News of the flooding presented this date coincided with coverage marking one year since the near-assassination of President Trump, including the death of a firefighter and two others wounded in Butler, Pennsylvania. Fortunately, a pre-positioned sniper made a miraculous one in 100 shot that killed the assassin.

Many people were surprised by the devastating flood on July 4, but historical data reveal that floods are the most frequent natural disaster and cause the most fatalities worldwide. The flood was caused by a mesoscale convective vortex (i.e., a mid-level low-pressure system with a circular shape) boosted by additional tropical moisture from the remnants of Tropical Storm Barry.

8.2 WATER MANAGER'S WORST NIGHTMARE

Municipal, industrial, and agricultural Water Managers today face a variety of challenges in their daily operations. They must ensure an adequate water supply for customers. They must ensure the water supplied is clean, safe, and healthy. They must work constantly to prevent leaks, waste, and contamination. They must continually maintain water infrastructure integrity. They must worry about terrorism, the poisoning of local water supplies, and/or the destruction of water distribution systems. They must ensure water personnel are trained, licensed if necessary, and dependable.

Water Managers must implement various strategies aimed at reducing water use within communities and industries. These strategies include Community Awareness

DOI: 10.1201/9781003655848-13

Programs, efficient irrigation systems, rainwater harvesting, and promoting water-efficient appliances.

But there is more.

All these concerns, duties, worries, and system upsets are part of the Water Managers' constant bag of worries. In their wide range of responsibilities, including planning, implementing, and evaluating, the Water Manager also must worry about too much water and flooding via weather events or human construction failures. The point is that the Water Manager worries about potential floodwater damage to the water system under their responsibility. Local water managers take flooding seriously, recognizing that humans cannot control nature or withstand catastrophic floodwaters and their destructive effects.

Water managers should act proactively and review measures regularly. They must develop and implement infrastructure development, such as building dams, levees, and flood channels, to divert water away from populated areas during heavy weather. They should use stormwater management systems like permeable pavement and green roofs to reduce surface runoff and lower flood risks. They are required to create and implement an Emergency Preparedness Plan and to communicate procedures for prompt response during flooding.

Water Managers must ensure compliance with all relevant water regulations at local, regional, and international levels, including regular monitoring to guarantee that water supplies meet safety standards. Water Managers must submit detailed reports to regulatory bodies. They must stay current not only by keeping informed of new regulations but also by keeping in tune with artificial intelligence (AI) applications in their water systems and ensuring that they are operable as per design and intent.

The bottom line here is twofold: A gallon of water weighs 8.34 lbs, and when trillions of gallons are on the loose, there is little in this world that is safe from such a rampage. Nothing is completely secure from water. Another crucial point is not to build, camp, or hang out in a flood alley.

8.3 THE ROAD TO RESILIENCE

Water managers need a comprehensive strategy to boost flood resilience. This strategy should prioritize land conservation, limit development in river corridors (again, do not build in a floodplain), protect vulnerable areas, encourage growth in safer locations, and coordinate stormwater management across the watershed.

The Water Manager should have a plan to boost resilience against future floods by implementing and following Hazard Mitigation Plans, local land use codes and regulations, and other necessary plans such as community planning, public works, and emergency services.

The first step the Water Manager must take is to develop a list of necessary checks to ensure the locality is ready for any flooding contingency, flood recovery, and long-term resilience. This listing can be used to develop a checklist.

8.3.1 Flood Resilience List of Pertinent Questions[1]

To ensure overall strategies to enhance flood resilience, the Water Manager should develop a list or checklist that includes key flood resilience elements. To achieve this, answer key questions; if any are negative, take corrective action. The Water Manager can use the following list of pertinent questions to devise their checklist.

The first question is: Does the community's comprehensive plan have a hazard element or flood planning section? Does the community's comprehensive plan reference the local Hazard Mitigation Plan and disaster recovery plans? Then the Water Manager needs to check if the comprehensive plan, if they have one, identifies flood- and erosion-prone areas, including river corridor and fluvial erosion hazard areas, if applicable. One of the key questions is whether the local government emergency response personnel, floodplain manager, and Department of Public Works participated in developing/updating the comprehensive plan. The Water Manager is advised to create emergency plans in collaboration with personnel involved in water operations. Including those responsible for implementing the plan can help ensure the plan is understood and executed according to established procedures and training.

Another important consideration for the Water Manager is whether the community has a Hazard Mitigation Plan that has been approved by both the Federal Emergency Management Agency (FEMA) and the state emergency management agency. You need these key players on board to effectively coordinate and manage any flooding event. It is recommended that your Flooding Plan be cross-referenced with the local comprehensive plan to ensure coordination of actions with other stakeholders. Team and local government planner input is key to gaining buy-in. They can provide valuable insight and recommendations for the Hazard Mitigation Plan—and at the same time, "know the plan" in case it needs to be enacted. Include emergency planners, responders, local businesses, hospitals, schools, landowners, watershed residents, and others affected by flooding in your planning. A large team makes it easier to promote green management, meaning getting everyone to buy in and participate in using green infrastructure to help prevent flooding.

SIDEBAR 8.1 GREEN STORMWATER INFRASTRUCTURE (GSI)

Section 502 of the Clean Water Act (CWA) defines green infrastructure as "the range of measures that use plant or soil systems, permeable pavement or other permeable surfaces or substates, stormwater harvest and reuse, or landscaping to store, infiltrate, or evapotranspirate stormwater and reduce flows to sewer systems or to surface waters."

Flooding is one of the biggest threats to water quality in the United States. It is a transporter of trash, bacteria, heavy metals, plastics, and other pollutants through storm sewers and into local rivers and other waterways. The heavier the storm, the more likely it is to cause flooding that can damage property and infrastructure. Traditionally, communities have used gutters, pipes, and tunnels to quickly carry stormwater downstream to wastewater treatment plants or untreated to local water bodies. This practice is not good because floodwater is not being managed at its source. To meet this challenge, many communities are installing green infrastructure to complement existing infrastructure and reduce flooding. Moreover, green infrastructure helps to comply with the CWA, treat stormwater, improve water quality, and mitigate flooding while also beautifying communities.

Green infrastructure uses filtration, infiltration, and evapotranspiration to treat and soak up rainwater where it falls, enabling multiple paybacks, including environmental, social, and economic benefits. Examples of green infrastructure practices are listed below.

- **Stormwater Wetlands**: Basins designed with water, soil, or sand substrate, and wetland plants. This technique stores floodwater during a storm and releases it slowly, reducing peak flows. However, it is limited in that it requires a lot of space and is not well-suited for urban areas.
- **Green and Blue Roofs**: Green roofs are gently sloping or flat roofs that contain a planting medium and vegetation on top of the roof material. Typically, a green roof has a filter membrane, support panel, thermal insulation, and vapor control layer, as well as a waterproof membrane. Blue roofs are non-vegetated roofs that detain water through a flow-restriction device around a roof drain. From there, the water travels to a storm sewer system or a green-infrastructure cistern or bioretention area. This technique is limited to functioning on flat or gently sloping roofs that can structurally withstand the necessary weight for these systems.
- **Existing Forests and Wetlands**: This is the most economical way to absorb and clean water by protecting existing forests and wetlands. This area should be protected in perpetuity and expanded where possible. The limitations of this technique are that it require the conservation of forests and wetland areas.
- **Tree Plantings and Conservation**: In addition to conserving existing forests, it is essential to plant native trees or conserve existing trees, which can slow down, capture, and store runoff in the canopy and release water into the atmosphere through evapotranspiration. Tree roots also create soil conditions that promote infiltration. The limitation is space; trees need adequate space and soil to ensure their health.
- **Stormwater Tree Trenches**: This is a row of trees that are connected by an underground trench engineered with layers of gravel and soil that store and filter stormwater runoff. This method requires careful selection of tree species, appropriate root zone areas, and the absence of conflicts with utility structures like electrical wires and signs.
- **Bioswales and Rain Gardens**: Bioswales convey stormwater at a slow, controlled rate, and the flood-tolerant vegetation and soil function as a filter medium, cleaning runoff and allowing infiltration. This practice can be expensive for large areas.
- **Permeable Pavement**: It consists of pavements and pavers with void spaces that enable runoff to pass through and be temporarily held in an underground stone base, before either infiltrating into the ground or being discharged through an underdrain. One problem with this setup is that snow removal by snowplows can damage the pavement surface.
- **Rain Harvesting with Rain Barrels or Cisterns**: Rain barrels and aboveground cisterns are closed containers that retain runoff for

non-potable reuse purposes such as landscaping and car washing. Roof runoff is directed into a downspout and then to a 50- to 100-gallon barrel or cistern. A limitation is that rain barrels can only collect a fixed volume of water and require draining after each rainfall. Also, the water collected is not suitable for drinking.

Okay, let's go back to the Water Manager's list of questions they should be asking.

In reviewing the Hazard Mitigation Plan, does it use green infrastructure techniques to help prevent flooding? Does the Hazard Mitigation Plan identify projects that could be included in pre-disaster grant applications? Regarding Community Plans, do they encourage green infrastructure techniques? Do community plans address potential flooding from climate change? Budget questions often arise, especially about funding necessary tasks. Are structural flood mitigation approaches, such as repairing bridges, culverts, and levees, and non-structural approaches, such as green infrastructure, that require significant investment of resources coordinated with local capital improvement plans and prioritized in the budget? Ensure that the community participates in the National Flood Insurance Program Community Rating System.

The community must discourage development in river corridors—again, *do not build* in a flood zone. Conserving land and limiting development in river corridors starts with acquiring property to support stormwater absorption, channel adjustment, and flood resilience. The Water Manager should ensure that buyouts are made on properties prone to frequent flooding. Are tax incentives available to property owners who conserve vulnerable land? Are incentives available for those who restore riparian and wetland vegetation in areas prone to erosion and flooding?

Again, building in a flood zone is not recommended and should be prohibited.

A forward-thinking Water Manager should encourage agricultural and landowners to adopt pre-disaster mitigation measures, such as storing hay bales and equipment in low-risk flood areas. Are installed ponds or swales in place to capture stormwater? Are those living in your area of concern planting vegetation that can tolerate flooding? Are land management practices used to improve the capability of the soil on local lands to retain water?

FEMA has minimum standards for Special Flood Areas; the Water Manager must ensure that their community follows these standards, and ideally, the community should adopt a plan that exceeds the FEMA requirements. Community planners and executives should prevent or limit new encroachment in flood-prone regions.

Again, it is not wise to build in a flood alley region or on a floodplain.

The Water Manager must ensure that the community enforces development regulations that protect vulnerable areas. These regulations should address fluvial erosion hazard zoning, agricultural or open space zoning, and conservation or cluster subdivision ordinances as appropriate. Are there any tools available to ensure limited development in river corridors and Special Flood Hazard Areas?

The Community Water Manager must ensure that people, buildings, and facilities in vulnerable settlements are protected from flooding. First, the areas or sites prone to flooding must be identified. Does the comprehensive plan and Hazard Mitigation

Plan identify developed areas prone to flooding? Does the plan discourage development in those areas or require strategies to reduce damage to buildings during floods? Have the HVAC systems been elevated and basements properly sealed for facilities located within the flood zone? The Hazard Mitigation Plan must list critical infrastructure in vulnerable areas—such as town facilities, bridges, roads, and wastewater treatment plants—that require protection, repair, or relocation.

Water Managers for public utilities and other entities need to check whether land development regulations and building codes promote safe building and rebuilding in flood-prone areas. As part of this process, the Water Manager must determine whether zoning or flood regulations mandate an elevation of two feet or more above the base flood elevation. Can the community implement a temporary post-disaster moratorium on new development? Moreover, the Water Manager needs to verify that non-conforming use and structure standards have been revised to encourage safer rebuilding in flood-prone areas. The city manager also needs to confirm that the community has adopted the International Building Code and the American Society of Civil Engineers standards that promote flood-resistant building. Another important question is whether the community plans for costs associated with follow-up inspections and enforcement of land development regulations and building codes.

Another key planning consideration is to verify if the community mandates developers rebuilding in flood-prone areas to include additional flood storage in redevelopment projects, such as creating parks, open spaces, or setting aside riverfront areas for high-water events.

The Water Manager should help plan parks and riverfront recreation that link the community to the river.

A significant concern for the Water Manager regarding flooding is to ensure that the comprehensive plan or Hazard Mitigation Plan discusses strategies to determine whether to relocate structures that have been repeatedly flooded, including identifying an equitable approach for community involvement in relocation decisions and potential funding sources.

Water managers in flood-prone regions should provide information to community planners and executives to support the consideration of new development in safer areas. Community planners need to identify safe growth areas for development. Has the community adopted policies to encourage development in these safe areas? Regarding these safe areas, has the community changed land use regulations and codes to allow for this type of development? The Water Manager must audit and verify that development in safe areas aligns with community needs for off-street parking, building height, density, and front-yard setbacks, ensuring that regulations do not unintentionally hinder development. Community planners are responsible for ensuring that budgeting and capital improvements, such as wastewater treatment plants and roads, are made to support development in designated growth areas. Community officers should encourage building flood-resistant structures in designated safe zones.

Water Managers need to study and then implement Stormwater Management Techniques throughout the whole watershed. The entire watershed can be quite extensive, which means that the Water Manager must determine if the community is coordinated with neighboring jurisdictions to explore a watershed-wide approach to stormwater management. The Water Manager should also help establish a stormwater utility to fund stormwater management. The Water Manager also needs to check on whether the community has implemented strategies to reduce stormwater impermeable surfaces like roads, driveways, and parking lots. Other questions or queries the Water Manager needs to ask may include the following: Do stormwater management regulations extend to areas not covered by federal or state requirements? Do stormwater management regulations promote the use of green infrastructure techniques? Has the community implemented measures to protect trees? Has the community adopted the steep slope strategy? Finally, has the community adopted riparian and wetland buffer requirements?

8.3.2 Low Impact Development (LID) Practices[2]

Frequently termed green infrastructure, Low Impact Development (LID) encompasses a range of practices designed to replicate or maintain natural drainage processes for effective stormwater management. Retention of stormwater and infiltration of the water into the ground to reduce runoff is what LID practices are all about, including the reduction of runoff, water pollution, and downstream flooding.

Protect unstable slopes from cleaning, grading, and development whenever possible. In practice, this is not always the case. In instances where development occurs on slopes, LID practices can be used when the proper precautions are followed. Keep in mind, however, that LID practices should not be a substitute for slope protection.

Implementing LID practices on or in proximity to slopes poses an increased risk of soil erosion and potential landslides, as these areas may experience heightened water saturation. Planners must consider the risks of water infiltration in hilly or valley areas when designing LID projects, as many LID practices promote soil absorption.

Not all the news is bad, however, because LID practices can be used successfully on sloped landscapes where site conditions are favorable, the correct practice is selected, and the design incorporates elements to prevent slope failure and blow-out of the LID practice.

So, what are these LID practices? What are the design features? These are just a couple of questions water management must ask and answer.

Regarding design, many LID practices can be implemented with features such as vegetative plantings, diversion berms, structural walls, check dams, and baffles. These features help slow down, retain, and infiltrate water on slopes.

- Slopes can be stabilized by planting trees and other vegetation that hold soil in place and absorb water.
- Diversion berms can be constructed across slopes to reduce runoff velocity and erosive flows and to promote infiltration and plant growth by retaining water in depressions.
- Terraces and weep gardens (aka bioretention, green infrastructure, or rain gardens) can be designed with structural walls on the downslope face that will discharge excess runoff when the system is saturated.

- Check dams can be incorporated on slopes to manage the flow volume, encourage retention and infiltration, and reduce erosion.
- Baffles can be constructed beneath permeable pavement to increase storage and promote infiltration.

8.3.3 GSI Design Elements

When a community decides to go green in handling stormwater runoff by employing certain design elements within the community, water management must be involved in the design process. Green Stormwater Infrastructure (GSI) facilities typically feature several key design elements. Since these elements are optional (e.g., the amount of water that can be stored in the region), and the GSI will differ depending on regional and site-specific factors, this is the type of information provided by water management, indeed, a key member of any GSI planning team and operation. The GSI planning team must understand the relevant terms and definitions to design the GSI project for their location effectively and to communicate with other team members. Several of these critical terms are listed and defined below.

- **Inflow**: For a GSI facility to capture and treat stormwater runoff, the runoff must enter the practice. Runoff entering GSI facilities, also called inflow, can travel in the form of sheet flow or concentrated flow that moves through engineered inlet structures (curb cuts, gutters, etc.). Sometimes other methods are used, such as tunnels and depressional channels.
- **Pretreatment components**: Forebays and vegetated filters function as pretreatment and are used for inflow entering a GSI facility as concentrated flow through inlets such as curb cuts, trench drains, and pipes. Pretreatment components prevent erosion by slowing inflow velocities and dissipating the energy. When the inflow speed is reduced, the suspended solids and trash settle out, preventing downstream clogging in the local area.
- **Energy dissipator**: Reducing runoff velocities and dissipating energy is accomplished in the pretreatment through the use of weirs, check dams, and rock rundowns. Installed weirs are the common choice of designers in reducing runoff velocity and energy dissipation. When energy dissipation is reduced, the likelihood of erosion and vegetation disturbance is also reduced. Moreover, slower flow velocities within the facility promote infiltration.

SIDEBAR 8.2 WEIRS

Note that weirs are used in other applications besides energy dissipation and reducing the velocity of flow. For example, weirs are widely used devices to measure open-channel flow. As can be seen in Figure 8.1, a weir is simply a dam or obstruction placed in the channel so that water backs up behind it and then flows over it. The sharp crest or edge allows the water to spring clear of the weir plate and to fall freely in the form of a *nappe*. As Nathanson (1997) points out, when the nappe discharges freely into the air, there is a hydraulic relationship between the height and depth of water flowing over the weir crest and the flow rate. This height, the vertical distance between the crest and the water surface, is called the head on the weir; it can be measured directly with a meter or yardstick or automatically by float-operated recording devices. Two frequently used types of weirs, rectangular and triangular, are shown in Figure 8.2.

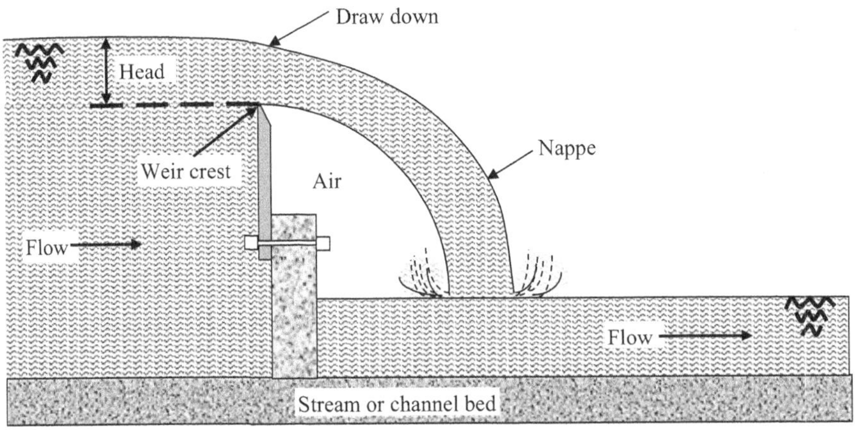

FIGURE 8.1 Side view of a weir.

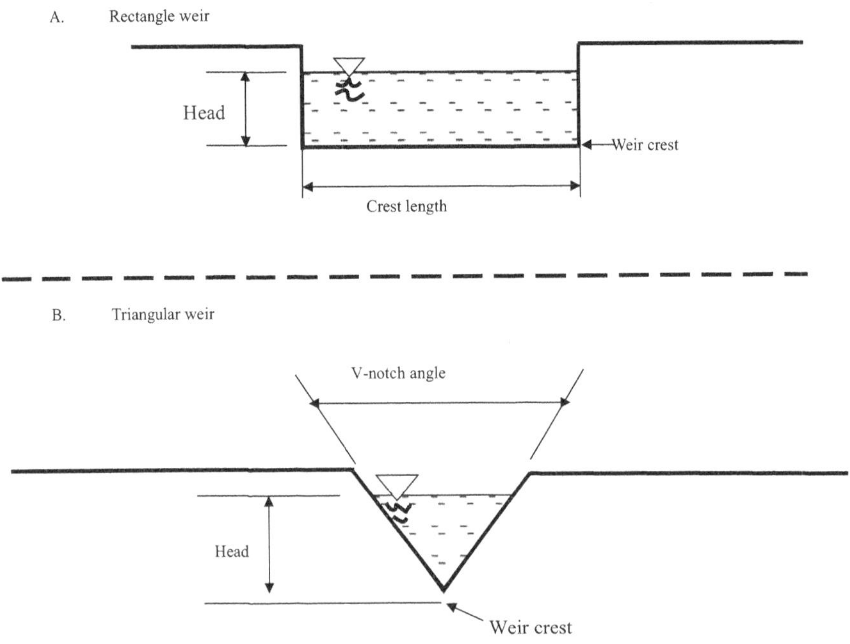

FIGURE 8.2 (A) Rectangular weir; (B) triangular V-notch weir.

Rectangular weirs are commonly used for large flows (see Figure 8.2A). The formula used to make rectangular weir computations is

$$Q = 3.33 \times L \times h^{1.5} \tag{8.1},$$

where
Q = flow
L = width of weir
h = head on weir (measured from the edge of the weir in contact with the water, up to the water surface).

EXAMPLE 8.1

Problem: A weir 4 ft high extends 15 ft across a rectangular channel in which 80 cfs is flowing. What is the depth just upstream from the weir?

Solution:

$$Q = 3.33 \times L \times h^{1.5}$$
$$80 = 3.33 \times 15\ h^{1.5}$$
$$h = 1.4\ \text{ft}\ \left(\text{w/calculator: 1.6 INV}\ y^{x1.5} = 1.36\ \text{or}\ 1.4\right)$$
$$4\ \text{ft height of weir} + 1.4\ \text{ft head of water} = 5.4\ \text{ft depth}$$

Triangular weirs, also known as V-notch weirs, can have notch angles ranging from 22.5° to 90°, but right-angle notches are the most common (see Figure 8.2B).

The formula used to calculate V-notch (90°) weir is

$$Q = 2.5 \times h^{2.5} \tag{8.2},$$

where
Q = flow
h = head on weir (measured from bottom of notch to water surface).

EXAMPLE 8.2

Problem: What is the minimum weir height needed to measure a 1,200 gpm flow using a 90-degree V-notch weir, given a 4 ft/sec flow rate in a 2.5 ft wide rectangular channel?

Solution:

$$\frac{1200\ \text{gpm}}{60\ \text{sec/min} \times 7.48\ \text{gal/cu ft}} = 2.67\ \text{cfs}$$

$$Q = A \times V$$
$$2.67 = 2.5 \times d \times 4$$
$$d = 0.27\ \text{ft}$$

$$Q = 2.5 \times h^{2.5}$$
$$2.67 = 2.5 \times h^{2.5}$$
$$h = 1.03 \ (\text{calculator: } 1.06 \text{ INV } y^{x2.5} = 1.026 \text{ or } 1.03)$$
$$0.27 \text{ ft } (\text{original depth}) + 1.03 \ (\text{head on weir}) = 1.3 \text{ ft}$$

It is essential to point out that weirs, aside from being operated within their flow limits, must also be operated within the available system head. In addition, the operation of the weir is sensitive to the approach velocity of the water, often necessitating a stilling basin or pound upstream of the weir. Weirs are not suitable for water that carries excessive solid materials or silt, which deposits in the approach channel behind the weir and destroys the conditions required for accurate discharge measurements.

Note: Accurate flow rate measurements with a weir cannot be expected unless the proper conditions and dimensions are maintained.

- **Mulch layer**: Mulch retains water, traps pollutants, prevents erosion, suppresses weed growth, and provides a favorable environment for beneficial soil organisms to thrive at the mulch/soil interface. However, mulch can be susceptible to suspension and washout during intense storms. It can leach dissolved organic carbon, which can stimulate unwanted biological activity in freshwater bodies (e.g., lakes, rivers, streams). As a result, other alternatives, such as expanded shale, gravel, or crushed rock, may be used.
- **Vegetation**: The planning team should be aware of the benefits of using native, noninvasive plants, such as grasses, perennials, shrubs, and trees, in bioretention. The goal is to incorporate vegetation in their plan and understand the need to design holistic bioretention for performance and longevity. This vegetation can reduce runoff velocities, improve infiltration rates, and enhance water quality through phytoremediation and nutrient uptake (via plant roots and biological processes). Limitations in using vegetation as a bioretention process include site conditions related to soil types, available water, light, climate, road salt application, and other environmental stressors. The team should also assess plant locations and purposes, such as aesthetics, nutrient removal, infiltration, or supporting wildlife.

8.4 SOIL INFILTRATION

Much of the world's freshwater is stored underground. The water begins its journey as precipitation, but through infiltration and seepage, large volumes of water seep into the ground. Groundwater sustains plant life and fulfills human needs. Plus, infiltration of stormwater helps to prevent or limit flooding.

Frank R. Spellman (2020)

Water Managers, community planners, and those responsible for stormwater management should have at least a fundamental understanding of soil infiltration. Personnel responsible for water management, ideally, should be generalists. This means that the Water Manager must possess widespread knowledge or should know a little bit about everything, dealing not only with water but also with its fate, especially ensuring not only freshwater is safe for consumption but also having knowledge of general science dealing with soil. Water Managers must understand soil water movement, local soil characteristics, and watershed conditions to assess infiltration. During storms, infiltration enables soil to store water for plants and organisms, while helping manage stormwater and reduce flooding.

8.4.1 We Take It for Granted

We take soil for granted. It has always been there, right?—With the implied corollary that it will always be there, right? But where does soil come from?

Of course, soil was formed, and in a never-ending process, it is still being formed. However, soil formation is a slow process—one at work throughout millennia, as mountains are worn away to dust through bare rock succession.

Any activity, human or natural, that exposes rock to air begins the process. Through the agents of physical and chemical weathering, through extremes of heat and cold, through storms, earthquakes, and entropy, bare rock is gradually worn away. As buildings deteriorate, plants accelerate their decay.

Lichens are often the initial organisms to colonize bare rock, growing on its surface and producing mild acids that slowly alter the rock. They also help collect and retain a thin layer of soil against the rock. As a result of the modified growth conditions, lichens do not persist and are instead succeeded by mosses.

The mosses establish themselves in the soil trapped and enriched by the lichens and collect even more soil. They hold moisture to the surface of the rock, setting up another change in environmental conditions.

Well-established mosses hold enough soil to allow herbaceous plant seeds to invade the rock. Grasses and small flowering plants move in, sending out fine root systems that hold more soil and moisture, and work their way into minute fissures in the rock's surface. More organisms join the growing community.

Weedy shrubs are the next invaders, with heavier root systems that find their way into every crevice. Each stage of succession affects the decay of the rock's surface and adds its own organic material to the mix. Mountains erode into soil over time due to plants and changing weather conditions. Over time, rock breaks down into smaller, weaker pieces, ultimately forming the mineral-rich soil we depend on.

The term "soil" often creates confusion among laypeople, soil scientists, engineers, and earth scientists. Soil is the upper layer of decomposed rock and organic material containing air, moisture, and nutrients that support life. Most people would have little difficulty in understanding and accepting this simple definition. Then why are various groups confused about the exact meaning of the word soil? Soil is complex, which often leads to confusion. In addition, the term soil has different meanings for diverse groups (such as pollution, the exact definition of soil is a personal judgment call). Let us look at how some of these diverse groups view soil.

The average person seldom gives soil a first or second thought. Why should they? Soil is not as big a deal. It is essential, but it does not impact their lives, pay their bills, or feed the bulldog.

Not exactly. Not directly.

The average person seldom thinks about soil as soil. He or she may think of soil in terms of dirt, but hardly ever as soil. Why is this? Having said the obvious about the confusion between soil and dirt, let us clear up this confusion.

First, soil is not dirt. Dirt is misplaced soil—soil where we do not want it, contaminating our hands, clothes, automobiles, and tracked in on the floor. We try to clean up dirt and keep our living environments clean.

Second, soil is too special to be called dirt. Why? Because soil is mysterious, and whether we realize it or not, essential to our existence. Because we think of it as common, we relegate soil to an ignoble position. As our usual course of action, we degrade it, abuse it, throw it away, contaminate it, and ignore it; we treat it like dirt, and only feces hold a lower status. Soil deserves better.

Why?

Again, because soil is not dirt—how can it be? It is not filth, grime, or squalor. Instead, soil is clay, air, water, sand, loam, organic detritus of former lifeforms, and, most importantly, the amended fabric of Earth itself. If water is Earth's blood and air is Earth's breath, then soil is its flesh and bone and marrow. Simply put, soil is the substance that most life depends on.

Soil scientists (or pedologists) are a group interested in soils as a medium for plant growth. Their focus is on the upper meter or so beneath the land surface (known as the weathering zone, which contains the organic-rich material that supports plant growth), directly above the unconsolidated parent material. Soil scientists have developed a classification system for soils based on the physical, chemical, and biological properties that can be observed and measured in the soil.

Soil engineers are typically soil specialists who look at soil as a medium that can be excavated using tools. Soil engineers focus on a soil's load-bearing capacity, not its suitability for plant growth. They attempt to determine (through examination and testing) a soil's particle size, particle-size distribution, and the plasticity of the soil.

Earth scientists (or geologists) have a view that typically falls between pedologists and soil engineers. They are interested in soils and weathering processes as past indicators of climatic conditions, as well as the geological formation of useful materials ranging from clay deposits to metallic ores.

Are you interested in learning more about soil? Take yourself out to a plowed farm field somewhere. Reach down and pick a handful of soil and look at it closely. What are you holding in your hand? Look at the two descriptions of that handful of soil that follow, and you may gain a better understanding of what soil is and why it is critically important to us all.

(1) A handful of soil is alive, a delicate living organism—as lively as an army of migrating caribou and as fascinating as a flock of egrets. Soil is full of diverse life and should be considered its ecosystem—or rather, multiple ecosystems.
(2) When we reach down and pick up a handful of soil, exposing Earth's stark bedrock surface, it should remind us (and maybe startle some of us) of the realization that without its thin living soil layer, Earth is a planet as lifeless as our moon (Spellman, 1999).

In our attempt to define soil, to differentiate soil from dirt, in our trying to make the point that soil is essential, critical, and vital to all of us, that it is arguably the most valuable of all mineral resources on Earth, we would be hard-pressed to do so in a more succinct but complete fashion than to quote E.L. Konigsburg (1996), who states, "Soil is the working layer of the earth."

This section provides comprehensive information about soil. And in so doing, we do not lose sight of the fact that we currently have a problem with soil. The problem? For example, the United States and most countries are seeing ongoing growth in both the volume and variety of discarded materials, reflecting our development into a throwaway society. This places a significant burden on disposal sites—landfills. It may surprise you to know that in the past three decades, a substantial amount of the material being landfilled is contaminated soil. This contaminated soil is only part of the problem, however. Soil directly contacts both water and air. Since water, air, and soil exist in a complex interrelationship, this contact causes issues with water and soil as well.

In the past, soil pollution problems have been exacerbated by the physical, economic, and technical limitations associated with the technologies currently available for the remediation of contaminated soil. Progress has been made in this vital area, however. While problems are still to be solved in the areas of soil remediation, technology is not necessarily one of them.

8.4.2 Definitions of Key Terms

Note: Every branch of science, including soil science, has its own language. The terminology used here is as different from that of astrophysics as that of astronomy is from botany. To address soil science, pollution, remediation, and water infiltration, Water Manager staff should understand the vocabulary in this text. From my experience conducting pre-OSHA and operational inspections at water and wastewater treatment plants and public utilities, I have observed that personnel often lack a complete understanding of their operations. When speaking with anyone in charge who does not know even the basics of operations, including flood control measures and so forth, there is an instant credibility problem. I define key terms at the beginning, not just in the glossary, because water operators and managers have told me they prefer seeing important definitions up front instead of having to keep checking the glossary.

Ablation till—a superglacial coarse-grained sediment or till, accumulating as the subadjacent ice melts and drains away, finally deposited on the exhumed subglacial surface.

Absorption—refers to the uptake of ions and water into plant roots, which occurs either through metabolic processes within the root (active absorption) or through diffusion along a concentration gradient (passive absorption).

Acid rain—atmospheric precipitation with pH values less than about 5.6, the acidity being due to inorganic acids such as nitric and sulfuric acids that are formed when oxides of nitrogen and sulfur are emitted into the atmosphere.

Acid soil—a soil with a pH value of less than 7.0 or neutral. Soils can become acidic due to their geological origin, leaching processes, the decomposition of plant material, or the addition of substances like aluminum sulfate (alum). The addition of lime products can neutralize acidic soils.

Adsorption—the attraction of ions or compounds to the surface of a solid.

Aeration, soil—the process by which air in the soil is replaced by air from the atmosphere. In a well-aerated soil, the soil air is similar in composition to the atmosphere above the soil. Poorly aerated soils usually contain more carbon dioxide and correspondingly less oxygen than the atmosphere above the soil.

Aerobic—growing only in the presence of molecular oxygen, as aerobic organisms.

Aggregates (soil)—soil structural units of various shapes, composed of mineral and organic material, formed by natural processes, and having a range of stabilities.

Agronomy—a specialization of agriculture concerned with the theory and practice of field crop production and soil management. It involves the scientific management of land.

Air capacity—percentage of soil volume occupied by air spaces or pores.

Air porosity—the proportion of the bulk volume of soil that is filled with air at any given time or under a given condition, such as a specified moisture potential, usually the large pores.

Alkali—a substance capable of liberating hydroxide ions in water, measured by a pH of more than 7.0, and possessing caustic properties. It can neutralize hydrogen ions, with which it reacts to form a salt and water, and is an essential agent in rock weathering.

Amendment, soil—any substance other than fertilizers (such as compost, sulfur, gypsum, lime, and sawdust) used to alter the chemical or physical properties of soil, generally to make it more productive.

Ammonification—the production of ammonia and ammonium nitrogen through the decomposition of organic nitrogen compounds in soil organic matter.

Anaerobic—without molecular oxygen.

Aspect (of slopes)—the direction that a slope faces in relation to the sun.

Assimilation—the uptake of plant nutrients and their conversion into actual plant tissues.

Atterberg limits—water contents of fine-grained soils at different states of consistency.

Autotrophs—plants and microorganisms capable of synthesizing organic compounds from inorganic materials through photosynthesis or oxidation reactions.

Available water—the portion of water in soil that plant roots can readily absorb. It is the amount of water released between field capacity and the permanent wilting point.

Bedrock—the solid rock underlying soils and regolith at depths ranging from zero (where exposed by erosion) to several hundred feet.

Biological function—the role played by a chemical compound or a system of chemical compounds in living organisms.

Biome—a major ecological community extending over large areas.

Blow-out—a deflation depression, eroded by wind from the face of a vegetated dune.

Breccia—a rock composed of coarse angular fragments cemented together.

Calcareous soil—containing sufficient calcium carbonate (often with magnesium carbonate) to effervesce visibly when treated with hydrochloric acid.

Caliche—a layer near the surface, cemented by secondary carbonates of calcium or magnesium precipitated from the soil solution. It may occur as a soft, thin soil horizon, as a hard, thick bed just beneath the solum, or as a surface layer exposed by erosion.

Capillary water—held within the capillary pores of soils, primarily available to plants.

Catena—the sequences of soils that occupy a slope transect, from the topographic divide to the bottom of the adjacent valley.

Cation—an atom that has lost one or more negatively charged electrons and is thus positively charged.

Class, soil—a group of soils having a definite range in a particular property such as acidity, degree of slope, texture, structure, land-use capability, degree of erosion, or drainage.

Clay—a soil separate consisting of particles <0.0002 mm in equivalent diameter.

Cohesion—holding together: force holding a solid or liquid together, owing to attraction between like molecules. It decreases with a rise in temperature.

Colloidal—matter of very fine particle size.

Convection—a process of heat transfer in a fluid involving the movement of substantial volumes of the fluid. Convection is very important in the atmosphere, and to a lesser extent, in the oceans.

Detritus—debris from dead plants and animals.

Diffusion—the movement of atoms in a gaseous mixture, or ions in a solution, primarily due to their own random motion.

Drainage—the removal of excess water, both surface and subsurface, from plants. All plants (except aquatics) will die if exposed to an excess of water.

Duff—the matted, partly decomposed organic surface layer of forest soils.

Erosion—the wearing away of the land surface by running water, wind, ice, or other geological agents, including processes such as gravitational creep.

Evapotranspiration—the combined loss of water from a given area, during a specified period by evaporation from the soil surface and by transpiration from plants.

Exfoliation—mechanical or physical weathering that involves the disintegration and removal of successive layers of rock mass.

Fertility, soil—the quality of soil that enables it to provide essential chemical elements in the right quantities and proportions for the growth of specific plants.

Fixation—the transformation of a plant nutrient in soil from an available to an unavailable state.

Fluvial—deposits of parent materials laid down by rivers or streams.

Friable—a soil consistency term about the ease of crumbling of soils.

Heaving—the partial lifting of plants, buildings, roadways, and fence posts, etc., out of the ground due to freezing and thawing of the surface soil during the winter.

Heterotroph—an organism capable of deriving energy for life processes only from the decomposition of organic compounds, and incapable of using inorganic compounds as the sole source of energy or for organic synthesis.

Horizon, soil—a layer of soil, parallel to the soil surface, differing in properties and characteristics from adjacent layers below or above it.

Humus—stable fraction of soil organic matter (usually dark in color) remaining after the major portions of added plant and animal residues have decomposed.

Hydration—the incorporation of water into the chemical composition of a mineral, converting it from an anhydrous to a hydrous form. The term is also applied to a form of weathering in which hydration swelling creates tensile stress within a rock mass.

Hydraulic conductivity—the rate at which water can move through a soil.

Hydrolysis—the reaction between water and a compound (commonly a salt). The hydroxyl from the water combines with the anion from the compound undergoing hydrolysis to form a base, while the hydrogen ion from the water combines with the cation from the compound to form an acid.

Hygroscopic coefficient—the amount of moisture in a dry soil when it is in equilibrium with a standard relative humidity close to a saturated atmosphere (approximately 98%), expressed as a percentage based on oven-dry soil.

Infiltration—the downward entry of water into the soil.

Ions—atoms that have lost or gained one or more negatively charged electrons.

Land classification—the arrangement of land units into various categories based on the properties of the land and its suitability for a particular purpose.

Leaching—the removal of materials in solution from the soil by percolating waters.

Loam—the textural class name for soil having moderate amounts of sand, silt, and clay.

Loess—an accumulation of wind-blown dust (silt) that may have undergone mild diagenesis.

Marl—an earthy deposit consisting of calcium carbonate, usually mixed with clay. It is used for liming acid soils and is slower-acting than most lime products used for this purpose.

Mineralization—the conversion of an element from an organic form to an inorganic state due to microbial decomposition.

Nitrogen fixation—the biological conversion of elemental nitrogen (N2) to organic combinations or forms readily utilized in biological processes.

Osmosis—the movement of a liquid across a membrane from a region of high concentration to an area of low concentration. Water and nutrients move into roots independently.

Parent material—the unconsolidated and chemically weathered mineral or organic matter from which pedogenic processes develop the solum of soils.

Ped—a unit of soil structure such as an aggregate, crumb, prism, block, or granule, formed by natural processes.

Pedogenic/pedagogical process—any process associated with the formation and development of soil.

pH—the degree of acidity or alkalinity of the soil. Also referred to as soil reaction, this measurement is based on the pH scale, where 7.0 is neutral. Values from 0.0 to 7.0 are acidic, and values from 7.0 to 14.0 are alkaline. The pH of soil is determined by a simple chemical test where a sensitive indicator solution is added directly to a soil sample in a test tube.

Photosynthesis—the process by which green leaves of plants, in the presence of sunlight, manufacture their own needed materials from carbon dioxide in the air, water, and minerals taken from the soil.

Porosity, soil—the volume percentage of the total bulk not occupied by solid particles.

Profile, soil—a vertical section of the soil through all its horizons and extending into the parent material.

Regolith—the unconsolidated mantle of weathered rock and soil material on the earth's surface; loose earth materials above solid rock.

Rock—the material that forms the essential part of the earth's solid crust, including loose incoherent masses such as sand and gravel, as well as solid masses of granite and limestone.

Salinization—the process of the accumulation of salts in soil.

Sand—a soil particle between 0.05 and 2.0 mm in diameter; a soil textural class.

Silt—a soil separate consisting of particles between 0.05 and 0.002 mm in equivalent diameter; a soil textural class.

Slope—the degree of deviation of a surface from horizontal, measured in a numerical ratio, percent, or degrees.

Soil—an assemblage of loose and normally stratified granular minerogenic and biogenic debris at the land surface; it is the supporting medium for the growth of plants.

Soil air—the soil atmosphere; the gaseous phase of the soil, being that volume not occupied by soil or liquid.

Soil horizon—a layer of soil, parallel to the soil surface, with distinct characteristics produced by soil-forming processes. These characteristics form the basis for the systematic classification of soils.

Soil profile—a vertical section of the soil from the surface through all its horizons, including C horizons.

Soil structure—the combination or arrangement of primary soil particles into secondary particles, units, or peds. These secondary units may be, but usually are not, arranged in the profile in such a manner as to give a distinctive characteristic pattern. The secondary units are characterized and classified based on size, shape, and degree of distinctness into classes, types, and grades, respectively.

Soil texture—the relative proportions of the various soil separates in a soil.

Soluble—will dissolve easily in water.

Solum (plural sola)—the upper and most weathered part of the soil profile, including the A, E, and B horizons.

Subsoil—that part of the soil below the plow layer.

Till—unstratified glacial drift deposited directly by the ice and consisting of clay, sand, gravel, and boulders intermingled in any proportion.

Tilth—the physical condition of soil as related to its ease of tillage, fitness as a seedbed, and its impedance to seedling emergence and root penetration.

Topsoil—the layer of soil moved in cultivation.

Weathering—all physical and chemical changes produced in rocks, at or near the earth's surface, by atmospheric agents.

8.4.3 Functions of Soil

We usually think of soil in relation to our backyards, farms, forests, or regional watersheds. We think of soil as the substance upon which plants grow. This is about as far as the average person's thoughts extend regarding soil and its usefulness to us. However, soils play other roles. They have five main functions important to us: (1) soil is a medium for plant growth; (2) soils regulate our water supplies; (3) soils are recyclers of raw materials; (4) soils provide a habitat for organisms; and (5) soils are used as an engineering medium.

Let us examine each soil function.

8.4.3.1 Soil: A Plant Growth Medium

We are all aware of the primary function of soil: Soil serves as a plant growth medium and is responsible for plant growth (a function that becomes more important with each passing day as Earth's population continues to grow). However, while it is true that soil is a medium for plant growth (thus critical to maintaining life), let us also point out that soil is alive as well. Soil is essential for supporting life on Earth, as the formation and maintenance of soil are closely connected to the presence of living plants and animals. What does this mean? Let us look at how one renowned environmental writer, whose elegant prose brought this point to the forefront, explained this paradox.

> The soil community . . . consists of a web of interwoven lives, each in some way related to the others—the living creatures depending on the soil, but the soil in turn is a vital element of the earth only so long as this community within it flourishes.
>
> —Rachel Carson, Silent Spring, p. 56, 1962

Rachel Carson's message is clear and essential. She points out what the soil might say to us if it could—"Don't kill off the life within me and I will do the best I can to provide life that will help to sustain your life." What we have here is a tradeoff—one vitally important both to soil and to ourselves. Remember that most of Earth's people are tillers of the soil—and the soil is their source of life—and those soil tillers provide food for us all.

Soil serves as an essential growth medium, supplying crucial resources and functions for plants. To grow in soil, plants must have water and nutrients—soil provides these. To develop and sustain growth, a plant must have a root system—soil provides pore spaces for roots. To grow and maintain growth, a plant's roots must have oxygen for respiration and carbon dioxide exchange, and ultimately diffusion out of the soil—soil provides the air and pore spaces (the soil's ventilation system) for this. To continue to grow, a plant must have support—soil provides this support.

Suppose a plant seed is planted in the soil and is exposed to the proper amount of sunlight for growth to occur. In that case, the soil must provide nutrients through a root system that has space to grow, a continuous stream of water (it requires about 500 grams of water to produce one gram of dry plant material) for root nutrient transport and plant cooling, and a pathway for both oxygen and carbon dioxide transfer. Just as importantly, soil water provides the plant with its normal fullness or tension (turgor) it needs to stand, the structural support it needs to face the sun for photosynthesis to occur.

In addition to the functions stated above, soil is also an essential moderator of temperature fluctuations. When digging in a garden on a sweltering day, you'll find the surface soil is warm, but it becomes much cooler just a few inches down.

8.4.3.2 Soil: Regulator of Water Supplies

Few realize that walking on land is similar to crossing a bridge. This bridge spans over a vast body of water lying beneath the earth's surface.

Consider what happens to rain. Where does the rainwater go? Some, falling directly over water bodies, become part of the water body again. But an enormous amount falls on land. Where does it go? Some of the water, obviously, runs off—always following the path of least resistance. In modern communities, stormwater runoff is a hot topic. Cities have made major efforts to manage runoff and prevent flooding.

This section will examine precipitation and the various sinks into which it is distributed, followed by a discussion of how this process relates to soil water. We will start with surface water and then cover groundwater beneath the soil.

Surface water (water on the Earth's surface as opposed to subsurface water—groundwater) is primarily a product of precipitation—rain, snow, sleet, or hail. Surface water is water open to the atmosphere, created by the movement of water across or just beneath the Earth's surface. This overland flow is the same thing as surface runoff, which is the amount of rainfall that passes over the Earth's surface. Specific sources of surface water include rivers, streams, lakes, impoundments, shallow wells, rain catchments, and tundra ponds or muskegs (peat bogs).

Most surface water is the result of surface runoff. The amount and flow rate of surface runoff are highly variable. This variability stems from two main factors: (1) human interference (influences) and (2) natural conditions. In some cases, surface water dashes off land. From a water resources perspective, this outcome is typically considered unfavorable as it limits the opportunity for water to percolate into the soil and replenish groundwater aquifers. Other problems associated with quick surface water runoff are erosion and flooding. The main advantage of rapid surface water runoff is that it typically does not have time to absorb high mineral content. Surface water running slowly off land may be expected to have the opposite effects.

From a water resources perspective, this outcome is typically considered unfavorable as it limits the opportunity for water to percolate into the soil and replenish groundwater aquifers.

Surface water travels over the land to a predetermined destination. What factors influence how surface water moves? Surface water's journey over the face of the earth typically begins at its drainage basin, sometimes referred to as its drainage area, catchment, and/or watershed. For a groundwater source, this is known as the recharge area—the area from which precipitation flows into an underground water source.

A surface water drainage basin is usually an area measured in square miles, acres, or sections, and if a city takes water from a surface water source, how large (and what lies within) the drainage basin is essential information for the assessment of water quality.

We know water does not flow uphill. Instead, surface water runs off (like the flow of electricity) and follows along the path of least resistance. Water within a drainage basin will naturally (by the geological formation of the area) be shunted toward one primary watercourse (a river, stream, creek, or brook) unless some manufactured distribution system diverts the flow.

Several factors directly affect surface water flow. The principal factors are as follows:

Rainfall Duration. The length of the rainstorm affects the amount of runoff. Even a light, gentle rain will eventually saturate the soil if it lasts long enough. Once the saturated soil can absorb no more water, rainfall builds up on the surface and begins to flow as runoff.

Rainfall Intensity. The harder and faster it rains, the more quickly the soil becomes saturated. With heavy rains, the surface inches of soil quickly become inundated, and during short, intense storms, most of the rainfall may end up as surface runoff, because the moisture is carried away before significant amounts of water are absorbed into the earth.

Soil Moisture. If the soil is already laden with water from previous rains, the saturation point will be reached sooner than if the soil were dry. Frozen soil also inhibits water absorption: Up to 100% of snow melt or rainfall on frozen soil will end up as runoff because frozen ground is impervious.

Soil Composition. Runoff amount is directly affected by soil composition. Hard rock surfaces will shed all rainfall, obviously, but so will soil with a heavy clay composition. Clay soils possess small void spaces that swell when wet. When the void spaces close, they form a barrier that prevents additional absorption or infiltration. On the opposite end of the spectrum, coarse sand allows easy water flow-through, even in a torrential downpour.

Vegetation Cover. Runoff is limited by ground cover. Roots of vegetation and pine needles, pinecones, leaves, and branches create a porous layer (a sheet of decaying natural organic substances) above the soil. This porous "organic" sheet (ground cover) readily allows water into the soil. Vegetation and organic waste also function as a cover to protect the soil from heavy rainfall. Hard rains can compact bare soils, close off void spaces, and increase runoff. Vegetation and ground cover work to maintain the soil's infiltration and water-holding capacity. Note that vegetation and groundcover also reduce the evaporation of soil moisture.

Ground Slope. Flat land water flow is usually so slow that copious amounts of rainfall can infiltrate the ground. Gravity works against infiltration on steeply sloping ground where up to 80% of rain may become surface runoff.

Human Influences. Various human activities have a definite impact on surface water runoff. Most human activities tend to increase the rate of water flow. Canals and ditches are built to ensure steady flow, while agriculture often strips ground cover that would otherwise slow runoff. On the opposite extreme, human-caused dams are generally built to retard the flow of runoff.

Human habitations, with their paved streets, tarmac, paved parking lots, and buildings, create surface runoff potential, since so many surfaces are impervious to infiltration. All these surfaces hasten the flow of water, and they also increase the possibility of flooding, often with devastating results. Because of urban increases in runoff, a whole new field (industry) has developed: Stormwater management.

Paving over natural surface acreage has another profound side effect. Without enough area available for water to infiltrate the ground and percolate through the soil to eventually reach and replenish—recharge—groundwater sources, those sources may eventually fail, with devastating impact on local water supply.

We will now discuss groundwater.

Water falling to the ground as precipitation normally follows three courses. Some of the water flows into rivers and streams, some seeps into underground reservoirs, and the remainder evaporates or is released through vegetation. The water in the ground (groundwater) is "invisible" and may be thought of as a temporary natural reservoir (ASTM, 1969). All groundwater is in constant motion toward rivers or other surface water bodies.

Groundwater is defined as water below the earth's crust, but above a depth of 2,500 feet. Thus, if water is located between the earth's crust and the 2,500-foot level, it is considered usable (potable) fresh water. In the United States, it is estimated "that at least 50% of total available freshwater storage is in underground aquifers" (Kemmer, 1979, p. 17).

This text focuses on the soil's water retention needed for plant growth and to minimize flooding. Having said this, recall that earlier we stated that producing one gram of dry plant material requires about 500 grams of water. Note that about five grams of this water becomes an integral part of the plant. When rainfall is infrequent, the capacity of soil to retain water against gravity becomes significant. This makes water regulation for plants one of the essential functions of soil.

8.4.3.3 Soils: Recycler of Raw Materials

Soil is Earth's top recycler. The simple fact is that if it were not for the soil's incredible recycling ability, plants and animals would have run out of nourishment long ago. Soil recycles in other ways. For example, consider the geochemical cycles (i.e., where soil is a key player in the chemical interactions among Earth's water, air, and living organisms).

Soil possesses the incomparable ability and capacity to assimilate significant quantities of organic wastes and turn them into beneficial organic matter (humus), converting the nutrients in the wastes to forms that can be utilized by plants and animals. In turn, the soil returns carbon to the atmosphere as carbon dioxide, where it will eventually become part of living organisms through photosynthesis. Soil performs diverse recycling roles through complex chemical interactions with water, air, and organisms, playing a fundamental role in Earth's systems.

Consider one recycling function of soil that may not be so good. Soils can store significant quantities of carbon as soil organic matter, which influences global phenomena such as the greenhouse effect.

8.4.3.4 Soil: Physical Properties

From the soil pollution technologist's point of view (regarding land conservation and methodologies for contaminated soil remediation through reuse and recycling), five major physical properties of soil are of interest. They are soil texture, slope, structure, organic matter, and soil color. **Soil texture**—the proportions of different soil separates—cannot be easily or significantly changed. It is determined by the size of the rock particles (sand, silt, and clay particles) or the soil separates within the soil. The largest soil particles are gravel, which consist of fragments larger than 2.0 mm in diameter.

Particles between 0.05 and 2.0 mm are classified as sand. Silt particles range from 0.002 to 0.05 mm in diameter, and the smallest particles (clay particles) are less than 0.002 mm in diameter. Though clays are composed of the smallest particles, those particles have stronger bonds than silt or sand. However, once broken apart, they erode more readily. Particle size has a direct impact on erodibility. Soil typically consists of various particle sizes.

The **slope** (or steepness of the soil layer) is another important factor because the erosive power of runoff increases with the steepness of the slope. Slope also allows runoff to exert increased force on soil particles, breaking them apart more readily and carrying them farther away.

Soil structure (tilth) should not be confused with soil texture—they are different. Soil structure can significantly alter the properties defined by soil texture in the field. Soil structure refers to the combination or arrangement of primary soil particles into secondary particles (units or peds). Simply stated, soil structure refers to the way various soil particles clump together. The size, shape, and arrangement of clusters of soil particles called aggregates form naturally formed larger clumps called peds. Sand particles do not clump because sandy soils lack structure. Clay soils tend to stick together in large clumps. Good soil develops small friable (easily crumbled) clumps. Soil develops a unique, stable structure in undisturbed landscapes, but agricultural practices break down the aggregates and peds, lessening erosion resistance.

Decomposed plant and animal matter in soil improve both fertility and structure, particularly enhancing water retention. Live organisms—protozoa, nematodes, earthworms, insects, fungi, and bacteria—are typical inhabitants of soil. These organisms work to either control the population of organisms in the soil or aid in the recycling of dead organic matter. All soil organisms, in one way or another, work to release nutrients from the organic matter, changing complex organic materials into products that plants can use.

Soil color often varies from place to place. Soil colors range from very bright to dull grays, with a wide range of reds, browns, blacks, whites, yellows, and even greens. Soil color is primarily dependent on the quantity of humus and the chemical form of iron oxides present.

Soil scientists use a set of standardized color charts (the Munsell Color Book) to describe soil colors. They compare soil colors to color chips based on hue, value, and chroma.

8.4.3.5 Soil Separates

Again, soil particles have been divided into groups based on their size, termed soil separates—sand, silt, and clay by the International Soil Science Society System, the United States Public Roads Administration, and the United States Department of Agriculture. In this text, we use the classification established by the U.S. Department of Agriculture (USDA, 1975). These size ranges mark significant shifts in particle behavior and their effects on soil properties.

In Table 8.1, the names of the separates are given, together with their diameters, and the number of particles in one gram of soil (according to USDA).

TABLE 8.1
Characteristics of Soil Separates (USDA)

Separate	Diameter (mm)	Number of Particles/Gram
Very coarse sand	2.00–1.00	90
Coarse sand	1.00–0.50	720
Medium sand	0.50–0.25	5700
Fine sand	0.25–0.10	46,000
Very fine sand	0.10–0.05	722,000
Silt	0.05–0.002	5,776,000
Clay	Below 0.002	90,260,853,000

Sand, one of the individual-sized groups of mineral soil particles, ranges in diameter from 2 mm to 0.05 mm and is divided into five classes (see Table 8.1). Sand grains are spherical (rounded) in shape, with variable angularity, depending on the extent to which they have been worn down by abrasive processes such as rolling around by flowing water during soil formation.

Sand constitutes the structural component of soil, providing stability when combined with finer particles. Sand has large particles, so the spaces between grains are also large. This promotes free drainage of water and the entry of air into the soil. Sand is usually composed of a high percentage of quartz because it is the most resistant to weathering, and its breakdown is extremely slow. Many other minerals are found in sand, depending on the rocks from which the sand was derived. In the short term (on an annual basis), sand contributes little to plant nutrition in the soil. However, in the long term (thousands of years of soil formation), soils with a lot of weatherable minerals in their sand fraction develop a higher state of fertility.

Another individual-sized group of mineral soil particles is silt (microsand). Silt is finer than sand, weathers more quickly, and releases plant nutrients faster; it is also too small to see without magnification. Too fine to be gritty, silt imparts a smooth feel (like flour) without stickiness. Silt has much smaller pores than sand, as both are finer fragments of parent rock crystals. In flowing water, silt is suspended until it drops out when the flow is reduced. Silt is carried by wind and deposited as loess. Silt is finer than sand, weathers more quickly, and releases plant nutrients faster; it is also too small to see without magnification.

The clay soil separates (for the most part) much differently from sand and silt. Clay is composed of secondary minerals that were formed by the drastic alteration of the original forms or by the recrystallization of the products of their weathering. Because clay crystals are plate-like (sheeted) in shape, they have a tremendous surface area-to-volume ratio, giving clay an immense capacity to adsorb water and other substances on its surfaces. Clay acts as a storage reservoir for both water and nutrients. There are many kinds of clay, each with different internal arrangements of chemical elements, which give them individual characteristics.

8.5 SOIL CHARACTERIZATION

When people become ill, they may go to a doctor to seek a determination (a diagnosis) of what is causing the illness, and hopefully, a forecast or outlook (prognosis) of how long it will take before they feel well again.

What do diagnosis and prognosis have to do with soil? Quite a lot. The diagnostic techniques used by a physician to identify the causative factors leading to an illness are analogous to the soil practitioner using diagnostic methods to identify a particular soil. Sound far-fetched? It will not, because it is not. Water Managers must be able to determine the type of soil they study or work with.

Determining the type of soil makes sense, but what does prognosis have to do with all this? Consider that soil practitioners not only need to be able to identify or classify a soil type but also that this information allows them to correctly predict how a pollutant will react or respond when spilled in that type of soil. The pollutant's fate is crucial in assessing environmental harm, as a spill can impact soil, groundwater, and air. Therefore, a soil practitioner should employ diagnostic tools to accurately identify soil types and possess comprehensive familiarity with them in order to evaluate how pollutants or contaminants may behave when introduced into specific soils.

From the time humans first advanced from hunter-gatherer status to cultivators of crops, they noticed differences in "good" (productive) soils and "bad" (unproductive) soils. This was the earliest attempt to classify soils as either good or bad. The ancient Chinese, Egyptians, Romans, and Greeks all recognized and acknowledged the differences in soils as media for plant growth. These early soil classification practices were based primarily on texture, color, and wetness.

Soil classification became established as a scientific practice in the late eighteenth and early nineteenth centuries alongside the rise of geology. Terms (with an obvious geological connotation) such as "limestone" soils, "lake-laid" soils, as well as "clayey" and "sandy" soils came into being. The Russian scientist Dokuchaev was the first to suggest a generic classification of soils, stating that soils are natural bodies. Dokuchaev's classification work was then further developed by Europeans and Americans. The system is based on the theory that each soil has a definite form and structure (morphology), related to a combination of soil-forming factors. This system was utilized until 1960, when the U.S. Department of Agriculture published Soil Classification, A Comprehensive System. This classification system places major emphasis on soil morphology and gives less emphasis to genesis or the soil-forming factors compared to previous systems. In 1975, Soil Classification, A Comprehensive System, was replaced by Soil Taxonomy, which classifies objects according to their natural relationships. Soils are classified based on measurable properties of soil profiles.

Note that no clear delineation or line of demarcation can be drawn between the properties of one soil and those of another. Rather than abrupt changes, soil properties typically exhibit a gradation—sometimes subtle, such as between shades of white—as one transitions from one soil type to another. Brady and Weil (1996) point out that "the gradation in soil properties can be compared to the gradation in the wavelengths of light as you move from one color to another. The change is gradual, and yet we identify a boundary that differentiates what we call 'green' from what we call 'blue'" (p. 58). To properly characterize the primary characteristics of a soil, a soil must be identified down to the smallest three-dimensional characteristic sample possible (which makes sense when considering the impossibility of studying an entire soil area at one time).

However, to accurately perform a soil sample characterization, a sampling unit must be large enough so that the nature of its horizons can be studied and the range of its properties identified. The Pedon (rhymes with head-on) is this unit. A Pedon is the smallest polygonal unit that meets the definition of soil.

Since pedons cover only about 1 to 10 m^2, they are too small to serve as practical units in field soil classification systems. To solve this problem, a group of pedons, termed a polypedon, is of sufficient size to serve as a basic classification unit (or, as commonly called, a soil individual). In the United States, these groupings have been called a soil series.

There is a distinction between "soil" and "the soil." This difference is essential in the soil classification scheme. A soil is characterized by a sampling unit (Pedon), which, as a group (polyhedrons), forms a soil individual. Soil consists of various natural components and is identified as distinct from other substances, including water, air, solid rock, and other portions of the Earth's crust. By incorporating the difference between a soil and the soil, a classification system has been developed that is effective and widely used.

8.5.1 Diagnostic Horizons and Temperature and Moisture Regimes

Soil taxonomy uses a strict definition of soil horizons called diagnostic horizons, which are used to define most of the orders. Two kinds of diagnostic horizons are recognized: surface and subsurface. The surface diagnostic horizons are called epipedons (Greek epi over; Pedon soil). The epipedon includes the dark (organic-rich) upper part of the soil, the upper eluvial horizons, and/or sometimes both. Those soils beneath the epipedons are called subsurface diagnostic horizons. Each of these layers is used to characterize different soils in Soil Taxonomy.

In addition to using diagnostic horizons to strictly define soil horizons, soil moisture regime classes can also be used. A soil moisture regime describes when plant-available water or groundwater is present at adequate levels. The control section of the soil (ranging from 10 to 30 cm for clay and from 30 to 90 cm for sandy soils) designates that section of the soil where water is present or absent during given periods in a year. The control section is divided into two parts: upper and lower portions. The upper portion is defined as the depth to which 2.5 cm of water will penetrate within 24 hours. The lower portion is the depth to which 7.5 cm of water will penetrate.

Six soil moisture regimes are identified:

Aridic: characteristic of soils in arid regions
Xeric: characteristic of having extended periods of drought in the summer
Ustic: where soil moisture is high enough to meet plant needs during the growing season
Udic: ordinary soil in humid climatic regions
Perudic: an extremely wet moisture regime annually
Aquic: soil saturated with water and free of gaseous oxygen

Table 8.2 lists the moisture regime classes and the percentage distribution of areas with different soil moisture regimes.

In soil taxonomy, several soil temperature regimes are used to define classes of soils. Based on mean annual soil temperature, mean summer temperature, and the difference between mean summer and winter temperatures, soil temperature regimes are shown in Table 8.3.

The diagnostic horizons and moisture/temperature regimes just discussed are the main criteria used to define the various categories in soil taxonomy.

8.5.2 Soil Taxonomy

The U.S. Soil Conservation Service's soil classification system, Soil Taxonomy (which is based on measurable properties of soil profiles), places soils in categories (see Table 8.4).

TABLE 8.2
Soil Moisture Regimes (Percent of Global Area Occupied by Soils)

Moisture Regime	Percent of Soils
Aridic	35.9
Xeric	3.5
Ustic	18.0
Udic	33.1
Perudic	1.0
Aquic	8.3

Source: Adaptation from Eswaran, H. "Assessment of global resources: Current status and future needs," Pedologie, XL111 pp. 19–39, 1993.

TABLE 8.3
Soil Temperature Regimes (Percent of Global Areas Occupied by Soils)

Soil Temperature Regimes/Mean Annual Temperature (°C)	Percent of Soils
Pergelic (0°C)	10.9
Cryic (0–8°C)	13.5
Frigid (0–8°C)	1.2
Mesic (8–15°C)	12.5
Thermic (15–22°C)	11.4
Hyperthermic (>22°C)	18.5
Isofrigid (0–8°C)	0.1
Isomesic (8–15°C)	0.3
Isothermic (15–22°C)	2.4
Isohyperthermic (>22°C)	26.0
Water (NA)	1.2
Ice (NA)	1.4

Source: Adaptation from Eswaran, H., "Assessment of global resources: Current status and future needs," Pedologie, XL111-1, pp. 19–39, 1993.

TABLE 8.4
Subdivision of Soil Taxonomy Classification System (Shown in Hierarchical Order)

Category	Number of Taxa
Order	11
suborder	55
Great group	Approximately 230
Subgroup	Approximately 1,200
Family	Approximately 7,500
Series	Approximately 18,500 in U.S.

Each category will be reviewed.

Order: soils not too dissimilar in their genesis. There are 11 soil orders in soil taxonomy. The names and major characteristics of each soil order are shown in Table 8.5.
Suborder: fifty-five subdivisions of the order that emphasize properties that suggest some common features of soil genesis.
Great group: diagnostic horizons are the major bases for differentiating approximately 230 great groups.
Subgroup: approximately 1,200 subdivisions of the great groups.
Family: approximately 7,500 soils with subgroups having similar physical and chemical properties.
Series: a subdivision of the family, and the most specific unit of the classification system. More than 18,000 soil series are recognized in the United States.

8.5.3 Soil Orders

As stated earlier, 11 soil orders are recognized, constituting the first category of the classification. The classification of the orders is illustrated in Table 8.5.

8.5.3.1 Soil Suborders

Soil orders are further divided into 55 suborders, based primarily on the chemical and physical properties that reflect either the presence or absence of waterlogging or genetic differences caused by climate and vegetation, to give the class the most

TABLE 8.5
Soil Orders (With Simplified Definitions)

Alfisol	Mild forest soil with a gray to brown surface horizon, medium to high base supply (refers to the number 4 of interchangeable cations that remain in the soil), and a subsurface horizon of clay accumulation
Andisol	Formed on volcanic ash and cinders and is lightly weathered
Aridisol	Dry soil with a Pedogenic (soil-forming) horizon, low in organic matter
Entisol	Recent soil without Pedogenic horizons
Histosol	Organic (peat or bog) soil
Inceptisol	Soil at the beginning of the weathering process with weakly differentiated horizons
Mollisol	Soft soil with a black, organic-rich surface horizon and a high base supply
Oxisol	Oxide-rich soil is principally a mixture of kaolin, hydrated oxides, and quartz.
Spodosol	Soil with an accumulation of amorphous materials in the subsurface horizons
Ultisol	Soil with a horizon of silicate clay accumulation and low base supply
Vertisol	Soil with high activity clays (cracking clay soil)

Source: Soil Survey Staff. Soil Classification, A Comprehensive System. Washington, DC: USDA Natural Resources Conservation Service, 1975. Soil Survey Staff. Keys to Soil Taxonomy. Washington, D.C.: USDA Natural Resources Conservation Service, 1994.

significant genetic homogeneity. Thus, the Equals (formed under wet conditions) are "wet" (aqu for aqua); Alfisols become saturated with water sometime during the year. The suborder names all have two syllables, with the first syllable indicating the order, such as "alf" for Alfisol and "oll" for Mollisol.

8.5.3.2 Soil Great Groups and Subgroups

Suborders are divided into great groups. Their classification depends on the presence and arrangement of diagnostic horizons. Effective group names are formed by adding prefixes to the suborder name. More than 230 great groups have been identified.

Subgroups are subdivisions of great groups. Subgroup names indicate the extent to which the characteristics of the great group are present. A Typic Fragiaqualf is a soil that is typical of the Fragiaqualf great group.

8.6 SOIL MECHANICS

By now, you know the basics of soil. In addition to soil composition, it is essential to understand the basic science, mechanics, and physics of soil; this and the following section address that need.

You may have wondered: Why does the Leaning Tower of Pisa lean? The tower leans because it was built on a nonuniform consolidation of a clay layer beneath the structure. This process is ongoing (about 1/25 of an inch per year) and may eventually lead to the failure of the tower.

The factors that caused the Leaning Tower of Pisa to lean (and affect the use of soil as foundational and building materials) are what this chapter is all about. Soil mechanics and physics help determine if a site is suitable for construction. Simply put, these two factors can help to answer the question: Will the soils present support buildings? This concerns us because wherever humans build, the opportunity for anthropogenic pollutants follows—and to clean up those pollutants, we must excavate below the surface of the soil again.

The mechanics of soil are physical factors important to engineers and water management personnel because their focus is on soil's suitability as a construction material and the need to possess knowledge of proper building codes and procedures that will not make the specific area subject to flooding or that will facilitate flooding. Simply put, water managers and others need an understanding of the response of a volume of soil to internal and external mechanical forces. This understanding is essential for assessing whether the soil can support various structures without causing harm to the surrounding area.

By studying soil survey maps and reports, Water Managers who consult with soil scientists and other engineers familiar with the region and its soil types to determine the suitability of a soil for any purpose. Conducting field sampling to ensure that the soil product possesses the necessary soil characteristics for its intended purpose is also essential.

The soil characteristics important for engineering purposes include soil texture, types of clay present, depth to bedrock, soil density, erodibility, corrosivity, surface geology, plasticity, content of organic matter, salinity, and depth to the seasonal water table. Engineers will also want to know the soil's spatial and volume properties (weight-volume relationships), stress-strain behavior, slope stability, and compaction. Because these concepts are also of paramount importance in determining the fate of materials that are carried through soil, we present them as concepts in this section.

As mentioned earlier, all-natural soil consists of at least three primary components (or phases), solid (mineral) particles, water, and/or air (within void spaces between the solid particles). Examining the physical relationships (for soils in particular) between these phases is essential. Remember that all water that is not chemically attached acts as a void filler. The relationship between free water and void spaces depends on the available water (moisture).

The volume of the soil mass is the sum of the volumes of the three components, or

$$VT = Va + Vw + Vs \quad (8.3).$$

Void volume equals Va plus Vw. However, since weighing air in the soil voids must be done within the Earth's atmosphere (as with other weighing), the weight of the solids is determined to compensate for that. Air weight in soil is considered zero. The total weight is expressed as the sum of the weights of the soil solids and water:

$$WT = Ws + Ww \quad (8.4)$$

The relationship between weight and volume can be expressed as

$$Wm = VmGmw \quad (8.5),$$

where

Wm = weight of the material (solid, liquid, or gas)
Vv = volume of the material
Gm = specific gravity of the material (dimensionless)
w = unit weight of water.

We can solve a few practical problems with the relationships described earlier. This soil information enables engineers to adjust the proportions of its main components by compacting or tilling to reorient mineral grains. In remediation, a decision to blend soil types to alter the proportions (such as increasing or decreasing the percentage of void space) may be part of a site cleanup process.

The void ratio, e, and porosity, H, describe the relationships between the volumes of soil and voids. We must first determine the void ratio, which is the ratio of the void volume to the volume of solids:

$$e = Vv/Vs \quad (8.6)$$

The first step is to determine the ratio of the volume of void spaces to the total volume. We do this by calculating the porosity (H) of the soil, which is the ratio of void volume to total volume. Porosity is usually expressed as a percentage.

$$H = Vv/Vt \times 100\% \quad (8.7)$$

The terms moisture content (w) and degree of saturation (S) relate to the water content of the soil and the volume of water in the void space to the total void volume:

$$w = Ww/Ws \times 100\% \tag{8.8}$$

and

$$S = Vw/Vv \times 100\% \tag{8.9}$$

The size, shape, density, and other characteristics of soil particles provide engineers with information about shear strength, compressibility, and various aspects of soil behavior. These index properties are used to create engineering classifications of soil. Simple classification tests are used to measure index properties (see Table 8.10) in the lab or the field.

From an engineering point of view, the separation of cohesive (fine-grained) soils from incohesive (coarse-grained) soils is an important distinction. Let's examine these two terms in further detail.

A soil's level of cohesion describes the tendency of soil particles to stick together. Cohesive soils contain silt and clay, which, along with water content, enable these soils to hold together through the attractive forces between individual clay and water particles. Clay particles enhance cohesion, leading to more complex index properties for cohesive soils compared to cohesionless soils. The soil's consistency—the arrangement of clay particles—describes the resistance of soil at various moisture contents to mechanical stresses or manipulations and is the most essential characteristic of cohesive soils.

TABLE 8.10
Index Property of Soils

Soil Type	Index Property
Cohesive (fine-grained)	Water content
	Sensitivity
	Type and amount of clay
	Consistency
	Atterberg limits
Incohesive (coarse-grained)	Relative density
	In-place density
	Particle-size distribution
	Clay content
	Shape of particles

Source: Adaptation from A.E. Kehew, Geology for Engineers and Environmental Scientists, 2nd ed., p. 284, 1995, Long Grove, Ill: Waveland Press.

Sensitivity (the ratio of unconfined compressive strength in the undisturbed state to strength in the remolded state) is another essential index property of cohesive soils. Soils with high sensitivity are highly unstable.

$$\mathrm{St} = \frac{\text{strength in an undi sturbed condition}}{\text{strength in remolded condition}} \tag{8.11}$$

As we described earlier, soil water content also influences soil behavior. Water content values of soil (the Atterberg limits, a collective designation of the so-called limits of consistency of fine-grained soils determined with simple laboratory tests) are usually presented as the liquid limit (LL), plastic limit (PL) [Note: Plasticity is exhibited over a range of moisture contents referred to as plasticity limits] and the plasticity index (PI). The PL is the lower water level at which soil begins to be malleable in the semi-solid state, but the molded pieces still crumble easily with a little applied pressure. The soil reaches the shrinkage state when its volume remains almost unchanged despite further water loss.

The upper plasticity limit (or LL) is reached when the water content in a soil-water mixture changes from a liquid to a semi-fluid or plastic state and tends to flow when jolted. A soil that exhibits a tendency to flow when wet can present distinct challenges for engineering applications as well as for contamination remediation efforts.

The range of water content over which the soil is plastic, called the PI, provides the difference between the LL and the PL. Soils with the highest plasticity indices are unstable in bearing loads—a key point to remember.

The Unified System of Classification is the most widely used system for classifying soil stability. This classification gives each soil type (14 classes) a two-letter designation based on particle-size distribution, LL, and plasticity index.

Cohesionless coarse-grained soils are classified by index properties, including the size and distribution of particles in the soil. Other index properties, such as particle shape, in-place density, and relative density, are essential in describing cohesionless soils because they relate to how closely particles can be packed together.

8.7 SOIL STRESS

Just as water pressure increases as you go deeper into the water, the pressure within the soil increases as the depth increases. A soil with a unit weight of 75 pounds per cubic foot exerts a pressure of 75 psi at one-foot depth—and 225 psi at three feet. Of course, as the pressure on a soil unit increases, soil particles reorient themselves to support the cumulative load. A soil sample from beneath a load may not reflect the material's properties at the surface. Representative samples are essential.

The response of soil to pressure (stress), such as when a load is applied to a solid object, is transmitted throughout the material. The load puts the material under pressure, which is equal to the amount of load divided by the surface area of the external face of the object over which it is applied. The response to this pressure or stress is called displacement or strain. Stress (like pressure) at any point within the object can be defined as force per unit area.

8.8 SOIL COMPRESSIBILITY

Compressibility, the tendency of soil to decrease in volume under load, is most significant in clay soils due to their inherently high porosity. The actual evaluation process for these properties is accomplished in the consolidation test. This involves testing a soil sample under increasing load. The change in thickness is measured after the application of each load increment.

8.9 SOIL COMPACTION

Compaction reduces the void ratio and increases the soil density, which affects how materials (including pollutants) travel through the soil. Compaction is accomplished by working the soil to reorient the soil grains into a more compact state. Water content sufficient to lubricate particle movement is critical to obtaining efficient compaction.

8.10 SOIL FAILURE

Soil structural implications are involved in natural processes such as frost heave, which could damage a septic system and disturb improperly set footings or shift soils under an improperly seated UST and its piping. They are also involved in changes applied to soils during remediation efforts, for example, when excavating to mitigate a hazardous materials spill in soil. When soil cannot support a load, soil failure occurs, which can include events as diverse as foundation overload, collapse of the sides of an excavation, or slope failure on the sides of a dike or hill. Structural stability of soil is essential for safety.

Classifying the type of soil to be excavated before an excavation can be accomplished includes determining the soil type. Stable Rock, Type A, Type B, or Type C soil are the classifications in which soils will be categorized, but finding a combination of soil types at an excavation site is common.

Stable rock: Stable (may lose stability when excavated) with natural solid mineral material that can be excavated with vertical sides and will remain intact while exposed.

Type A soil: Most stable soil. Includes clay, silty clay, sandy clay, clay loam, and sometimes silty clay loam and sandy clay loam.

Type B soil: Moderately stable soil. Includes silt, silt loam, sandy loam, and sometimes silty clay loam and sandy clay loam.

Type C soil: Least stable soil. Includes granular soils like gravel, sand, loamy sand, submerged soil, as well as soil from which water is freely seeping, and unstable submerged rock.

Both visual and manual tests are used to classify soil for excavation.

Visual soil testing concerns soil particle size and type. In a mixture of soils, if the soil clumps when dug, it could be clay or silt. The presence of cracks in walls and spalling (breaks up in chips or fragments) may indicate Type B or C soil. Standing water or water seeping through trench walls classifies the soil as Type C, automatically.

Manual soil testing includes the sedimentation test, wet shaking test, thread test, and ribbon test. Evaluate soil samples immediately to preserve moisture. Soil can be evaluated either on-site or off-site.

A sedimentation test determines the amount of silt and clay in sandy soil. Saturated sandy soil is placed in a straight-sided jar with about five inches of water. After the sample is thoroughly mixed (by shaking it) and allowed to settle, the percentage of sand is visible. A sample containing 80% sand, for example, will be classified as Type C.

The wet shaking test is another way to determine the amount of sand, clay, and silt in a soil sample. This test is accomplished by shaking a saturated sample by hand to gauge soil permeability based on the following facts: (1) Shaken clay resists water movement through it; and (2) water flows freely through sand and less freely through silt.

The thread test is used to determine cohesion (remember, cohesion relates to stability—how well the grains hold together). A representative soil sample is rolled between the palms of the hands to about 1/8″ diameter and several inches in length. The rolled piece is placed on a flat surface and then picked up. If the sample remains intact for two inches, it is classified as cohesive.

The ribbon test is used as a backup for the thread test and also determines cohesion. A representative soil sample is rolled out (using the palms of your hands) to a diameter of 3/4″, and several inches in length. The sample is then squeezed between the thumb and forefinger into a flat, unbroken ribbon 1/8″ to 1/4″ thick, which can fall freely over the fingers. If the ribbon does not break off before several inches are squeezed out, the soil is considered cohesive.

Once the soil has been properly classified, the necessary measures for safe excavation can be chosen based on both soil classification and site restrictions. The two standard protective systems include sloping or benching and shoring or shielding.

8.11 SOIL PHYSICS

A dynamic, heterogeneous body, soil is non-isotropic—it does not have the same properties in all directions. Because soil properties vary directionally, various physical processes are always active in soil.

Soil practitioners must understand the factors involved in the physical processes that are active in the soil. These include physical interactions related to soil water, soil grains, organic matter, soil gases, and soil temperature.

8.11.1 WATER AND SOIL

Water is not only a vital component of every living being but also essential to plant growth, microorganisms that live in the soil, and the weathering process, which involves the breakdown of rocks and minerals to form soil and release plant nutrients. In this section, we focus on soil water and its importance in soil. Let us review the physical properties of water.

Water exists as a liquid between 0° and 100°C (32°–212°F), as a solid at or below 0°C (32°F), and as a gas at or above 100°C (212°F). One gallon of water weighs 8.33

pounds (3.778 kg) and equals 3.785 liters. One cubic foot of water equals 7.5 gallons (28.35 liters), one ton of water equals 240 gallons, and one acre-foot of water equals 43,560 cubic feet (325,900 gallons). Earth's rate of rainfall equals 340 cubic miles per day (16 million tons per second). Finally, water is dynamic, constantly in motion, evaporating from the sea, lakes, and the soil; being transported through the atmosphere; falling to earth; running across the land; and filtering downward into and through the soil to flow along rock strata.

Water is often assumed to be one of the simplest compounds known on Earth. It is rarely found in its pure form in nature. On Earth, water, dating back three to five billion years, contains various constituents even in its purest form. Along with H_2O molecules, hydrogen (H+), hydroxyl (OH–), sodium, potassium, and magnesium, other ions and elements are present.

Water contains additional dissolved compounds, including various carbonates, sulfates, silicates, and chlorides. Rainwater (often assumed to be the equivalent of distilled water) is not immune to contamination, as it collects impurities as it descends through the atmosphere. The movement of water across the land contributes to its contamination, as it takes up dissolved gases such as carbon dioxide and oxygen, along with a multitude of organic substances and minerals leached from the soil.

In soil physics, the physical properties (which are also a function of water's chemical structure) of water that concern us are density, viscosity, surface tension, and capillary action. Let's examine these physical properties.

Density is a measure of the mass per unit volume. The number of water molecules occupying the space of a unit volume determines the magnitude of the density. As the temperature (which measures internal energy) increases or decreases, the molecules vibrate strongly and frequently (which changes the distance between them), expanding or diminishing the volume occupied by the molecules.

As discussed previously, liquid water reaches its maximum density at 4°C and its minimum at 100°C. In soil science work, the density may be a unit weight (62.4 pounds per cubic foot or 1 gram per cubic centimeter).

Viscosity is the measure of the internal flow resistance of a liquid or gas. Stated differently, viscosity is the ease of flow of a liquid, or the capacity of a fluid to convert energy of motion (kinetic energy) into heat energy. Viscosity is the result of the cohesion between fluid particles and the interchange of molecules between layers of different viscosities. High viscosity fluids flow slowly, while low viscosity fluids flow freely. Liquid viscosity drops as the temperature increases.

Have you ever wondered why a needle can float on water? Or why some insects can stand on water? The reason: surface tension. Surface tension (or cohesion) is the property that causes the surface of a liquid to behave as if it were covered with a weak elastic skin. It is caused by the exposed surface's tendency to contract to the smallest possible area due to unequal cohesive forces between molecules at the surface.

What does surface tension have to do with soil? The surface tension property of water significantly influences the behavior of water in soils. Consider an example you may be familiar with, one that will help you understand surface tension and the other important physical properties of the water-soil interface. Water frequently rises in clays, fine silts, and other soils due to their specific properties. But why? How does this happen? Surface tension plays a significant role here: The rise of water through clays, silts, and other soils is termed capillarity or capillary action (the property of the interaction of the water with a solid). The two primary factors of capillary rise are surface tension (cohesion) and adhesion (the attraction of water to the solid walls of channels through which it moves).

Why does the water rise? Because the water molecules are attracted to the sides of the tube (or soil pores) and start moving up the tube in response to this attraction. The cohesive force between individual water molecules ensures that water not directly in contact with the side walls is also pulled up the tube (or soil pores). This action continues until the weight of water in the tube counterbalances the cohesive and adhesive forces.

Keep in mind that for water in soil, the rate of movement and the rise in height of soil water are less than one might expect based on soil pore size, because soil pores are not straight like glass tubes, nor are the openings uniform. Also, many soil pores are filled with air, which may prevent or slow down the movement of water by capillarity.

A final word on capillarity—keep in mind that capillarity means movement in any direction, not just upward. Since the attractions between soil pores and water are as effective with horizontal pores as with vertical ones, water movement in any direction occurs.

Water is always moving. This occurs due to the water cycle. In simple terms, the water cycle can be explained as follows: The sun helps transfer water from lakes and oceans to the land. When the sun shines on the earth, the surface water is heated and evaporates, forming an invisible gas that mixes with the air. This gas is water vapor, which is pure water without any minerals or bacteria. Water vapor rises into the air, then cools and condenses into tiny drops of water that form clouds. Further cooling may form drops large enough to fall as rain. In this way, water is brought from the oceans to the land, where it reappears in springs and wells, soaks into the ground, or runs off again through streams and rivers back to the sea. Of course, the actual movement of water on Earth is much more complex.

Water moves through evaporation, precipitation, and runoff. Evaporation of water is a significant factor in hydrologic systems. Evaporation is a function of temperature, wind velocity, and relative humidity. Evaporation (or vaporization) is, as the name implies, the formation of vapor. Dissolved constituents (such as salts) remain behind when water evaporates. Evaporation of the surface water of oceans provides most of the water vapor, though water can also vaporize through plants, especially from leaf surfaces. This process is called evapotranspiration. Ice can also vaporize without melting first. However, this sublimation process is slower than the vaporization of liquid water.

Precipitation includes all forms in which atmospheric moisture descends to earth—rain, snow, sleet, and hail. Before precipitation can occur, the water that enters the atmosphere through vaporization must first condense into liquid (clouds and rain) or solid (snow, sleet, and hail) before it can fall. This vaporization process absorbs energy, which is released in the form of heat when the water vapor condenses. This phenomenon is similar to water evaporating from your skin, which absorbs heat and makes you feel cold.

Note: The annual evaporation from ocean and land areas is the same as the annual precipitation.

Runoff is the flow back to the oceans of the precipitation that falls on land. This journey to the oceans is not always unobstructed—flow back may be intercepted by vegetation (from which it later evaporates), a portion is held in depressions, and some infiltrates into the ground. A part of the infiltrated water is taken up by plant life and returned to the atmosphere through evapotranspiration. At the same time, the remainder either moves through the ground or is held by capillary action. Eventually, water drips, seeps, and flows its way back into lakes, ponds, rivers, streams, and the oceans.

Have you ever wondered what happens to water after it enters the soil? Most people don't need to know this, but if you work in water management, you must thoroughly understand it. Water that enters the soil has (in simple terms) four ways it may go: It may move on through the soil and percolate out of the root zone, where it may eventually reach the water table. It may be drawn back to the surface and evaporate. It may be taken up (transpired—used) by plants. Finally, it may be "saved" in storage in the water profile.

What determines how much water ends up in each of these categories? It depends. Climate, soil properties, and plant needs all influence how much water is distributed among distinct categories. But don't forget the impact of anthropogenic actions (what we like to call the heavy hand of humans)—people alter the movement of water, not only by irrigation and stream diversion practices and by building but also by choosing which crops to plant and the types of tillage practices employed.

8.12 SOIL CHEMISTRY

Most current and future environmental issues stem from chemicals. In short, in environmental science, studying such problems as the greenhouse effect, ozone depletion, groundwater contamination, toxic wastes, air pollution, stream pollution, and acid rain in environmental science would be difficult (if not impossible) without a fundamental understanding of basic chemical concepts. Of course, an environmental practitioner who must solve environmental problems and understand environmental remediation clean-up processes (such as emission control systems or waste treatment facilities) must be well-grounded in chemical principles and the techniques of chemistry in general, because many of them are used to solve these problems.

This text assumes that the reader may or may not have some fundamental knowledge of chemistry. Thus, the chemistry topics have been selected to review, providing only the essential chemical principles required to understand the nature of environmental problems we face, and the chemistry involved in scientific and technological approaches to their solutions.

So, what is chemistry?

Chemistry is the science concerned with the composition of matter (gas, liquid, or solid) and the changes that take place in it under certain conditions.

Everything in the environment—every material, substance, and object—is a chemical substance or a mixture of chemical substances. Your body is made up of thousands of chemicals. The chemical reactions and interactions we are a part of or cause daily—the food we eat, the clothes we wear, the fuel we burn, the vitamins we take in from natural or synthetic sources—are all products of chemistry created by nature or our own technology. Chemistry is about matter, its constituents, consistency, and measuring and quantifying matter.

What is the matter? All matter can exist in three states: gas, liquid, or solid. Matter, composed of constantly moving, minute particles termed molecules, may be further divided into smaller units called atoms.

Molecules containing atoms of only one kind are called elements; those containing atoms of more than one kind are known as compounds.

Chemical compounds, produced by chemical reactions that alter the arrangements of the atoms in the reacting molecules, are changed when they experience an environmental change. Changes that could make a chemical change include heat, light, vibration, catalytic action, radiation, pressure, and moisture (for ionization). Analyzing a compound's components through examination and breakdown is called analysis, while building compounds from the components is known as synthesis. Bringing substances together without changing their molecular structures results in mixtures.

Key Point: All compounds that contain carbon are organic substances. All other substances are inorganic substances.

Material that has been separated from all other materials, pure substances are indistinguishable from each other, no matter what procedures were used to purify them or their origin. Copper metal, aluminum metal, distilled water, table sugar, and oxygen are all pure substances. No matter the original source, all samples of table sugar are alike and are chemically indistinguishable from all other samples.

A substance is characterized as a material having a fixed composition, usually expressed in terms of percentage by mass. Distilled water is a pure substance consisting of approximately 89% oxygen and 11% hydrogen by mass. A substance that is not pure will have varying constituents—a lump of coal's carbon content may vary from 35% to 90% by mass, for example. Materials that are not pure substances are mixtures. Substances that can be broken down into two or more simple substances are called compounds.

Substances that cannot be broken down or decomposed into simpler forms of matter are called elements, which are the basic substances of which all matter is composed. Presently, we know only 100+ elements, but well over a million compounds. Only 88 of the 100+ elements are present in detectable amounts on Earth, and many of these 88 are rare. Only ten elements make up approximately 99% (by mass) of the Earth's crust, including the surface layer, the atmosphere, and bodies of water (see Table 8.6). As you can see from this table, the most abundant element on Earth is oxygen, which is found in the free state in the atmosphere, as well as in combined form with other elements in numerous minerals and ores.

TABLE 8.6
Elements Making Up 99% of Earth's Crust, Oceans, and Atmosphere

Element	Symbol	% of Composition	Atomic Number
Oxygen	O	49.5%	8
Silicon	Si	25.7%	14
Aluminum	Al	7.5%	13
Iron	Fe	4.7%	26
Calcium	Ca	3.4%	20
Sodium	Na	2.6%	11
Potassium	K	2.4%	19
Magnesium	Mg	1.9%	12
Hydrogen	H	1.9%	1
Titanium	Ti	0.58%	22

Table 8.6 includes symbols and atomic numbers for the ten listed elements. The symbols consist of either one or two letters, with the first letter capitalized. The atomic number of an element is the number of protons in the nucleus.

8.13 TIME TO SHIFT GEARS

Water Managers familiar with this material are more informed than users who only care that water comes out when they turn on the tap. However, those unfamiliar should recognize that water is essential for life and presents its own challenges. Water is necessary for life; however, it may be scarce during droughts or cause damage in situations of flooding.

If water itself were able to look in a mirror, it would be shocked at the many faces that it mirrors back. After years of research, it has been observed that water exhibits a variety of characteristics and faces. Water is not a single substance; it is not a specific thing. Water is a mystery. However, there are a few things we know about water. We know it is rain, it's hydration, it's irrigation, it's flood, it's nature empyreal, but its personality is best described as being Jekyll and Hyde, meaning it exhibits dual or contradictory behavior and often characterized or defined as the opposite of fire-gazing or waterfall gazing, meaning that watching either can mesmerize us and put us into a trance-like state where we are at ease with ourselves and the world around us. However, when the Mr. Hyde-type personality manifests itself, water can cause either floods or droughts, reflecting its unpredictable nature.

Earlier, flooding was discussed and shown to be the Water Manager's nightmare. But while too much water is not a good thing, dying of thirst is even worse and becomes the Water Manager's worst nightmare.

While floods, hurricanes, and earthquakes are easy to define, drought is more subjective. When resources are depleted, definitions lose significance, and reality takes precedence. Note that floods are immediate and, in our face, in our houses, in our cars, and destroy everything they touch. Drought, however, is insidious and, like the fictional Mr. Hyde, poses a significant risk. Although droughts do not have the immediate effects of floods, sustained droughts can cause economic stress throughout an area. Due to its subjective nature, the word "drought" has various meanings, depending on a person's perspective.

The shifting of gears occurs when a local water issue shifts from catastrophic flooding to the opposite, where water is scarce and a drought manifests itself. Water floods and droughts each pose unique, unpredictable challenges. Drought is a serious event, depending on its severity and length.

Drought affects different sectors of life and the economy. To a farmer, a drought is a period of moisture deficiency that affects the crops under cultivation—a dry spell of even a few weeks without water can stress many crops during specific periods of the growing cycle. To a meteorologist, a drought is a prolonged period when precipitation is less than usual. For hydrologists, a drought is defined as an extended period of reduced rainfall and streamflow. To a Water Manager, drought means insufficient water supply affecting both availability and quality. The transition from periods of flooding to drought presents significant challenges for water managers. It is essential to recognize that even when there is an adequate water supply without flooding, the water may not necessarily be potable or safe for consumption. These issues are why a Water Manager's responsibilities and worries are never-ending. However, with proper knowledge about floods and droughts and correct response plans, one can alleviate concerns to a point.

A drought is a period of drier-than-normal conditions that results in water-related problems for water managers wherever it occurs. The problem is precipitation as rain or snow falls in uneven patterns across the country. Although annual rainfall fluctuates, long-term averages remain stable. In the U.S. deserts of the Southwest, the average precipitation is less than three inches per year. In contrast, the average yearly rainfall in Atlanta is about 50 inches. This might make one wonder why they call Seattle the rain city with its 39.34 inches per annum, while Atlanta receives more.

The amount of rain and snow also varies with the seasons. In Georgia, for example, most of the yearly precipitation falls during winter, early spring, and in July. Even if the total amount of rainfall for a year is about average, a rainfall shortage can occur during a period when moisture is critically needed for plant growth, such as in early summer.

When little or no rain falls, soils dry out and plants can die. When several weeks, months, or years pass with little to no precipitation, stream and river flow declines, water levels of lakes and reservoirs fall, and the depth of water in wells increases. A dry period can become a drought if dry weather persists and water-supply problems develop—a Water Manager's head-knocker, for sure (Spellman, 2020; Moreland, 1993).

Several weeks, months, or even years may pass before Water Managers and others know that a drought is occurring. Also, a drought can end as gradually as it began. Dry periods can last for a decade or more.

In the United States, in the 1930s, during the Dust Bowl era some might argue that the famous Dust Bowl had little to do with drought. This is untrue. As a case in point, consider Sidebar 8.2.

SIDEBAR 8.3 THE DUST BOWL[3]

That renowned literary genius, Charles Dickens, once stated in one of his classic works: "It was the best of times, it was the worst of times." This statement applies to many times and situations in the human history, including times of conflict between family and love, hatred and oppression, good and evil, light and darkness, and wisdom and folly. The truth is, it is quite impossible to write a classic comparison and contrast work that can match Dickens' classic tale. However, human events have a way of paralleling literature and vice versa. This can certainly be the case, and is the case, when comparing and contrasting the Twenties (when good times rolled) with the "Dirty Thirties" (aka the Dust Bowl era—salted with the Great Depression).

It is not outside the boundaries of reason to make the connection between the Dust Bowl of the thirties and the human migration it generated—the great human migration from America's High Plains to greener pastures out west. In his masterpiece, *The Grapes of Wrath*, John Steinbeck gave voice to those who fled the Dust Bowl. Moreover, it is not outside the boundaries of reason to connect the migration from the Dust Bowl to climate: Climate Migration.

DID YOU KNOW?

Dryland farming encompasses specific agricultural techniques for the non-irrigated cultivation of crops. As the name implies, dryland farming is associated with drylands, areas characterized by a cool wet season followed by a warm dry season. Dryland farming has evolved as a set of techniques and management practices used by farmers to continually adapt to the lack of moisture in each crop cycle.

Okay, it is definition time. I mentioned the Dust Bowl and the Dirty Thirties, and those two terms are not hieroglyphics needing the Rosetta Stone to decipher. But for clarity (in science clarity is boss—or should be, must be), the Dust Bowl is a colloquial term for the Dust Bowl era, and we can say the same for the term Dirty Thirties. So, what does the colloquial term "Dust Bowl" consist of; that is, what are its boundaries? Well, what is certain is that the Black Sunday dust storm of April 14, 1935, drifted across Oklahoma, Texas, Kansas, and Colorado. The exact delineation of the Dust Bowl remains a subject of debate, and its boundaries do not cover a significant area of the Great Plains. However, the worst or "most" severe events between 1935 and 1938 occurred in the Oklahoma-Texas panhandles and the southwestern corner of Kansas, the northeast corner of New Mexico, and the southeast corner of Colorado. The events not listed as "most" severe but at the severe level occurred in 1935–1936 and 1938, and in these same states, but to a more extensive, wide-ranging degree, reaching southern Nebraska and more profoundly into Texas and New Mexico.

The Great Plains is an extensive semi-arid region 2,000 miles (3,200 km) in length, 500 miles (800 km) in width, and 1,100,000 square miles (2,800,000 km^2) in total area, stretching from southern Texas to beyond Moose Jaw, Saskatchewan, into parts of Alberta and Manitoba, Canada.

Research on topics related to the Dust Bowl era centers on global environmental change, drought modeling, atmospheric circulation, land management practices, institutional responses, adaptive processes, and the phenomenon of human migration—referred to in this context as climate refugees. Study of the Dust Bowl era helps us gain understanding about human migration drivers; that is, drivers that convert local inhabitants into becoming climate refugees.

When someone, anyone, uses the old idiom, "the icing on the cake," this is a term that many of us interpret to mean something that is an additional benefit or as a positive aspect to something that is already considered positive or beneficial. The point is, you may not think that my choice of describing the Great Depression as the icing on the cake of the Dust Bowl era is appropriate. However, I use this terminology, this phraseology, to paint a picture of both tragic events as being enhancements of the other; that is, I view the Great Depression as an enhancement of the Dust Bowl era and vice versa. During the 1930s, the two were synergistic, or, as stated, coactive.

This enhancement (coactiveness) of the situation is not that far off base when you consider that during the worst years of the Great Depression large areas of the North American Great Plains experienced uncompromising, multi-year droughts that led to soil erosion, dust storms, farm abandonments, personal hardships, dearth, scarcity, famine, and death (caused mainly by dust pneumonia or suicide). For those able to gravitate to greener pastures, they became part of the distress migration on scales not previously seen—to this very day and time—the U.S. climate refugees.

Okay, what about the causes and consequences of the tragic Dust Bowl events of the 1930s? Water when it is not enough to sustain life and is too much to float all boats, houses, schools, hospitals, and you name it. . . . I am talking about drought . . . little to no water or water that is unfit for even a mosquito to indulge in.

Throughout the globe, drought and its reoccurring frequency are of increasing interest, not only to scientists but also to those dying of thirst. We should study the past to learn from it and prevent similar events in the future. Keep in mind that drought has not only ecological consequences but also socio-economic impacts. Note that after years of study, I have found that our knowledge of the physical causes and human impact of droughts and the Dust Bowl era is incomplete. However, one thing seems inevitable: The Dust Bowl era has much to teach us about climate change and its enduring implications.

So, what exactly happened to make the 1930s the Dust Bowl era in certain parts of the Great Plains region of the United States? Drought and poor farming practices led to severe dust storms that ruined crops and forced many residents to migrate west, with most settling in California. When local farmers lost their farms due to drought, dust storms, or bank foreclosure, they abandoned the land. Note that the massive migration from the Dust Bowl region did not include all the residents within the Dust Bowl; three out of four farmers stayed in their homes and retained their land. Although approximately 75% of the population stayed in the Dust Bowl region, population numbers declined in specific areas due to migration.

The Dust Bowl migration was the largest in American history. By 1940, more than two million people had moved out of the Dust Bowl states, of whom approximately 200,000 moved to California. In the realist novel *The Grapes of Wrath*, Steinbeck focused on the Joads, a poor family of tenant farmers driven from their Oklahoma home by drought, economic hardship, changes in the agricultural industry, and bank foreclosures, which forced tenant farmers out of work. Because of their desperate situation and being trapped in the Dust Bowl, Joad leads his family out of Oklahoma using Route 66 and goes to California. Why did the Joads and thousands of other Dust Bowl residents head for California? Thousands of Dust Bowl residents thought of that golden state as a paradise, of sorts, certainly better than the Dust Bowl. The climate migrants were seeking jobs, land, self-esteem, and a future.

For those who have read or studied Steinbeck's realist novel or who have studied Dust Bowl climate migration, or, better yet, for those alive today who were driven from their homes (keep in mind that today we are talking about very few survivors of the Dust Bowl, and those still around were babies in 1935) because they wanted a better life somewhere else, they might refer you to a 1935 excerpt from Collier's magazine where the climate migrants' welcome was described. "Very erect and primly severe, [a man] addressed the slumped driver of a rolling wreck that screamed from every hinge, bearing, and coupling, 'California relief rolls are overcrowded now. No use to come farther,' he cried. The weakened driver ignored him, only glancing back to check on his family. They were packed so closely together that leaving the area was not possible. 'There is nothing for you here,' the neat trooperish young man went on. 'Nothing, really nothing.' And the forlorn man on the moaning car looked at him, dull, emotionless, incredibly weary, and said: 'So? Well, you ought to see what they got where I come from.'"

Migrants who moved to California discovered that life there presented challenges similar to those they had previously experienced.

8.13.1 Black Sunday, April 14, 1935

The presentation has sufficiently covered the Dust Bowl. But to gain knowledge of something (anything), it is important to get different or differing views on topics, including historic happenings. Thus, I have chosen the U.S. National Weather Service's (NWS's) 01/31/2022 view to relate here.

As previously noted, the 1930s were a period of significant hardship throughout the Great Plains region. The settlers dealt with the double whammy—the Great Depression and years of drought that plunged an already suffering society into an onslaught of relentless dust storms for days, weeks, and months on end. They were known as dirt storms, sandstorms, "dusters," and black blizzards. Just when it seemed things could not get worse, Sunday, April 14, 1935, proved they could—far beyond words. The day is known in history as "Black Sunday," when the wind did "scoop the dust like hands" and send the mountain of blackness sweeping across the high plains and instantly turn a warm, sunny afternoon into a horrible blackness that was darker than the darkest night. Famous songs were written about it, and on the following day, the world would hear the region referred to for the first time as "The Dust Bowl" (NWS, 2022).

At first, it was a wall of blowing sand that blasted its way into and through the eastern Oklahoma panhandle around 4 p.m. Then it raced to the south and southeast across the main body of Oklahoma that evening, accompanied by heavy blowing dust, winds of 40 MPH or more, and rapidly falling temperatures. But the worst conditions were in the panhandle, where the rolling mass raced more toward the southwest—accompanied by a massive wall of blowing dust that resembled a land-based tsunami. Winds in the panhandle exceeded 60 MPH, and at times the darkness was so dense that visibility was reduced to zero (NWS, 2022).

The blowing dust that blasted the High Plains was attributed not only to dry soil conditions but also to poor soil conservation techniques that were in use during the 1930s. It was one of President Roosevelt's advisors, Hugh Hammond Bennett, who sounded the alarm bell before Congress about the poor soil conservation techniques. The irony is that Bennett testified weeks before Black Sunday about the soil conservation problem, and it was the dust cloud that reached the capital and blotted out the sun that convinced Congress to pass the Soil Conservation Act.

8.13.2 The Great Plains Committee

The U.S. government established the Great Plains Committee in the early 1930s to identify the causes, impacts, and necessary remedies for the crisis in the region. Table 8.7 summarizes the key findings—which put much of the blame on land settlement patterns and land use practices of the late nineteenth century—and the recommendations for action.

TABLE 8.7
Summary of Key Findings of the Great Plains Committee (1936)

Outcome	Root Causes	Suggested Farm-Level Action
Soil erosion	High rates of farm tenancy and absentee landlords mean overproduction of crops relative to livestock, soil mining, lack of farm improvement/long-term planning, expansion of farming into marginal areas, over-cultivation of small landholdings, and failure to recognize diversity of soil conditions across the region.	Plow along contours; list and furrow fields at right angles to prevailing winds; plant crops in strips; terrace slopes; till soil roughly and leave high stubble after harvest; avoid bare summer fallow in wind-exposed areas and instead rotate in cover crops like clover; plant windbreaks.
Loss of forage cover for grazing	Overstocking of rangelands; expansion of farming into marginal areas	Reduce herd sizes or keep herds off fragile lands.
Inefficient use of water	Poor farming technologies and practices fail to conserve soil moisture; inadequate capacity for irrigation	Create deeper, better water ponds for livestock; use supplemental irrigation where cost-effective to do so.
Highly variable farm incomes; high rates of farm indebtedness	Undue dependence on wheat as a cash crop; high rates of tenancy; family farm landholdings too small in size; mechanization in the 1920s financed on credit during a period of good rainfall and favorable crop prices	Maintain a higher ratio of fodder and livestock to cash crops; reduce the proportion of wheat and corn on farms; create diversified operational plans; keep larger feed and seed reserves.

Source: Adaptation from McLeman, R, Dupre, J, Berang Ford, L, Ford, J, Gajewski, K, Marchildon (2014). What we learned from the Dust Bowl: lessons in science, policy, and adaptation. Accessed 01/30/2022 @ https://www.ncbi.nim.nih.gov/pmc/articles/PMC4015056.

DID YOU KNOW?

For those who lived through the Dust Bowl, that is, those who not only migrated to other locations but also those who stayed put, they were exposed to the dust in a manner whereby they inhaled it. Inhalation of dust can lead to dust pneumonia. This form of respiratory disorder affected a substantial number of people during the 1930s Dust Bowl era.

8.13.3 Bottom Line

When migrants from the Dust Bowl states arrived in California, they found themselves in unfamiliar territory for more than one reason. Many corporations own California farms. They were larger and more modernized than anything the migrants were used to or had ever seen, and the crops were unfamiliar. The rolling fields of wheat were replaced by crops of fruit, nuts, and vegetables. In Steinbeck's classic, "The Grapes of Wrath," he pointed out that about 40% of the migrant farmers ended up in the San Joaquin Valley, picking grapes and cotton. The migrants replaced the Mexican migrant workers, a legion of workers numbering around 120,000. To say that the life of the migrants was hard is to make a gross understatement (see Figure 8.3). They were paid by the quantity of fruit and cotton picked and earned seventy-five cents to $1.25 a day. And out of that, they had to pay for their board and room.

The effects of the Dust Bowl and the experiences of the migrants are difficult to convey fully through words alone. However, Figure 8.3 may convey significant information, as visual data can often communicate more than text. This photo provides insight into natural events that can impact individuals who may not be fully aware of their actions or the resulting consequences. Natural processes are ultimately designed, and when we interfere, it may be a good idea to wonder what we have done to the ecological balance before it smacks us in the face, so to speak.

8.13.4 References

Cook, B., R. Miller, R. Seager. Did dust storms make the Dust Bowl worse? Accessed 01/30/2022 @ http://ocp.ideo.columbia.edu/res/res/div/ocp/drought/dust_storms.shtml.

McLeman, R, Dupre, J, Berang Ford, L, Ford, J, Gajewski, K, Marchildon (2014). What we learned from the Dust Bowl: lessons in science, policy, and adaptation. Accessed 01/30/2022 @ https://www.ncbi.nim.nih.gov/pmc/articles/PMC4015056.

National Weather Service (NWS) (2022). The Black Sunday Dust Storm of April 14, 1935. Accessed 02/01/2022 @ https://www.weather.gov/oun/events-19350414.

The first evidence of a drought is usually seen in records of rainfall. Within a short period of time, the amount of moisture in the soil can begin to decrease. Drought is insidious; its effects on flow in streams and reservoirs may not be noticed for several

FIGURE 8.3 Dust Bowl children.

weeks or months. Water levels in wells may not reflect a shortage of rainfall for a year or more after a drought begins (Spellman, 2020; Moreland, 1993).

It is important to note that a period of below-normal rainfall does not necessarily result in drought. Some rain returns to the air as water vapor when water evaporates from water surfaces and from moist soil. Plant roots draw some of the moisture from the soil and return it to the air through a process called transpiration.

Regarding the effect of drought on groundwater, note that the groundwater levels that are sources of water do not always remain the same. Along with drought, seasonal variations in rainfall, and the piping effect, the height of underground water levels is affected. One location of groundwater can be affected by massive pumping at other nearby locations when water withdrawal is excessive.

8.14 WATER RESOURCE OVERUSE

Many may feel that the Earth has abundant water. Oceans cover 70% of Earth, making water everywhere, right?

No, not exactly.

Of all Earth's water, less than 1% is available for human use. The rest is salt water found in oceans, frozen freshwater in polar ice caps, or water that is inaccessible for practical usage. The supply of potable freshwater is decreasing, and population and demand for freshwater are increasing. This is becoming a massive problem because we have a finite quantity of freshwater available for use and consumption—the water supply on Earth is constant, no more or less than we have right now, the same amount we have always had. It is true that the hydrologic (water) cycle continuously returns water to Earth, but it is not always deposited where needed (where accessible) or in the same quantity and quality. Even if we locate freshwater, if its quality or quantity is inadequate or contaminated, it becomes a problem.

Water Managers have numerous issues that can turn a good scenario into a nightmare. We have discussed two elements of the nightmare manifested by flooding and drought, but there are several other issues to consider.

First, is there sufficient water in the utility? Other water-related issues are less significant when water is unavailable or insufficient. The public water utility or independent groundwater user via wells must ensure and secure adequate sources of water.

Second, if there is a sustainable amount of fresh water accessible, is it of a quality where it can be safely consumed or used for whatever purpose?

Third, if there is a sustainable amount of potable fresh water, is it overused? In this book, water overuse is synonymous with waste. Water waste persists at a significant rate.

Okay, straightforward and understandable to this point, but more information is needed. Where do we use potable fresh water? This is an essential question if we are to understand not only water's use but also where it is wasted. We cannot solve a problem unless we recognize the problem.

Wasting water is a significant problem that cannot be underestimated. Water management at all levels must be knowledgeable about the fate of water. In the United States, Water Managers benefit from readily available, safe, treated drinking water. In most locations in the United States, all one needs to do is turn on the tap, and the water flows.

As pointed out earlier, in Sidebar 5.1 dealing with Sick Water, a family of four wakes up in the morning, takes a shower, brushes their teeth, grabs a cup of coffee, and then heads out the door for the day. This is routine behavior. The only time this behavior is interrupted is when they turn on the water faucets, and either there is no flow or the flow delivers water that is disgusting in appearance, odor, and quality. We rely on water daily but often overlook how much we use or waste.

- The average American family uses more than 300 gallons of water per day at home. Note that most of this use, 70%, is indoors with 24% for toilet(s), 20% for shower, 19% for faucets, 17% for washing machine, 12% for leaks, and 8% for other uses.
- Outdoor water use makes up 30% of household consumption, but this can be higher in dry regions or areas with water-intensive landscaping. For example, the arid West has some of the highest per capita residential water use because of landscape irrigation (USEPA, 2024).

Before assessing water distribution, a Water Manager should first develop a personal understanding of water use. Understand your water usage before assessing or tracking it. Looking at your water bill is a good place to begin understanding your water usage. It is not the cost that is important, but instead how much water was used. Once you gain an understanding of how much water you use, enacting efficient water conservation measures can follow.

Attempting to determine water usage on a larger scale can be more complicated because there are uses of water that are basically invisible—we do not see their use as readily as we see when the water is used for personal needs, to grow food, manufacture goods, and keep many businesses running smoothly.

Managing water is a growing concern not only for Water Managers but also for communities. Currently and in the future, challenges related to water supply must be addressed, and there is an increasing need to update aging water infrastructure, water treatment, and delivery systems. Competition for water supply is ongoing and increasing due to population growth, where per capita water use is increasing. Communities may encounter higher water prices, stricter summer watering restrictions, loss of seasonal recreation areas, and costly water treatment or transport projects as water supplies dwindle and systems age. The water demand is expected to remain constant, as it is essential for human survival and supports other forms of life. Water has numerous applications, making it a fundamental resource in various contexts.

If Water Managers are not worried about the growing concerns in the U.S. communities across the country, while they are starting to face challenges regarding water supply and a need to update aging water treatment and delivery systems, then they need to find some other line of work; maybe selling widgets would be more suitable for them. This bit of sarcasm is based on personal knowledge of several water operations in the United States, where I was invited to lecture on operations or to inspect the plant for compliance before an OSHA or EPA audit. The two challenges, proper water management and care of water distribution, are lacking based on my observations in more than a few operations throughout the country.

I say this because the primary attribute of a competent Water Manager is leadership. Without leadership, there is no management.

CASE STUDY 8.1 SWIFT: THE FIX

Water Managers in U.S. communities are constantly trying to figure out how to obtain safer potable drinking water. The problem is, of course, that supplying water from a water resource that is not growing to meet the growing population and increased water uses is a significant challenge. This is not to say that this challenge cannot be abated by far-thinking leadership. One example of how to secure safe drinking water is in Hampton Roads, Virginia, where the far-thinkers at Hampton Roads Sanitation District (HRSD) have implemented what they call the Sustainable Water Initiative For Tomorrow (SWIFT). The (SWIFT) program is an ongoing, long-term project spanning a decade. Do not confuse the acronym "SWIFT" with the adjectives for fast, speedy, rapid, hurried, immediate, or quick. SWIFT is a long-term project scheduled to complete installation and operations by 2030.

What is HRSD's SWIFT? SWIFT is a program to inject treated wastewater into the subsurface; specifically, it is designed to inject treated wastewater to improve the drinking water quality in the Potomac Aquifer. Injection of water into the subsurface is expected to raise groundwater pressures, potentially expanding the aquifer system, raising the land surface, and counteracting land subsidence occurring in the Virginia Coastal Plain. In 2016, a pilot project at the HRSD Nansemond Wastewater Treatment Plant in Suffolk, Virginia, began testing aquifer injection. HRSD has asked USGS to prepare a proposal for the installation of an extensometer monitoring station at the test site to monitor groundwater levels and aquifer compaction and expansion.

SWIFT is designed to counter land subsidence at various locations in the Hampton Roads area of southern Chesapeake Bay, where land subsidence rates of 1.1 to 4.8 millimeters per year have been observed (Eggleston and Pope, 2013; Holdahl and Morrison, 1974).

Injecting treated wastewater (at drinking water quality) is expected to reduce land subsidence or raise ground levels in the area. Careful monitoring of aquifer system compaction and groundwater levels can be used to optimize the injection process and improve the fundamental understanding of the relationship between groundwater pressures and aquifer system compaction and expansion.

There is more to HRSD's treated wastewater injection project, SWIFT, than just arresting or mitigating land subsidence and relative sea-level rise in the Hampton Roads region. An additional objective of the project is to discontinue the discharge of treated wastewater from seven out of 14 plants, which would reduce the annual release of nitrogen, phosphorus, and sediment into the bay by 18 million pounds. Assuming SWIFT works as designed, this is a huge benefit to Chesapeake Bay in that it may help to prevent or reduce the formation of algal bloom dead zones. Success from treated wastewater injection would benefit both the bay and HRSD ratepayers. To meet regulatory guidelines, removing nutrients from discharged treated wastewater would cost hundreds of millions of dollars and require almost non-stop retrofitting at the treatment plants to keep up with advances in treatment technology and regulatory requirements. Another goal of HRSD's SWIFT project is to restore or restock potable groundwater supplies in the local aquifers. The drawdown of water from the groundwater supply has contributed not only to land subsidence but also to a reduction of water available for potable use.

HRSD's planned restocking of the Hampton Roads groundwater supply with injected wastewater treated to potable water quality is not without its critics. The critics state that HRSD's wastewater injection project would contaminate potable water aquifers. For critics and others, this is where the so-called yuck factor comes into play. In this case, the "yuck factor" refers to concerns about groundwater being tainted with recycled toilet water. This is the standard view of many of the critics who feel HRSD's SWIFT project is nothing more than a direct reuse of wastewater; that is, a pipe-to-pipe connection of toilet water to their home water taps.

What the critics and others do not realize is that we are already using and drinking treated and recycled toilet water. Regarding concerns about HRSD's SWIFT project contaminating aquifers, the water is treated—currently at the Nansemond Treatment Plant in Suffolk, Virginia—to drinking water standards. This is the key phase here.

Concerns about consuming treated toilet water are overstated, as explained below.

WASTEWATER YUCK FACTOR OVER-STATED

Hercules, often considered the first environmental engineer, was tasked by Eurystheus with cleaning King Augeas' stables as his fifth labor. Hercules, faced with a mountain of horse and cattle waste piled high in the stable area, had to devise some method to dispose of the waste; he did. He diverted a couple of river streams to the inside of the stable area so that all the animal waste could simply be deposited into the river streams: Out of sight, out of mind. The waste flowed downstream. Hercules understood the principal point in pollution control technology that is pertinent to this very day and this discussion: that is, *dilution is the solution to pollution.*

When people say they would never drink toilet water, they have no idea what they are saying. As pointed out in my textbook, *The Science of Water*, fifth edition, the fact is that we drink recycled wastewater every day. In Hampton Roads, for example, HRSD's wastewater treatment plants outfall (discharge) treated water to the major rivers in the region. Many of the region's rivers are sources of local drinking water supplies. Even local groundwater supplies are routinely infiltrated with surface water inputs, which, again, are commonly supplied by treated wastewater (and sometimes infiltrated by raw sewage that is accidentally spilled).

My compliments to Mr. Henifin, former General Manager of HRSD, who stated in a local newspaper article that he would be the first to drink the treated wastewater effluent from the unit treatment processes at York River Treatment Plant, which he did. My only contention with his statement is that because of Mother Nature's Water Cycle, which we learned about in grade school, we have been drinking toilet water all along. I have yet to find anything yucky about it or its taste.

8.15 WHEN WATER AND AI MIX

AI is a developing technology that has been applied in various sectors. These sectors include healthcare, finance, retail, transportation, education, manufacturing, and water management, which can be enhanced by improving efficiency, accuracy, data analysis capabilities, and decision-making capabilities. In water supply management, AI preserves water supplies by:

- Detecting potential water supply failures before they occur.
- Enhancing water distribution networks and interceptor lines in wastewater treatment.
- Advancing water quality monitoring.
- Improving the use of energy in wastewater treatment.
- Stronger climate resilience and adaptation.
- Analyzing data through AI algorithms to predict floods and droughts, manage water supply efficiently, and ensure efficient water usage. AI also helps reduce pollution, improve water quality, and enhance flood protection.
- Allowing for more accurate flood predictions and adaptation.

Regarding the last item listed, flooding, AI applications have revolutionized the way communities prepare for and respond to flood risks. Some of the key applications include providing flood prediction tools. These tools generate satellite images to show potential flooding scenarios using generative AI combined with physics-based models. AI can also analyze substantial amounts of data quickly and relatively cheaply to inform adaptation plans and disaster management. When AI is combined with advanced water resource analytics, it is used to monitor and model systems.

8.16 THE BOTTOM LINE

The slow, inexorable progress of rock changing into soil gives us an ongoing demonstration of nature's unrelenting power. While we move around on the surface of the earth and push and shove soil into various stages and conditions, we forget one vital fact: We can do little to either slow down or speed up the natural process of soil creation.

Nature is a powerful and omnipresent force in our environments that we must both depend on and contend with. Human beings present another powerful force, and when you put their anthropogenic achievements into the equation (the heavy hand of man and what we can do (and do) to our environment), the repercussions of both the destruction that Nature can dish out and the impact to the environment that humans contribute can become overwhelming to comprehend. Thankfully, we rarely consider this way of thinking. Is this a good thing? Is it good to ignore Nature and what she can do, what she does always—around us? Is it essential to think about the impact man has on his surroundings?

We answer yes to both questions. Many individuals are indifferent. They think that while humans impact the environment, it will eventually recover or sustain itself without help. And besides, why should we be concerned with what man does to the environment today—soon it will all be forgotten? No matter what we do, we leave our mark on the land. We do impact the environment we use and inhabit more than we are sometimes aware. Have you ever walked through an old cemetery, one with crumbling and unreadable headstones? This can be a sobering reminder to all of us of our fragility—not only of human life but also of the shortness of memory. Our point: We must not forget; we must not only increase our awareness of our impact on our environment but also increase our memory spans—our sense of history. For long-range planning, we need long-range vision—and the ability to remember what has gone on before.

And this philosophy certainly is in tune with water management. To remain in tune with water management operations, the tuning fork is being aided by AI. AI improves water treatment, optimizes distribution, and aids in water management for both employers and employees.

NOTES

1 Adaptation from NOAA (2015). *Green Infrastructure to Reduce Flooding*. Washington, DC: National Oceanic and Atmospheric Administration.
2 Adaptation from USEPS (2015). *Slopes and Low Impact Development*. Washington, DC: United States Environmental Protection Agency.
3 Adaptation from F.R. Spellman (2020). *Climate Migration*. Boca Raton, FL: CRC Press.

REFERENCES

ASTM. 1969. *Manual on Water*. Philadelphia: American Society for Testing and Materials.
Brady, N.C., and Weil, R.R. 1996. *The Nature and Properties of Soils*, 11th ed. Upper Saddle River, NJ: Prentice-Hall.
Eggleston, J. and Pope, J. 2013. *Land Subsidence and Relative Sea Level Rise in the Southern Chesapeake Bay Region. Report No. 1392*. Washington, DC: U.S. Geological Survey.
Holdahl, S.R. and Morrison, N.L. 1974. *Regional Investigation of Vertical Crust Movements, Using Releveling and Mareograph Data*. Washington, DC: U.S. Geological Survey.
Kemmer, F.N. 1979. *Water: The Universal Solvent*. Oak Ridge, IL: NALCO Chemical Company.
Konigsburg, E.M. 1996. *The View from Saturday*. New York: Scholastic.
Moreland, J.A. 1993. *Drought*. Washington, DC: United States Geological Survey.
Nathanson, J.A. 1997. *Basic Environmental Technology: Water Supply, Waste Management, and Pollution Control*. Upper Saddle River, NJ: Prentice Hall.
NWS. 2022. *The Dust Bowl*. Washington, DC: National Weather Service.
Spellman, F.R. 1999. *Environmental Science and Technology: Concepts and Applications*. Rockville, MD: Government Institutes.
Spellman, F.R. 2020. *The Science of Water*, 4th ed. Boca Raton, FL: CRC Press.
USDA. 1975. *Soil Taxonomy: A Basic System of Soil Classification for Making and Interpreting Soil Surveys*. Washington, DC: USDA Natural Resources Conservation Service.
USEPA. 2024. *Outdoors*. Washington, DC: United States Environmental Protection Agency.

9 Stream Ecology and AI Synergy

9.1 INTRODUCTION

> The "control of nature" is a phrase conceived in arrogance, born of the The Neanderthal age of biology and the convenience of man.
>
> **—Rachel Carson (1962)**

What is ecology? Why is ecology critical? Why study ecology? Why study stream ecology? These questions are straightforward, but giving clear answers can be difficult. However, for Water Managers whose water supply is drawn from streams and rivers, these are important questions augmented by many more.

In this chapter, a convergence paradigm between ecology and AI is discussed. Artificial intelligence (AI) assists Water Managers in stream water supply management. The Water Manager benefits when the AI application is used to enhance ecological research and management. AI can analyze environmental data to identify patterns and trends, which helps in understanding stream health and ecosystem dynamics. Generative AI is being studied for its ability to reduce environmental impacts, marking a move toward sustainable stream ecology practices. The point is that integration of AI into stream ecology not only enables the Water Manager to enhance data analysis but also promotes more sustainable environmental management practices.

This chapter simplifies complex science to provide Water Managers, who are generalists (hopefully), with essential stream ecology knowledge. This chapter aims to summarize key information on ecology's functions, values, links with other sciences, and its impact on our lives. In doing so, the author hopes not only to dispel the common misconception that ecology is too tricky for the average person to understand but also to instill the concept of ecology as an asset that can not only be learned but also cherished.

The bottom line: Water Managers benefit from research in both ecology and AI in providing predictive understanding of complex stream ecology systems.

9.1.1 Key Terms

- *Abiotic factor*—the nonliving part of the environment composed of sunlight, soil, mineral elements, moisture, temperature, topography, minerals, humidity, tide, wave action, wind, and elevation.

Important Note: Every community is influenced by a particular set of abiotic factors. While it is true that the abiotic factors affect the community members, it is also true that the living biotic factors may influence the abiotic factors. For example, the

DOI: 10.1201/9781003655848-14

amount of water lost through the leaves of plants may add to the moisture content of the air. Also, the foliage of a forest reduces the amount of sunlight that penetrates the lower regions of the forest. The air temperature is therefore much lower than in non-shaded areas (Tomera, 1989).

- *Autotroph*—(green plants) fix the energy of the sun and manufacture food from simple, inorganic substances.
- *Biogeochemical cycles*—are cyclic mechanisms in all ecosystems by which biotic and abiotic materials are constantly exchanged.
- *Biotic Factor (Community)*—the living part of the environment composed of organisms that share the same area, are mutually sustaining, interdependent, and constantly fixing, utilizing, and dissipating energy.
- *Community*—in an ecological sense, community includes all the populations occupying a given area.
- *Consumers and decomposers*—dissipate energy fixed by the producers through food chains or webs. The available energy decreases by 80%–90% during transfer from one trophic level to another.
- *Ecology*—is the study of the interrelationship of an organism or a group of organisms and their environment.
- *Ecosystem*—is the community and the nonliving environment functioning together as an ecological system.
- *Environment*—is everything necessary to an organism in its surroundings.
- *Heterotrophs*—(animals) use food stored by the autotroph, rearrange it, and finally decompose complex materials into simple inorganic compounds. Heterotrophs may be carnivorous (meat-eaters), herbivorous (plant-eaters), or omnivorous (plant- and meat-eaters).
- *Homeostasis*—is a natural occurrence during which an individual population or an entire ecosystem regulates itself against negative factors and maintains an overall stable condition.
- *Niche*—is the role that an organism plays in its natural ecosystem, including its activities, resource use, and interaction with other organisms.
- *Pollution*—is an adverse alteration to the environment by a pollutant.

9.2 BASIC ECOLOGY

Ecology can be defined in various ways. It is the scientific study of how living organisms are distributed and how their abundance is influenced by interactions with each other and their environment. The German biologist Haeckel coined the term "ecology" in 1866, meaning the study of nature's household. Odum (1983) explains that the word "ecology" is derived from the Greek oikos, meaning home. Ecology then implies the study of organisms in their natural habitats. It refers to studying an organism in its natural habitat. Ecology is the study of the relationship of an organism or a group of organisms to their environment. In a broader sense, ecology is the study of the relationship of organisms or groups to their environment.

Important Point: No ecosystem can be studied in isolation. If we were to describe ourselves, our histories, and what made us the way we are, we could not leave the world around us out of our description! So it is with streams: They are directly tied to the world around them. They take their chemistry from the rocks and dirt beneath them as well as from a great distance around them (Spellman, 1996).

Charles Darwin (1998) explained ecology in a famous passage in the *Origin*, a passage that helped establish the science of ecology. According to Darwin, a "web of complex relations" binds all living things in any region. Darwin writes that adding or subtracting even a single species causes waves of change that race through the web, "onwards in ever-increasing circles of complexity." The simple act of adding cats to an English village would reduce the number of field mice. Killing mice would benefit the bumblebees, whose nest and honeycombs the mice often devour. Bumblebees are essential for fertilizing heartsease and red clover. Including cats in the village could eventually result in adding flowers. For Darwin, the whole of the Galapagos archipelago argues this fundamental lesson. The island volcanoes are much more diverse in their ecology than in their biology. The contrast suggests that in the struggle for existence, species are shaped at least as much by the local flora and fauna as by the local soil and climate. "Why else would the plants and animals differ radically among islands that have the same geological nature, the same height, and climate?" (Darwin, 1998). An effective way to grasp ecology is to read the following excerpt by Rachel Carson (1962):

> When caddis flies in a stream are exposed to toxic substances, salmon population subsequently decline and may disappear. We poison the gnats in a lake and the poison travels from link to link of the food chain and soon the birds of the lake margins become victims. We spray our elms and the following springs are silent of robin song, not because we sprayed the robins directly but because the poison traveled, step by step, through the now familiar elm leaf-earthworm-robin cycle. These are matters of record, observable, and part of the visible world around us. They reflect the web of life—or death—that scientists know as ecology.

As Carson pointed out, what we do to any part of our environment affects other parts. In other words, there is an interrelationship between the parts that make up our environment. Ecology, as defined here, is the science of the specific interactions between organisms and their living and nonliving environment (Tomera, 1989).

When the environment is mentioned in the preceding text and as it is discussed throughout this text, it includes everything important to the organism in its surroundings. The organism's environment can be divided into four parts:

(1) Habitat and distribution—its place to live
(2) Other organisms—whether friendly or hostile
(3) Food
(4) Weather—light, moisture, temperature, soil, etc.

Ecology has four main subdivisions:

- Behavioral Ecology
- Population Ecology (Autecology)

- Community Ecology (Synecology)
- Ecosystem Ecology

Behavioral ecology is the study of the ecological and evolutionary basis for animal behavior. *Population ecology* (or Autecology) is the study of the individual organism or a species. It focuses on life history, adaptation, and behavior. It is the study of communities, ecosystems, and the biosphere. A classic example of autecology is biologists dedicating their careers to studying salmon ecology. *Community ecology* (or Synecology), on the other hand, is the study of groups of organisms associated together as a unit and deals with the environmental problems caused by humankind. For example, the effect of discharging phosphorus-laden effluent into a stream involves several organisms. The activities of human beings have become a significant component of many natural areas. As a result, it is essential to realize that the study of ecology must involve people. *Ecosystem ecology* is the study of how energy flows and matter interact with biotic elements of ecosystems (Odum, 1971).

Important Point: Ecology is generally categorized according to complexity, the primary kinds of organisms under study (plant, animal, insect ecology), the biomes principally studied (forest, desert, benthic, grassland, etc.), the climatic or geographic area (e.g., Artic or tropics), and/or the spatial scale (macro or micro) under consideration.

9.2.1 Why Is Ecology Important?

Ecology, in its true sense, is a holistic discipline that does not dictate what is right or wrong. Ecology matters because it helps us understand life on Earth. Ecology shows us that each living organism has an ongoing and continual relationship with every other element that makes up our environment. Simply put, ecology is all about interrelationships, intraspecific and interspecific, and how important it is to maintain these relationships to ensure our very survival.

There are numerous examples that highlight the importance of ecology and interrelationships. However, to not only demonstrate the significance of ecology but also to point out that an ecological principle can be a double-edged sword, depending on point of view (ecological problems along with pollution can be a judgment call—that is, they are a matter of opinion), a famous parable (*The Keeper of the Spring*) adapted from Peter Marshall's (1950) *Mr. Jones: Meet the Master* is used here to demonstrate that many misconceptions exist in ecology.

SIDEBAR 9.1 THE KEEPER OF THE SPRING

This is the story of the keeper of the spring. He lived high in the Alps above an Austrian town and had been hired by the town council to clear debris from the mountain springs that fed the stream flowing through the town. The man did his work well, and the village prospered. Graceful swans floated in the stream, and the surrounding countryside was irrigated. Several mills used the water for power, and restaurants flourished for townspeople and a growing number of tourists.

Years went by. One evening at the town council meeting, someone questioned the money being paid to the keeper of the spring. Nobody knew his identity or whether he was still working in the mountain heights. Before the evening was over, the council decided to dispense with the old man's services.

Weeks went by, and nothing seemed to change. Then autumn came. The trees began to shed their leaves. Branches broke and fell into the pools high up in the mountains. The villagers noticed the water darkening below. A foul odor appeared. The swans disappeared, and so did the tourists. The disease began to circulate in the town.

When the town council reassembled, they realized that they had made a costly error. They rehired the old keeper of the spring. Within a few weeks, the stream cleared up and life returned to the village as it had been before (Marshall, 1950).

After reviewing Marshall's parable and the restoration of the spring, the average person might say to him/herself, "Gee, all is well with the town again." Residents are satisfied that the stream's restoration benefits swans, irrigation, hydropower, and scenic views. The trained ecologist, however, would take a different view of this same stream. The ecologist would go beyond the hype (as portrayed in the popular media, including literature) about what a healthy stream is. For example, a trained ecologist would know that a perfectly clean stream, clear of all terrestrial plant debris (woody debris and leaves), would not be conducive to ensuring diverse, productive invertebrates and fish, preserving natural sediment and water regimes, and ensuring overall stream health (Dolloff and Webster, 2000).

9.2.2 Why Study Ecology?

Does anyone need to study ecology to observe, relish, feel, and/or sense the real thing? We might be talking about the clear, sunny days and cool, crisp autumn nights that attract outdoor enthusiasts of all backgrounds. To take in the splendor and delight of autumn's color display, many head for the hills, mountains, countryside, lakes, streams, and recreation areas of the national forests. The more adventurous can enjoy a horseback ride or backpack through Nature's glory and solitude on trails winding deep into the forest tranquility—just being outdoors in those golden days rivals any thrill in life. Even those of us fettered to the chains of city life are often exposed to city streets with their columns of life ablaze in color.

No, one does need to study ecology to witness, appreciate, and/or understand the enchantment of Autumn's annual color display—summer extinguished in a blaze of color. It is a different story, however, for those involved in trying to understand all the complicated actions—and even more complex interactions—involving pigments, sunlight, moisture, chemicals, temperatures, site, hormones, length of daylight, genetic traits, and so on that make for a perfect autumn color display (USDA, 1999). An ecologist's job is to explore nature's fundamentals and continuously seek answers. To find the answers, the ecologist must be a synthesis scientist; that is, he/she must be a generalist well-versed in botany, zoology, physiology, genetics, and other disciplines like geology, physics, and chemistry.

Earlier, we used Marshall's parable to make the point that a clean stream and other downstream water bodies can be a good thing, depending on one's point of view—pollution is a judgment call. It was also pointed out that this view may not be shared by a trained ecologist, especially a stream ecologist. The ecologist recognizes that terrestrial plant debris is essential, not just beneficial. Why? Consider the following explanation (Spellman, 1996):

A stream has two sources of primary energy: (1) instream photosynthesis by algae, mosses, and higher aquatic plants, and (2) imported organic matter from streamside vegetation (e.g., leaves and other parts of vegetation). A considerable amount of the food consumed by aquatic organisms originates within the stream itself, including algae, diatoms, nymphs, larvae, and fish. This food that originates from within the stream is called autochthonous (Benfield, 1996).

Most food in a stream, however, comes from outside the stream. This is especially true in small, densely wooded streams with limited light, where energy comes from external sources due to minimal instream photosynthesis. Leaves provide a substantial portion of this imported energy. Worms drown in floods and get washed in. Leafhoppers and caterpillars fall from trees. Adult mayflies and other insects mate above the stream, lay their eggs in it, and then die in it. All this food from outside the stream is called allochthonous.

9.2.3 Leaf Processing in Streams

Autumn leaves entering streams are nutrition-poor because trees absorb most of the sugars and amino acids (nutrients) that were present in the green leaves (Suberkoop et al., 1976). Leaves that fall into streams often travel only briefly before being trapped by streambed structures and forming leaf packs. These leaf packs are then processed in place by components of the stream communities in a series of well-documented steps (Figure 9.1) (Peterson and Cummins, 1974).

Within 24 to 48 hours of entering a stream, many of the remaining nutrients in leaves leach into the water. After leaching, leaves are composed chiefly of structural

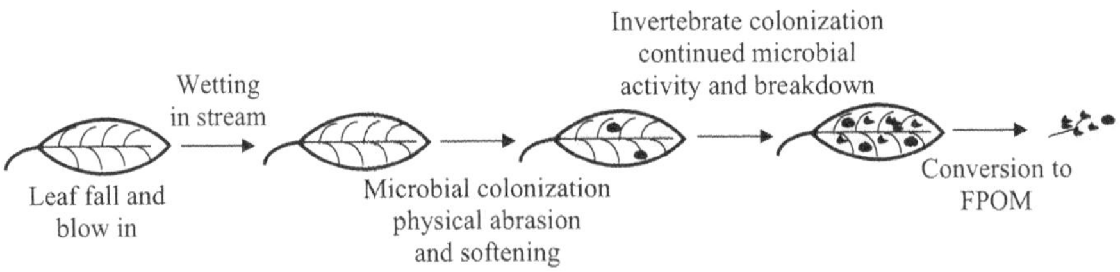

FIGURE 9.1 Leaf processing in streams.

Source: Adapted from Allen, J. D., *Stream Ecology: Structure and Functions of Running Waters*, Tandem: Chapman & Hall, p. 114, 1996.

materials like non-digestible cellulose and lignin. Within a few days, fungi (especially Hyphomycetes), protozoa, and bacteria process the leaves by microbial processing (see Figure 9.1) (Barlocher and Kendrick, 1975). Two weeks later, microbial conditioning leads to structural softening of the leaf and, among some species, fragmentation. Reduction in particle size from whole leaves (coarse particulate organic matter) to fine particulate organic matter is accomplished mainly through the feeding activities of a variety of aquatic invertebrates collectively known as "shredders" (Cummins, 1974; Cummins and Klug, 1979). Shredders (e.g., stoneflies) help produce fragments that are shredded but not ingested, as well as fecal pellets that reduce the particle size of organic matter. These particles are then collected (by mayflies, e.g.) and serve as a food resource for a variety of micro- and macroconsumers. Collectors eat what they want and send even smaller fragments downstream. These tiny fragments may be filtered out of the water by a true fly larva (i.e., a filterer). Leaves may also be fragmented by a combination of microbial activity and physical factors such as water current and abrasion (Benfield et al., 1977; Paul et al., 1978).

Leaf-pack processing by all the elements mentioned earlier (i.e., leaf species, microbial activity, physical and chemical features of the stream) is essential. The crucial point is that these ecosystem processes break down leaves into fine particles, which are carried downstream and serve as energy for various consumers.

The bottom line on allochthonous material in a stream: Insects that have fallen into a stream are ready to eat, and may join leaves, exuviae (castoff coverings, such as crab shells or the skins of snakes), copepods, dead and dying animals, rotifers, bacteria, and dislodged algae and immature insects in their float (drift) downstream to a waiting hungry mouth.

9.2.4 Levels of Organization

Odum (1983) suggested that the best way to delimit modern ecology is to consider the concept of *levels of organization*. Levels of organization can be simplified as shown in Figure 9.2. In this relationship, organs form an organism; organisms of a particular species form a population; populations occupying a specific area form a community. Communities, interacting with nonliving or abiotic factors, separate into a natural unit to create a stable system known as the ecosystem (the central ecological unit); and the part of the Earth in which the ecosystem operates is known as the *biosphere*. Tomera points out "every community is influenced by a particular set of abiotic factors" (Tomera, 1989). Inorganic substances such as oxygen, carbon dioxide, several other inorganic substances, and some organic substances represent the abiotic part of the ecosystem.

The physical and biological environment in which an organism lives is referred to as its *habitat*. For example, the habitat of two common aquatic insects, the "backswimmer" (Notonecta) and the "water boatman" (Corixa), is the littoral zone of

Organs → Organism → Population → Communities → Ecosystem → Biosphere

FIGURE 9.2 Levels of organization.

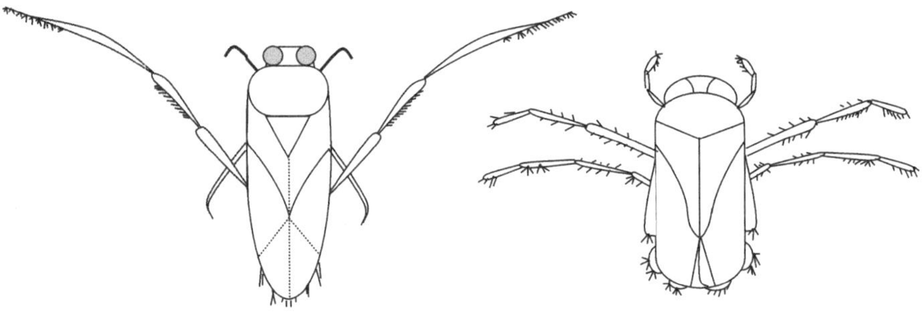

FIGURE 9.3 *Notonecta* (left) and *Corixa* (right).

Source: Adapted from Odum, E.P., Basic Ecology, Saunders, Philadelphia, PA, 1983, p. 42.

ponds and lakes (shallow, vegetation-choked areas) (see Figure 9.3) (Odum, 1983). Within each level of organization of a particular habitat, each organism has a special role. The role the organism plays in the environment is referred to as its *niche*. A niche might be that the organism is food for some other organism or is a predator of other organisms. Odum refers to an organism's niche as its "profession" (Odum, 1971). In other words, each organism has a specific job or role to fulfill in its environment. Even if two species share a habitat, differences in their food habits distinguish their niches (Odum, 1983). A comparison of the niches of the water backswimmer and water boatman illustrates niche separation. The backswimmer hunts actively, but the water boatman eats decaying plants (McCafferty, 1981).

9.3 ECOSYSTEMS

An *ecosystem* denotes an area that includes all organisms therein and their physical environment. The ecosystem is the major ecological unit in nature. Living things and their environment interact closely, forming a self-regulating, self-sustaining system. To create a self-regulating and self-maintaining system, ecosystems are homeostatic, that is, they resist any change through natural controls. These natural controls are necessary in ecology. This is especially the case because people, through their complex activities, tend to disrupt natural controls.

As stated earlier, the ecosystem encompasses both the living and nonliving factors in a particular environment. The living or biotic part of the ecosystem is formed by two components: *autotrophic* and *heterotrophic*. The autotrophic (self-nourishing) component does not require food from its environment but can manufacture food from inorganic substances. For example, some autotrophic components (plants) manufacture needed energy through photosynthesis. Heterotrophic components, on the other hand, depend on autotrophic components for food.

The nonliving or abiotic part of the ecosystem is formed by three components: inorganic substances, organic compounds (linking the biotic and abiotic parts), and the climate regime. Figure 9.4 is a simplified diagram showing a few of the living and nonliving components of an ecosystem found in a freshwater pond.

An ecosystem is a cyclic mechanism in which biotic and abiotic materials are constantly exchanged through *biogeochemical cycles.* Biogeochemical cycles involve living organisms (*bio*) and elements of the environment, such as water, air, rocks, or soil (*geo*). *Chemistry* is concerned with the chemical composition of the earth. Biogeochemical cycles are driven by energy, directly or indirectly from the sun.

Figure 9.4 depicts an ecosystem where biotic and abiotic materials are constantly exchanged. Producers construct organic substances through photosynthesis and chemosynthesis. Consumers and decomposers use organic matter as their food and convert it into abiotic components. That is, they dissipate energy fixed by producers through food chains. The abiotic component of the pond depicted in Figure 9.4 consists of both inorganic and organic compounds, including elements such as carbon, oxygen, nitrogen, sulfur, calcium, and hydrogen, as well as humic acids, all dissolved within the sediment. Producers such as rooted plants and phytoplankton represent the biotic part. Fish, crustaceans, and insect larvae make up the consumers. Mayfly nymphs represent detritivores, which feed on organic detritus. Decomposers make up the final biotic part. They include aquatic bacteria and fungi, which are distributed throughout the pond.

As stated earlier, an ecosystem is a cyclic mechanism. From a functional viewpoint, an ecosystem can be analyzed in terms of several factors. The factors important in this study include the biogeochemical cycles, energy, and food chains.

9.3.1 Energy Flow in the Ecosystem

Simply defined, *energy* is the ability or capacity to do work. For an ecosystem to exist, it must have energy. All activities of living organisms involve work, which is the

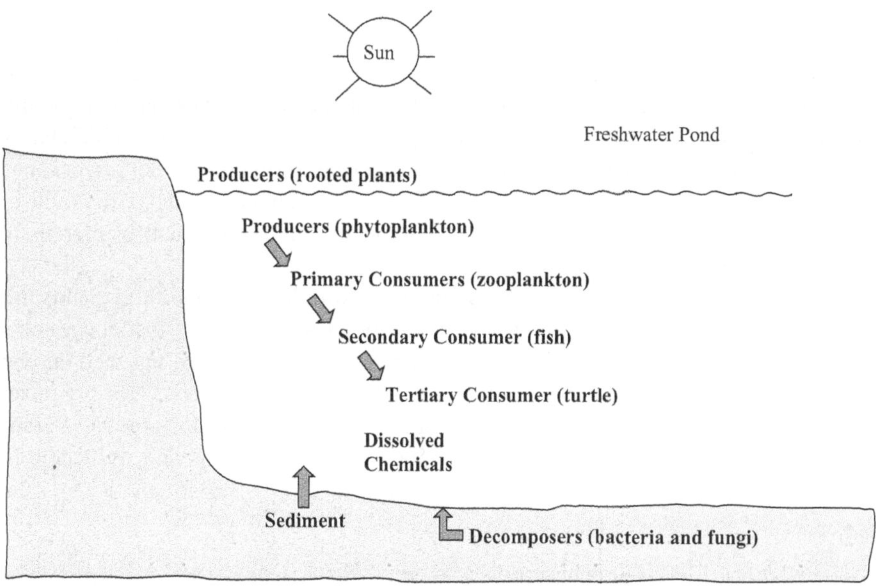

FIGURE 9.4 Major components of a freshwater pond ecosystem.

expenditure of energy. This means the degradation of a higher state of energy to a lower state. Two laws govern the flow of energy through an ecosystem: *The First and Second Laws of Thermodynamics*. The first law, known as the conservation law, states that energy cannot be created or destroyed. The second law states that no energy transformation is 100% efficient. In every energy transformation, some energy is dissipated as heat. The term *entropy* is used as a measure of the unavailability of energy to a system. Entropy increases with an increase in dissipation. Due to entropy, energy input always exceeds useful output, so efficiency is always below 100%.

The interaction of energy and materials in the ecosystem is essential. Energy powers biogeochemical cycles. Note that energy does not cycle like nutrients do in biogeochemical cycles. For example, when food passes from one organism to another, the energy contained in the food is systematically reduced until all the energy in the system is dissipated as heat. Price (1984) refers to this process as "a ***unidirectional flow*** of energy through the system, with no possibility for recycling of energy." When water or nutrients are recycled, energy is required. The energy expended in this recycling process is not recyclable.

As mentioned, the principal source of energy for any ecosystem is sunlight. Green plants, through the process of photosynthesis, transform the sun's energy into carbohydrates, which are consumed by animals. This transfer of energy, again, is unidirectional—from producers to consumers. Often, this transfer of energy to different organisms is called a *food chain*. Figure 9.5 shows a simple aquatic food chain.

All organisms, alive or dead, are potential sources of food for other organisms. All organisms that share the same general type of food in a food chain are said to be at the same *trophic level* (nourishment or feeding level). Since green plants use sunlight to produce food for animals, they are called the *producers*, or the first trophic level. The herbivores, which eat plants directly, are called the second trophic level or the *primary consumers*. The carnivores are flesh-eating consumers; they include several trophic levels from the third level and up. At each transfer, a large amount of energy (about 80% to 90%) is lost as heat and waste. Thus, nature typically limits food chains to four or five links. In aquatic ecosystems, however, food chains are commonly longer than those on land. The marine food chain is longer because several predatory fish may be feeding on the plant consumers. Even so, the built-in inefficiency of the energy transfer process prevents the development of extremely long food chains.

Only a few simple food chains are found in nature. Most simple food chains are interlocked. This interlocking of food chains forms a *food web*. Most ecosystems support a complex food web. A food web involves animals that do not feed on one trophic level. For example, humans feed on both plants and animals. An organism in a food web may occupy one or more trophic levels. An organism's role in its community, not its species, determines its trophic level. Food chains and webs help to explain how energy moves through an ecosystem.

FIGURE 9.5 Aquatic food chain.

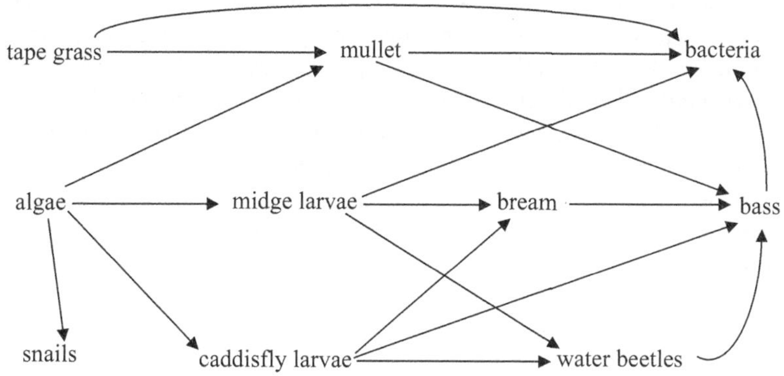

FIGURE 9.6 Aquatic food web.

A critical trophic level of the food web is comprised of the *decomposers*. Decomposers consume dead organisms and recycle nutrients in ecosystems. Simply put, there is no waste in ecosystems. All organisms, dead or alive, are potential sources of food for other organisms. An example of an aquatic food web is shown in Figure 9.6.

9.4 ECOLOGICAL PYRAMIDS

In a food chain, a natural community typically has many small organisms at the base and fewer large organisms at higher levels. A grassy field, for example, has a larger number of grasses and other small plants, a smaller number of herbivores like rabbits, and an even smaller number of carnivores like foxes. The practical significance of this is that we must have several more producers than consumers.

This pound-for-pound relationship, where it takes more producers than consumers, can be demonstrated graphically by building an *ecological pyramid*. An ecological pyramid displays the number of organisms at each trophic level, with producers at the base and final consumers at the top. The pyramid shape is formed due to a significant amount of energy loss at each trophic level. The same is true if the corresponding biomass or energy substitutes are numbers. Ecologists typically use three ecological pyramids: number, biomass, and energy. Obviously, there will be differences among them. Some generalizations:

1. Energy pyramids must always be larger at the base than at the top (because of the second law of thermodynamics, which has to do with dissipation of energy as it moves from one trophic level to another).
2. Likewise, biomass pyramids (in which biomass is used as an indicator of production) are usually pyramid-shaped. This is particularly true of terrestrial systems and aquatic ones dominated by large plants (marshes), in which consumption by heterotrophs is low and organic matter accumulates with time.

However, biomass pyramids can sometimes be inverted. This is common in aquatic ecosystems, in which the primary producers are microscopic planktonic organisms that multiply very rapidly, have very short life spans, and are heavily grazed by herbivores. At any single point in time, the amount of biomass in primary producers is less than that in larger, long-lived animals that consume primary producers.

3. Number pyramids can have various shapes (and not be pyramids at all) depending on the sizes of the organisms that make up the trophic levels. In forests, large trees are the primary producers, while insects make up most herbivores, resulting in a producer base smaller than the herbivore level above it. In grasslands, the number of primary producers (grasses) is much larger than that of the herbivores above (large grazing animals) (Ecosystems Topics, 2000).

9.5 PRODUCTIVITY

As mentioned, the flow of energy through an ecosystem starts with the fixation of sunlight by plants through photosynthesis. Photosynthesis measurement is crucial for ecosystem evaluation. Ecosystems can be classified as high or less productive. Therefore, the study of ecosystems must involve some measure of the productivity of that ecosystem. Primary production is the rate at which the ecosystem's primary producers capture and store a given amount of energy in a specified time interval. In simpler terms, primary productivity is a measure of the rate at which photosynthesis occurs. The four successive steps in the production process are as follows:

(1) *Gross primary productivity*—the total rate of photosynthesis in an ecosystem during a specified interval.
(2) *Net primary productivity*—the rate of energy storage in plant tissues in excess of the rate of aerobic respiration by primary producers.
(3) *Net community productivity*—the rate of storage of organic matter not used.
(4) *Secondary productivity*—the rate of energy storage at consumer levels.

When attempting to comprehend the significance of the term *productivity* as it relates to ecosystems, it is wise to consider an example. Consider the productivity of an agricultural ecosystem such as a wheat field. Often, its productivity is expressed as the number of bushels produced per acre. This is an example of the harvest method for measuring productivity. For a natural ecosystem, several one-square-meter plots are marked off, and the entire area is harvested and weighed to give an estimate of productivity as grams of biomass per square meter per given time interval. From this method, a measure of net primary production (net yield) can be determined.

Productivity can differ widely between and within natural and cultivated ecosystems. Several factors influence year-to-year productivity within an ecosystem. Such factors as temperature, availability of nutrients, fire, animal grazing, and human cultivation activities are directly or indirectly related to the productivity of a particular ecosystem.

Productivity can be measured in several different ways in the aquatic ecosystem. For example, the production of oxygen may be used to determine productivity. Oxygen content may be measured in several ways. One way is to measure it in the water every few hours for 24 hours. During daylight, when photosynthesis is occurring, the oxygen concentration should rise. At night, the oxygen level should drop. The oxygen level can be measured by using a simple *x-y* graph. The oxygen level can be plotted on the *y*-axis with time plotted on the *x*-axis, as shown in Figure 9.7.

Another method of measuring oxygen production in aquatic ecosystems is to use light and dark bottles. Biochemical oxygen demand bottles (300 ml) are filled with water to a particular height. One bottle undergoes testing for initial dissolved oxygen (DO), and the remaining two bottles—one clear and one dark—are placed back at their original depth in the water. After 12 hours, the bottles are collected, and the DO values for each bottle are recorded. After determining oxygen production, productivity (g/m/day) can be calculated. In the aquatic ecosystem, pollution can have a profound impact on the system's productivity.

9.6 POPULATION ECOLOGY

The Water Manager needs basic information about key ecological sectors, including populations that use water for drinking and other purposes. Water management personnel should also possess basic knowledge of population ecology related to stream or river ecology. Over the years, I have received several questions from students and water operators as to why stream ecology and population are important. Water management personnel must keep abreast of the ecological conditions of source water from streams and rivers because these conditions are direct indicators of source water health.

Webster's Third New International Dictionary defines population as all entities in an area, the organisms of a specific region or biotype, or an interbreeding group where speciation begins. The concept of population is interpreted differently in various sciences. In *human demography*, a population refers to all humans within a specific area. In *genetics*, a population is a group of interbreeding individuals of the same species, which is isolated from other groups. In *population ecology*, a population is a group of individuals of the same species inhabiting the same area.

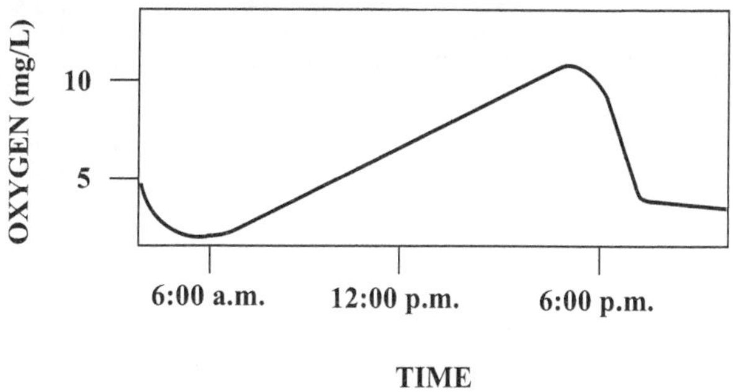

FIGURE 9.7 Diurnal oxygen curve for an aquatic ecosystem.

If we wanted to study the organisms in a slow-moving stream or stream pond, we would have two options. We could examine every fish, plant, crustacean, and insect individually. In that case, we would be studying individuals. It would be easier to do this if the subject were trout, but it would be difficult to separate and study each aquatic plant. The second option is to study every trout, each insect species, and all aquatic plants present in the stream or pond during the study period. Ecologists studying a group of similar individuals in one place and time are examining a *population*. When attempting to determine the population of a particular species, it is important to remember that time is a factor. Time is important because populations change, whether it is at various times during the day, during different seasons, or from year to year.

Population density may change dramatically. For example, if a dam is closed off in a river midway through the spawning season, with no provision allowed for fish movement upstream (a fish ladder), it would drastically decrease the density of spawning salmon upstream. Along with the swift and sometimes unpredictable consequences of change, it can be difficult to draw exact boundaries between various populations. The population density or abundance of a species is determined by factors such as birth, mortality, immigration, and emigration. Changes in population density are the result of both births and deaths. The birth rate of a population is called *natality,* and the death rate is called *mortality*. In aquatic populations, two factors besides birth and mortality can affect density. For example, in a run of returning salmon to their spawning grounds, the density could vary as more salmon migrated in or as others left the run for their own spawning grounds. The arrival of new salmon to a population from other places is termed *immigration* (ingress). The departure of salmon from a population is called *emigration* (egress). Thus, natality and immigration increase population density, whereas mortality and emigration decrease it. The net increase in population is the difference between these two sets of factors.

Each organism occupies only those areas that can provide for its requirements, resulting in an irregular distribution. How a particular population is distributed within a given area has a considerable influence on density. As shown in Figure 9.8, organisms in nature may be distributed in three ways. In a random distribution, there is an equal probability of an organism occupying any point in space, and "each individual is independent of the others." In a regular or uniform distribution, organisms are spaced more evenly; they are not distributed by chance. Animals defend their

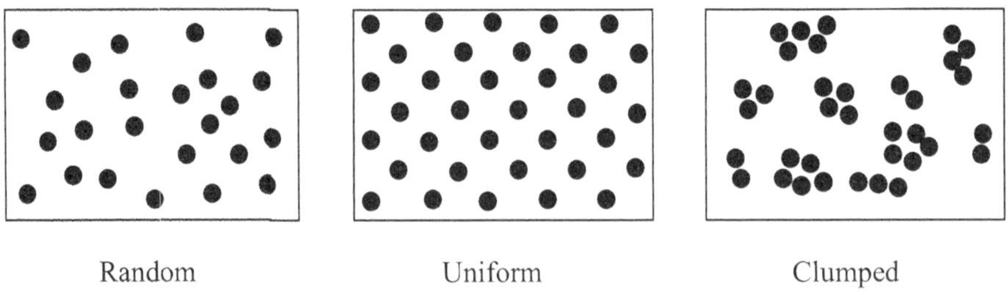

FIGURE 9.8 Basic patterns of distribution.

Source: Adapted from Odum, E.P., *Fundamentals of Ecology*, Saunders, Philadelphia, PA. 1971, p. 205.

territory by excluding others of their species. In regular or uniform distribution, the competition between individuals can be quite severe and antagonistic to the point where the spacing generated is quite even. The most common distribution is the contagious or clumped distribution, where organisms are found in groups; this may reflect the heterogeneity of the habitat. Organisms with clumped distributions often form social hierarchies to coexist efficiently. Animals within the same species have evolved many symbolic aggressive displays that carry meanings that not only are mutually understood but also prevent injury or death within same species.

The size of animal populations is constantly changing due to birth, mortality, emigration, and immigration. As mentioned, the population size will increase if the birth and immigration rates are high. On the other hand, it will decrease if the mortality and emigration rates are high. Each population has an upper limit on size, often called the *carrying capacity*. Carrying capacity is the optimum number of individuals of a species that can survive in a specific area over time. Stated differently, the carrying capacity is the maximum number of species that can be supported in a bioregion. A pond may be able to support only a dozen frogs, depending on the food resources for the frogs in the pond. If 30 frogs shared one pond, likely over half would die due to insufficient food. Carrying capacity is based on the quantity of food supplies, the physical space available, the degree of predation, and several other environmental factors.

The carrying capacity is of two types: ultimate and environmental. *Ultimate carrying capacity* is the theoretical maximum density; that is, it is the maximum number of individuals of a species in a place that can support itself without rendering the place uninhabitable. The *environmental carrying capacity* is the actual maximum population density that a species can maintain in an area. The ultimate carrying capacity is always higher than the environmental carrying capacity. Ecologists have concluded that a major factor affecting population stability or persistence is *species diversity*. Species diversity is a measure of the number of species and their relative abundance.

If the stress on an ecosystem is small, the ecosystem can usually adapt quite easily. Moreover, even when severe stress occurs, ecosystems have a way of adapting. Severe environmental change to an ecosystem can result from natural occurrences such as fires, earthquakes, and floods, as well as from human-induced changes such as land clearing, surface mining, and pollution. One of the most important applications of species diversity is in the evaluation of pollution. Stress of any kind will reduce the species diversity of an ecosystem to a significant degree. For example, domestic sewage pollution reduces DO, stressing aquatic life.

Ecosystems can and do change; for example, if a fire devastates a forest, it will grow back, eventually, because of *ecological succession*. Ecological succession is the observed process of change (a normal occurrence in nature) in the species structure of an ecological community over time. Succession usually occurs in an orderly, predictable manner and involves the entire system. Ecologists can now predict future changes in ecosystems years in advance. These predictions about the water ecosystem are important for Water Managers. Scientists know that if a burned-out forest region receives light, water, nutrients, and an influx or immigration of animals and seeds, it will eventually develop into another forest through a sequence of steps or stages. Ecologists recognize two types of ecological succession: primary and secondary. The type that occurs depends on the site's initial conditions. *Primary succession*, sometimes called *bare-rock succession*, occurs on surfaces such as hardened volcanic lava, bare rock, and sand dunes, where no soil exists, and where nothing has ever grown before (see Figure 9.9). Plants require soil to grow. Thus, soil

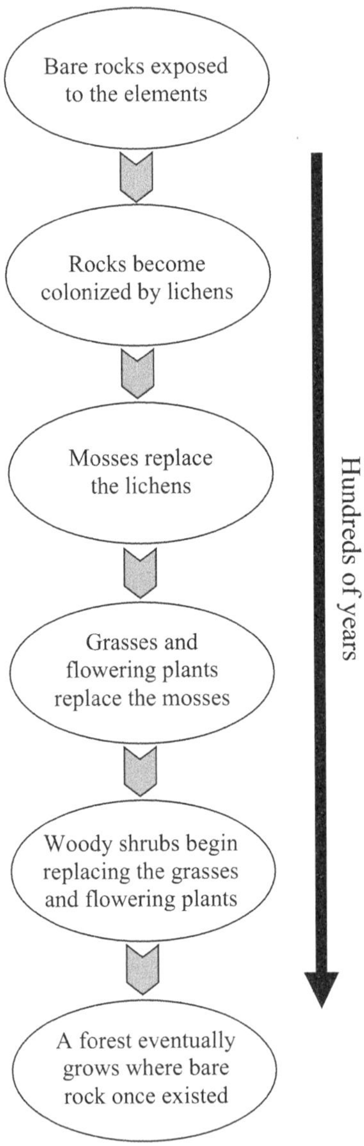

FIGURE 9.9 Bare-rock succession.

Source: Adapted from Tomera, A.N., *Understanding Basic Ecological Concepts*, J. Weston Walch, Publisher, Portland, ME, 1989, p. 67.

must form on the bare rock before succession can begin. Usually, this soil formation process results from weathering. Atmospheric exposure—weathering, wind, rain, and frost—forms tiny cracks and holes in rock surfaces. Water collects in the rock fissures and slowly dissolves the minerals out of the rock's surface. A pioneer soil layer is formed from the dissolved minerals and supports plants such as lichens. Lichens gradually cover the rock surface and secrete carbonic acid, which dissolves additional minerals from the rock. Eventually, mosses replace the lichens. Organisms called *decomposers* move in and feed on dead lichen and moss. A few small animals, such as mites and spiders, arrive next. The result is what is known as a *pioneer community*. The pioneer community is defined as the first successful integration of plants, animals, and decomposers into a bare-rock community.

After several years, the pioneer community builds up enough organic matter in its soil to support rooted plants like herbs and shrubs. Eventually, the pioneer community is crowded out and replaced by a different environment. This, in turn, works to thicken the upper soil layers. The progression continues through several other stages until a mature or climax ecosystem is developed, several decades later. In bare-rock succession, each stage in the complex succession pattern dooms the stage that existed before it. *Secondary succession* is the most common type of succession. Secondary succession occurs in an area where the natural vegetation has been removed or destroyed but the soil is not destroyed. For example, succession that occurs in abandoned farm fields, known as *old-field succession*, illustrates secondary succession. An example of secondary succession can be seen in the Piedmont region of North Carolina. Early settlers of the area cleared away the native oak-hickory forests and cultivated the land. In the ensuing years, the soil became depleted of nutrients, reducing its fertility. As a result, farming ceased in the region a few generations later, and the fields were abandoned. Some 150 to 200 years after abandonment, the climax oak-hickory forest was restored.

In a stream ecosystem, growth is enhanced by biotic and abiotic factors, including:

- Ability to produce offspring
- Ability to adapt to new environments
- Ability to migrate to new territories
- Ability to compete with species for food and space to live
- Ability to blend into the environment so as not to be eaten
- Ability to find food
- Ability to defend itself from enemies
- Favorable light
- Favorable temperature
- Favorable DO content
- Sufficient water level

The biotic and abiotic factors in an aquatic ecosystem that reduce growth include:

- Predators
- Disease
- Parasites

- Pollution
- Competition for space and food
- Unfavorable stream conditions (i.e., low water levels)
- Lack of food

In freshwater ecosystems, greater species diversity increases both inertia and resilience. At the same time, when species diversity is high within a stream ecosystem, a population within the stream can become out of control due to an imbalance between growth and reduction factors, while the ecosystem remains stable. Regarding instability in a freshwater ecosystem, remember that imbalance occurs when growth and reduction factors are out of balance. For example, when sewage is accidentally dumped into a stream, the stream ecosystem, through the self-purification process (discussed later), responds and returns to normal. This process can be described as follows:

- Raw sewage is dumped into the stream.
- It decreases the oxygen available as the detritus food chain breaks down the sewage.
- Some fish die at the pollution site and downstream.
- Sewage is broken down and washed out to sea, and is finally broken down in the ocean.
- Oxygen levels return to normal.
- Fish populations that were deleted are restored as fish about the sewage spill reproduce and the young occupy the real estate formerly occupied by the dead fish.
- Populations all return to "normal."

Shifts in ecosystem balance, such as in streams, commonly occur. In this case, the stream corrected the sewage imbalance on its own and returned to normal through self-purification. Recall that succession is the method by which an ecosystem either forms itself or heals itself. Thus, we can say that a type of succession has occurred in the polluted stream described earlier, because, in the end, it healed itself. More importantly, this healing process is a good thing; otherwise, long ago there would have been few streams on Earth suitable for much more than the dumping of garbage.

In summary, through research and observation, ecologists have found that the succession patterns in different ecosystems usually display common characteristics. First, succession brings about changes in the plant and animal members present. Second, organic matter increases from stage to stage. Finally, as each stage progresses, there is a tendency toward greater stability or persistence. Remember, succession is usually predictable—this is the case, unless humans interfere.

9.7 STREAM GENESIS AND STRUCTURE

Consider the following: Early in the spring, on a snow- and ice-covered high alpine meadow, is the time and place where the water cycle continues. Water, the main component of the cycle, stays frozen throughout the winter but begins to thaw as longer, warmer spring days bring more direct sunlight. The melt starts with a single drop, then two, then increasingly. As the snow and ice melt, the drops join a chorus that continues unending; they fall from their ice-bound lip to the bare rock and soil terrain below.

The terrain where the snow melts differs from glacial till, which is a loose mix of clay, sand, gravel, and boulders left by slowly moving glaciers. Instead, this soil and rock ground are exposed to the falling drops of snow-melt because of a combination of wind and the tiny, enduring force exerted by drops of water as over season after season they collide with the thin soil cover, exposing the intimate bones of the earth.

Gradually, the single drops increase to a small rush—they join to form a splashing, rebounding, helter-skelter cascade, many separate rivulets that trickle then run their way down the face of the granite mountain. A clear, cold, beautiful pool forms on an indented ledge midway down the hill, offering visitors a rare and remarkable experience.

The mountain pool fills slowly, tranquil under the blue sky, reflecting the pines, snow, and sky around and above it, an open invitation to lie down and drink, and to peer into the glass-clear, deep phantom blue-green eye, so clear that it seems possible to reach down over 50 feet and touch the very bowels of the mountain. The pool has no transition from shallow margin to depth; it is simply deep and pure. As the pool fills with meltwater, we wish we could freeze this rare, perfect moment forever. However, this cannot be—Mother Nature calls, prodding, urging—and for a brief instant, the water laps in the breeze against the outermost edge of the ridge, and then a trickle flows over the rim. Gravity exerts a force on the excess melt, directing it downward along the path of least resistance toward its next destination, located several thousand feet below.

When the overflowing, still high in altitude, rock-strewn bed bends downward toward the sea, it meets the angled broken rocks below, bounces, bursts, and mists its way against steep, V-shaped walls that form a small valley carved out over time by water and the forces of the earth. In the valley, melted water has increased from drops to small streams. It flows through what is at first a narrow opening, gaining strength, speed, and power as the V-shaped valley widens to form a U-shape. The water speeds up over large boulders and then slows down again.

At a larger but shallow pool, waters from higher elevations have joined the main body—from the hillsides, crevices, springs, rills, and mountain creeks. At the influent poolside, all appears peaceful, quiet, and restful, but not far away, at the effluent end of the pool, gravity takes control again. The overflow is flung over the jagged lip and cascades downward several hundred feet, where the waterfall again brings its load to a violent, mist-filled meeting.

The water separates and joins repeatedly, forming a deep, furious, wild stream that calms gradually as it continues to flow over less steep lands. The waters widen into pools overhung by vegetation, surrounded by tall trees. The once-clear waters have turned brown and muddy with humic acid and suspended sediments during their descent.

The water mass divides and flows in different directions over various landscapes. Small streams divert and flow into open country. Soils can slow or speed up water flow, sometimes creating shallow swamps, bogs, marshes, fens, or mires. Other streams pause long enough to fill deep depressions in the land and form lakes. For a while, the water remains and pauses on its journey to the sea. However, this is only a short-term pause, because lakes are only a short-term resting place in the water cycle. The water will eventually move on, by evaporation, or seepage into groundwater. Other portions of the water mass stay with the main flow, and the speed of flow changes to form a river, which braids its way through the landscape, heading for the sea. As it changes speed and slows, the river bottom changes from rock and stone to silt and clay. Plants begin to grow, stems thicken, and leaves broaden. The river is now full of life and the nutrients needed to sustain life. However, the river courses onward; its destiny is met when the rich flowing mass slows its last and finally spills into the sea.

Freshwater systems are divided into two broad categories: running waters (lotic systems) and standing waters (lentic systems). We concentrate on lotic systems (although many of the principles described herein apply to other freshwater surface bodies as well, which are known by common names). Some examples include seeps, springs, brooks, branches, creeks, streams, and rivers. Again, because it is the best term to use in freshwater ecology, it is the stream we are concerned with here. Although there is no standard scientific definition of a stream, it is usually subjectively distinguished as follows: A stream is of intermediate size that can be waded from one side to the other.

Physical processes involved in the formation of a stream are important to the ecology of the stream because stream channel and flow characteristics directly influence the functioning of the stream's ecosystem, and the biota found therein. Thus, in this section, we discuss the pathways of water flow contributing to stream flow; namely, we discuss precipitation inputs as they contribute to flow. We also cover streamflow, material transport, channel characteristics, profiles, sinuosity, floodplains, pool-riffle sequences, and depositional features, all of which influence stream ecology.

9.7.1 Water Flow in a Stream

Most elementary students learn early in their education process that water on Earth flows downhill—from land to the sea. However, they may or may not be told that water flows downhill toward the sea by various routes. The route (or pathway) that we are primarily concerned with is the surface water route taken by surface water runoff. A range of contributing factors influences surface runoff. For example, climate, vegetation, topography, geology, soil characteristics, and land use determine how much surface runoff occurs compared with other pathways.

The primary source (input) of water for total surface runoff, of course, is precipitation. This is the case even though a substantial portion of all precipitation input returns directly to the atmosphere by evapotranspiration. *Evapotranspiration* is a combination process, as the name suggests, whereby water in plant tissue and the soil evaporates and transpires to water vapor in the atmosphere. A substantial portion of precipitation input returns directly to the atmosphere by evapotranspiration. It is also important to point out that when precipitation occurs, some rainwater is intercepted by vegetation, where it evaporates, never reaching the ground or being absorbed by plants. Much of the rainwater that reaches the ground, lakes, and streams evaporates back into the atmosphere.

Although plants display special adaptations to minimize transpiration, plants they still lose water to the atmosphere during the exchange of gases necessary for photosynthesis. While much precipitation evaporates, the water that reaches the ground travels through various routes to streams or groundwater.

Soil can absorb rainfall up to its *infiltration capacity* (i.e., its maximum rate). During a rain event, this capacity decreases. Rainfall that exceeds the infiltration capacity collects on the surface. When this surface water exceeds the depression storage capacity of the surface, it moves as an irregular sheet of overland flow. In arid areas, overland flow is likely due to the low permeability of the soil. Overland flow is also likely when the surface is frozen and/or when human activities have rendered the land surface less permeable. In humid areas, where infiltration capacities are high, overland flow is rare.

In rain events, where the infiltration capacity of the soil is not exceeded, rain penetrates the soil. Eventually, it reaches the groundwater, from which it slowly discharges to the stream over a long period. This phenomenon helps to explain why stream flow through a dry weather region remains constant; the flow is continuously augmented by groundwater. This type of stream is known as a *perennial stream*, as opposed to an *intermittent* one, because the flow continues during periods of no rainfall.

When a stream flows through a humid region, it is fed water via the water table, which slopes toward the stream channel. Discharge from the water table into the stream accounts for flow during periods without precipitation and explains why this flow increases, even without tributary input, as one proceeds downstream. Such streams are called *gaining* or *effluent*, opposed to *losing* or *influent streams* that lose water into the ground (see Figure 9.10). The same stream can shift between gaining and losing conditions along its course due to changes in underlying strata and local climate.

9.7.2 Stream Water Discharge

Water velocity in a channel differs across the stream's cross section due to friction with the bed, banks, sediment, the atmosphere, and factors like bends and obstructions. The highest velocities typically occur near the surface and center of the channel, where friction is minimal. In deep streams, the current is fastest just below the surface; in shallow streams, it is fastest at the surface. Velocity decreases as a function of depth, approaching zero at the substrate surface.

9.7.3 Transport of Material

Water flowing in a channel may exhibit *laminar flow* (parallel layers of water shearing over one another vertically) or *turbulent flow* (complex mixing). In streams, laminar flow is uncommon, except at boundaries where flow is very low and in groundwater. Stream flow is typically turbulent. Turbulence creates a shearing force that moves particles along the streambed through processes such as pushing, rolling, and skipping, which is known as the *bed load*. This same shear causes turbulent

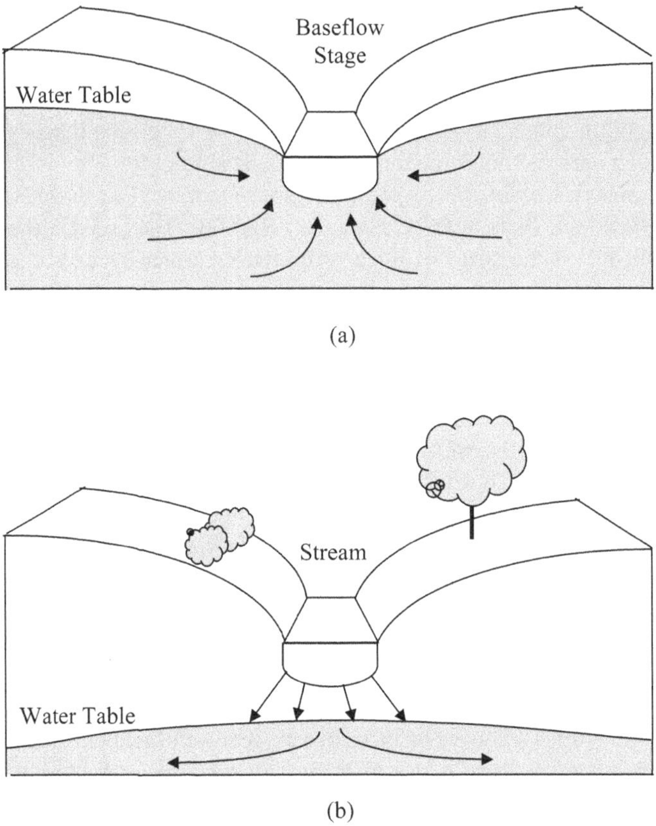

FIGURE 9.10 (A) Cross section of a gaining stream. (B) Cross section of a losing stream.

eddies that entrain particles in suspension (called the *suspended* load—particles with a size under 0.06 mm).

Entrainment is the incorporation of particles when the stream velocity exceeds the *entraining velocity* for a particular particle size. The entrained particles in suspension (suspended load) also include fine sediment, primarily clays, silts, and fine sands that require only low velocities and minor turbulence to remain in suspension. These are referred to as *wash* load (under 0.002 mm). Thus, the suspended load includes the wash load and coarser materials (at lower flows). The combination of suspended load and bed load is collectively referred to as the *solid load*. It is important to note that in bedrock streams, the bed load will be a lower fraction than in alluvial streams, where channels are composed of easily transported material.

A substantial amount of material is also transported as the *dissolved load*. Solutes mainly come from chemical weathering of bedrock and soils, especially in subsurface flows and limestone areas. The relative amount of material transported as a solute rather than a solid load depends on basin characteristics, lithology (i.e., the physical character of rock), and hydrologic pathways. In areas of very high runoff, the contribution of solutes approaches or exceeds the sediment load, whereas in dry regions, sediments make up as much as 90% of the total load.

Deposition occurs when *stream competence* (i.e., the largest particle that can be moved as bedload) and the critical erosion-competent velocity (the lowest velocity at which a particle resting on the streambed will move) fall below a given velocity. Simply stated, the size of the particle that can be eroded and transported is a function of current velocity.

Sand particles are the most easily eroded. The greater the mass of larger particles (e.g., coarse gravel), the higher the initial current velocities must be for movement. However, smaller particles (silts and clays) require even greater initial velocities because of their cohesiveness and because they present smaller, streamlined surfaces to the flow. During transport, particles move at lower speeds than needed to start their movement and settle when the velocity drops further.

Size, flow conditions, and mode of entrainment determine particle movement. Particles over 0.02 mm (medium-coarse sand size) tend to move by rolling or sliding along the channel bed as a *traction load.* When sand particles fall out of the flow, they move by *saltation,* or repeated bouncing. Particles under 0.06 mm (silt) move as *suspended* load, and particles under 0.002 (clay) move indefinitely as *wash* load. If sediment supply remains available, both the concentration and quantity of transported solids increase. However, discharge is usually too low throughout most of the year to scrape or scour, shape channels, or move significant quantities of sediment in all but sand-bed streams, which can experience change more rapidly. During extreme events, the greatest scour occurs, and the amount of material removed increases dramatically.

Sediment inflow into streams can be both increased and decreased due to human activities. Poor farming methods and deforestation sharply boost erosion. Dams and channel diversions can significantly reduce sediment inflow.

9.7.4 Characteristics of Stream Channels

Rivers and streams carve their own channels, following physical laws that are not yet fully understood. The development of stream channels and entire drainage networks, as well as the existence of various regular patterns in the shape of channels, indicate that streams are in a state of dynamic equilibrium between erosion (sediment loading) and deposition (sediment deposit) and governed by common hydraulic processes. However, because channel geometry is four-dimensional with a long profile, cross section, depth, and slope profile, and because these factors mutually adjust over a time scale as short as years and as long as centuries or more, cause-and-effect relationships are difficult to establish. Other variables that are presumed to interact as the stream achieves its graded state include width and depth, velocity, size of sediment load, bed roughness, and the degree of braiding (sinuosity).

9.7.5 Stream Profiles

Mainly due to gravity, most streams exhibit a downstream decrease in gradient along their length. Beginning at the headwaters, the steep gradient becomes less so as one proceeds downstream, resulting in a concave longitudinal profile. Though diverse geography provides for almost unlimited variation, a lengthy stream that originates in a mountainous area typically comes into existence as a series of springs and rivulets; these coalesce into a fast-flowing, turbulent mountain stream, and the addition of tributaries results in a large and smoothly flowing river that winds through the lowlands to the sea.

Studying any stream quickly reveals that its flowing water changes significantly along its course. For instance, when discharge rises, the stream's width, depth, and velocity also increase. In addition to physical changes that occur from location to location along a stream's course, there are a legion of biological variables that correlate with stream size and distance downstream. The main changes are a steeper slope and a shift from a shallow, rocky stream to a deeper one with a sandy bottom. The particle size of bed material at various locations is also variable along the stream's course. Particles become finer downstream.

9.7.6 Sinuosity

Unless forced by humans in the form of heavily regulated and channelized streams, straight channels are uncommon. Stream flow creates distinctive landforms composed of straight (usually in appearance only), meandering, and braided channels, channel networks, and flood plains. Simply put, flowing water will follow a sinuous course. The main measure is the *sinuosity index* (SI). Sinuosity equals 1 in straight channels and more than 1 in sinuous channels. *Meandering* is the natural tendency for alluvial channels and is usually defined as an arbitrarily extreme level of sinuosity, typically an SI greater than 1.5. Many variables affect the degree of sinuosity.

Even in many natural channel sections of a stream course that appear straight, meandering occurs in the line of maximum water or channel depth (known as the *thalweg*). Remember, streams meander to renew themselves. By meandering, they wash plants and soil from the land into their waters, and these serve as nutrients for the plants in the rivers. If rivers are not allowed to meander and are *channelized*, the amount of life they can support will gradually decrease. This results in fewer fish and, consequently, a decline in bald eagles, herons, and other fishing birds (Spellman, 1996).

Meander flow exhibits a consistent pattern, resulting in systematic regions of erosion and deposition (refer to Figure 9.11). The streamlines of maximum velocity and the deepest part of the channel lie close to the outer side of each bend and cross over near the point of inflection between the banks (see Figure 9.11). A huge elevation of water outside a bend causes a helical flow of water toward the opposite bank. Surface flow separation also creates a back eddy. The result is zones of erosion and deposition, which explains why point bars develop in a downstream direction in depositional zones.

9.7.7 Bars, Riffles, and Pools

Implicit in the morphology and formation of meanders are *bars*, *riffles*, and *pools*. Bars develop through deposition in slower, less competent flow on either side of the sinuous mainstream. Onward-moving water, depleted of bed load, regains competence and shears a pool in the meander—reloading the stream for the next bar. Alternating bars

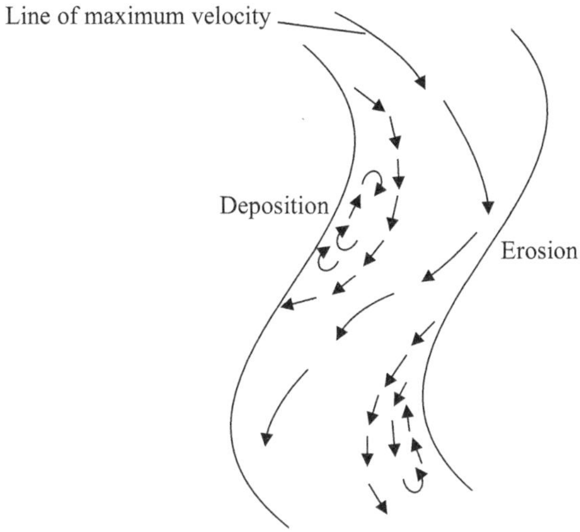

FIGURE 9.11 A meandering reach.

migrate to form riffles (see Figure 9.12). As stream flow continues along its course, a pool-riffle sequence is formed. A riffle is a raised area, and a pool is a depression.

9.7.8 The Floodplain

A stream channel influences the shape of the valley floor through which it courses. The self-formed, self-adjusted flat area near the stream is the *flood plain*, which loosely describes the valley floor prone to periodic inundation during over-bank discharges. What is not commonly known is that valley flooding is a regular and natural behavior of the stream. A stream's aquatic community has several unique characteristics. The marine community operates under the same ecological principles as terrestrial ecosystems, but the physical structure of the community is more isolated and exhibits limiting factors that are very different from those of a terrestrial ecosystem. Certain materials and conditions are necessary for the growth and reproduction of organisms. If, for instance, a farmer plants wheat in a field containing too little nitrogen, it will stop growing when it has used up the available nitrogen, even if the wheat's requirements for oxygen, water, potassium, and other nutrients are met. In this particular case, nitrogen is said to be the limiting factor. A *limiting factor* is a condition or a substance (the resource in shortest supply) that limits the presence and success of an organism or a group of organisms in an area.

Even the smallest mountain stream provides an astonishing number of different places for aquatic organisms to live, or *habitats.* If it is a rocky stream, every rock of the substrate offers several different habitats. On the side facing the upriver, organisms with special adaptations that are very good at clinging to rock do well here. On the side that faces the downriver, a certain degree of shelter is provided from the current, but it still allows organisms to hunt for food. The top of a rock, if it contacts

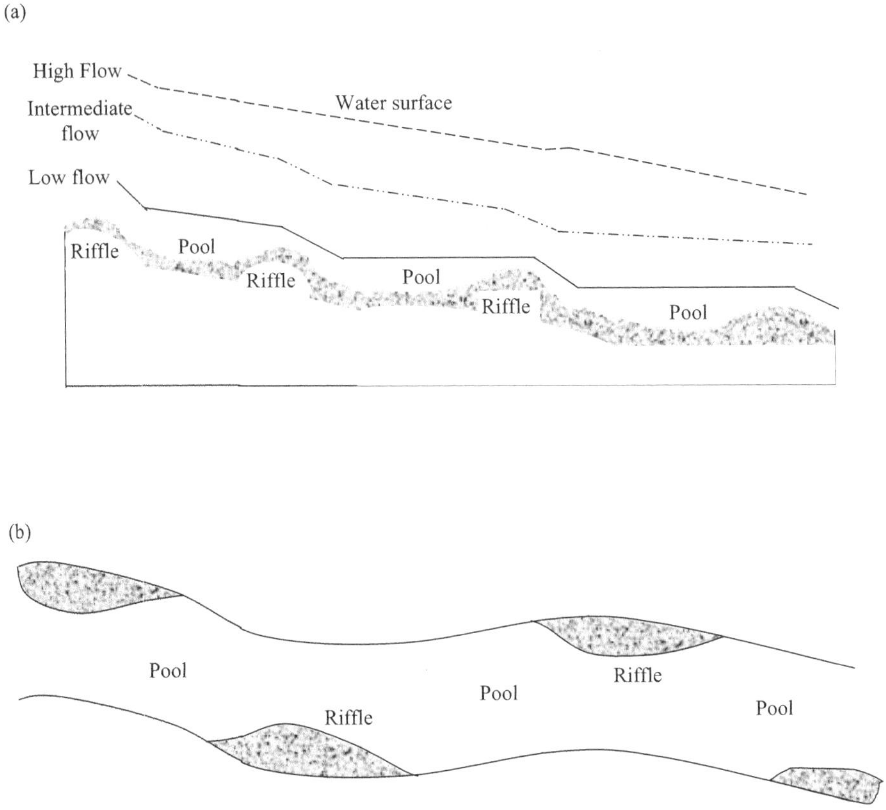

FIGURE 9.12 (A) Longitudinal profile of a riffle-pool sequence; (B) plain view of riffle-pool sequence.

air, is a good place for organisms that cannot breathe underwater and need to surface now and then. Underneath the rock is a popular place for organisms that hide to prevent predation. Normal stream life can be compared to that of a "balanced aquarium" (ASTM, 1969); that is, nature continuously strives to provide clean, healthy, normal streams. This is accomplished by maintaining the stream's flora and fauna in a balanced state. Nature balances stream life by maintaining both the number and the type of species present in any one part of the stream. Such a balance ensures that there is never an overabundance of one species compared to another. Nature structures the stream environment so that both plant and animal life are dependent on the existence of others within the stream.

As mentioned, lotic (washed) habitats are characterized by continuously running water or current flow. Running water bodies usually have three zones: riffle, run, and pool. The *riffle zone* contains faster-flowing, well-oxygenated water with coarse sediments. In the riffle zone, the velocity of the current is significant enough to keep the bottom clear of silt and sludge, thus providing a firm bottom for organisms. This zone contains specialized microorganisms, which are adapted to live in running water. For example, organisms adapted to live in fast streams or rapids (trout) have streamlined bodies, which aid in their respiration and in obtaining food (Smith, 1996). Stream organisms that live under rocks to avoid the strong current have flat or streamlined bodies. Others have hooks or suckers to cling to or attach to a firm substrate to prevent the washing-away effect of the strong current. The *run zone* (or intermediate zone) is the slow-moving, relatively shallow part of the stream with moderately low velocities and little or no surface turbulence. The *pool zone* of the stream is usually a deeper water region where the velocity of water is reduced, and silt and other settling solids provide a soft bottom (more homogeneous sediments), which is unfavorable for sensitive bottom dwellers. Decomposition of some of these solids causes a lower amount of DO. Some stream organisms spend some of their time in the rapids part of the stream, and others can be found in the pool zone (trout, for example). Trout typically spend about the same amount of time in the rapid zone pursuing food as they do in the pool zone pursuing shelter.

Organisms are sometimes classified based on their mode of life:

(1) *Benthos* (Mud Dwellers): The term originates from the Greek word for "bottom" and broadly includes aquatic organisms living on the bottom or on submerged vegetation. They live under and on rocks and in the sediments. A shallow sandy bottom contains sponges, snails, earthworms, and various insects. A deep, muddy bottom will support clams, crayfish, and nymphs of damselflies, dragonflies, and mayflies. A firm, shallow, rocky bottom supports mayfly nymphs, stoneflies, and water beetle larvae.

(2) *Periphytons or Aufwuchs*: The first term usually refers to microfloral growth upon substrata (i.e., benthic-attached algae). The term *aufwuchs* (pronounced OWF-vooks; German for "growth upon") describes the fuzzy, slimy green coating that clings to stems, leaves, or objects above the bottom without penetrating their surfaces. It consists of not only algae like Chlorophyta but also diatoms, protozoans, bacteria, and fungi.

(3) *Plankton (Drifters)*: They are small, mostly microscopic plants and animals that are suspended in the water column; movement depends on water currents. They typically follow the current's direction. There are two types of plankton: (1) *Phytoplankton* are assemblages of small plants (algae) and have limited locomotion abilities; they are subject to movement and distribution by water movements. (2) *Zooplankton* are animals that remain suspended in water and possess limited locomotion abilities. Examples of zooplankton include crustaceans, protozoans, and rotifers.

(4) *Nektons or Pelagic Organisms* (capable of living in open waters): They are distinct from other plankton in that they are capable of swimming independently of turbulence. They swim and can move against the current. Examples of nektons include fish, snakes, diving beetles, newts, turtles, birds, and large crayfish.

(5) *Neustons*: They are organisms that float or rest on the surface of the water (never break water tension). Some varieties can spread out their legs so that the surface tension of the water is not broken, for example, water striders (see Figure 9.13).

(6) *Madricoles*: Organisms that live on rock faces in waterfalls or seepages.

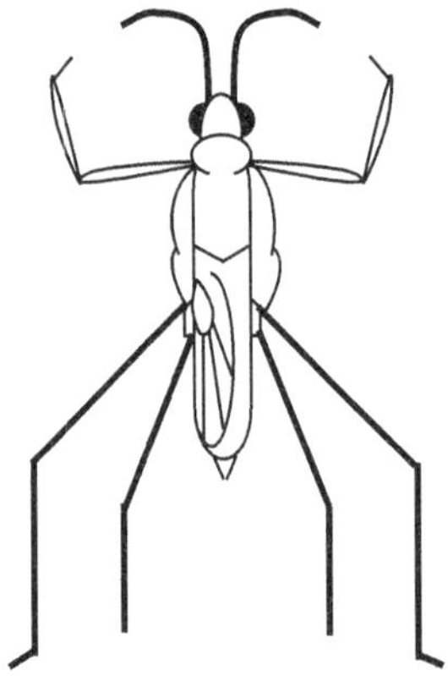

FIGURE 9.13 Water strider.

Source: Adapted from *Standard Methods*, 15th Edition. Copyright © 1981 by the American Public Health Association, the American Water Works Association, and the Water Pollution Control Federation.

In a stream, the rocky substrate is home to many organisms. Thus, we need to know something about the particles that make up the substrate. Namely, we need to understand how to measure the particles so we can classify them by size.

Substrate particles are measured with a metric ruler, in centimeters (cm). Because rocks can be long and narrow, we measure them twice: first the width, then the length. By adding the width to the length and dividing it by two, we obtain the average size of the rock.

It is important to randomly select the rocks we wish to measure. Otherwise, we might tend to select larger rocks, more colorful rocks, or those with unusual shapes. Instead, we should just reach down and collect the rocks in front of us and within easy reach. Then measure each rock. Upon completion of the measurement, each rock should be classified. Ecologists use the Wentworth scale to categorize substrate rock and mineral materials by size. These are:

Boulder	>256 mm
Cobble	64–256 mm
Pebble	16–64 mm
Gravel	2–16 mm
Sand	0.0625–2 mm
Silt	0.0039–0.0625 mm
Clay	<0.0039 mm

Organisms that live in/on/under rocks or small spaces occupy what is known as a *microhabitat*. Some organisms make their microhabitats: Many of the caddisflies build a case about themselves and use it for their shelter.

Rocks are not the only physical features of streams where aquatic organisms can be found. For example, fallen logs and branches (commonly referred to as Large

Woody Debris) provide an excellent place for some aquatic organisms to burrow into and surfaces for others to attach themselves to, as they might to a rock. They also create areas where small detritus, such as leaf litter, can pile up underwater. These piles of leaf litter are excellent shelters for many organisms, including large, fiercely predaceous larvae of dobsonflies.

Another key aquatic habitat is the drifting matter carried downstream. Drift serves as a primary food source for many fish. It may include insects such as mayflies (*Ephemeroptera*), some true flies (*Diptera*), and some stoneflies (*Plecoptera*) and caddisflies (*Trichoptera*). In addition, dead or dying insects and other small organisms, terrestrial insects that fall from the trees, leaves, and other matter, are common components of drift. Among the crustaceans, amphipods (small crustaceans) and isopods (small crustaceans including sow bugs and gribbles) have also been reported in the drift.

9.7.9 Adaptation to Stream Current

Stream current is the key factor shaping organism distribution in streams. The current is determined by the steepness of the bottom gradient, the roughness of the streambed, and the depth and width of the streambed. The current in streams has promoted many special adaptations by stream organisms. Odum (1971) listed these adaptations as follows (see Figure 9.14):

(1) *Attachment to a firm substrate:* Attachment is to stones, logs, leaves, and other underwater objects such as discarded tires, bottles, and pipes.

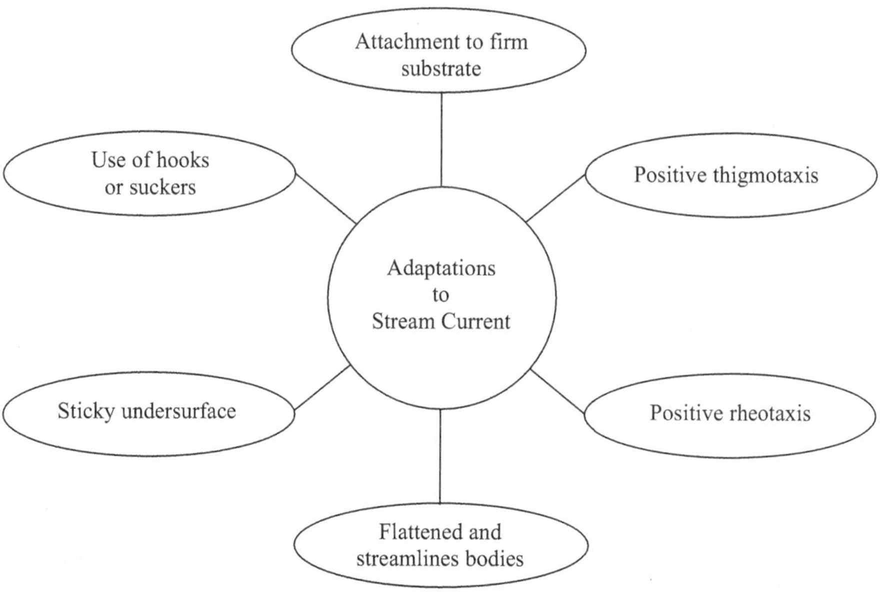

FIGURE 9.14 Adaptations to stream current.

Organisms in this group are primarily composed of primary producers, such as green algae, diatoms, aquatic mosses, caddisfly larvae, and freshwater sponges.

(2) *The use of hooks and suckers:* These organisms have the unusual ability to remain attached and withstand even the strongest rapids. Two Diptera larvae, *Simulium* and *Blepharocera,* are examples.

(3) *A sticky undersurface:* Snails and flatworms use their adhesive undersides to stick to underwater surfaces.

(4) *Flattened and streamlined bodies:* All macroconsumers have streamlined bodies; that is, the body is broad in front and tapers posteriorly to offer minimum resistance to the current. All nektons, such as fish, amphibians, and insect larvae, exhibit this adaptation. Some organisms have flattened bodies, which enable them to stay under rocks and in narrow places. Examples include the water penny (a beetle larva), mayfly nymphs, and stonefly nymphs.

(5) *Positive rheotaxis* (*rheo*: current; *taxis*: arrangement): An inherent behavioral trait of stream animals (especially those capable of swimming) is to orient themselves upstream and swim against the current.

(6) *Positive thigmotaxis* (*thigmo:* touch, contact): Another inherent behavior pattern for many stream animals is to cling close to a surface or keep their bodies in close contact with the surface. Stonefly nymphs, when relocated, instinctively cling to any available surface, even to each other.

Describing the many adaptations that aquatic organisms have made to their environments in streams would require a detailed discussion. Instead, we focus on the specific adaptations relevant to this discussion. The key point to remember is that an aquatic organism can adapt to its environment in several fundamental ways.

9.7.10 Types of Adaptive Changes

Adaptive changes are classed as genotypic, phenotypic, behavioral, or ontogenic:

(1) *Genotypic changes*: Tend to be great enough to separate closely related animals into species, such as mutations or recombination of genes. A salmonid is an example that has evolved a subterminal mouth (i.e., beneath the snout) to feed from the benthos.

(2) *Phenotypic changes*: Are the changes that an organism might make during its lifetime to better utilize its environment (e.g., a fish that changes sex from female to male because of an absence of males).

(3) *Behavioral changes*: Have little to do with body structure or type. A fish might spend more time under an overhang to hide from predators.

(4) *Ontogenetic change*: That which takes place as an organism grows and matures (e.g., a coho salmon that inhabits streams when young and migrates to the sea when older, changing its body chemistry to allow it to tolerate saltwater).

9.7.11 Specific Adaptations

Specific adaptations observed in aquatic organisms include mouth, shape, color, aestivation, and schooling.

(1) *Mouths*: Aquatic organisms such as fish change mouth shape (morphology) depending on the food the fish eats. Fish adapt their jawbones, head bones, gill rakers, teeth, and barbels to consume a wide variety of stream foods.
(2) *Shape*: Changes to allow fish to do different things in the water. Some organisms have body shapes that push them down in the water, against the substrate, and allow them to hold their place against even strong currents (e.g., chubs, catfish, dace, and sculpins). Certain organisms, particularly predators, have developed specific fin arrangements and shapes that enable them to remain motionless before quickly surging forward to capture prey. Examples include bass, perch, pike, trout, and sunfish.
(3) *Color*: May change within hours, to camouflage, or within days, or may be genetically predetermined. Fish tend to turn dark in clear water and pale in muddy water.
(4) *Aestivation:* Helps fish survive in arid desert climates, where streams may dry up from time to time. *Aestivation* refers to the ability of some fish to burrow into the mud and wait out the dry period.
(5) *Schooling*: Serves as protection for many fish, particularly those that are subject to predation.

9.7.12 Benthic Life

The benthic habitat exists in the streambed. As mentioned, the streambed is comprised of various physical and organic materials where erosion and/or deposition are continuous characteristics. Erosion and deposition can occur concurrently or in succession at multiple points along a single streambed. Habitats can develop on channel slopes where the channels are intense and gradually merge with the flat streambed. These habitats are referred to as littoral habitats. Shallow channels may dry up periodically due to weather changes. The streambed is then exposed to the open air and may take on the characteristics of a wetland.

Silt and organic materials settle and accumulate in the streambed of slowly flowing streams. These materials decay and become the primary food resource for the invertebrates inhabiting the streambed. Productivity in this habitat depends upon the breakdown of these organic materials by herbivores. Bottom-dwelling organisms use not only organic materials; a substantial amount becomes part of the streambed in the form of peat.

In faster-moving streams, organic materials do not accumulate so easily. Primary production occurs in a different type of habitat found in the riffle regions, where there are shoals and rocky regions for organisms to adhere to. Therefore, plants that can root themselves into the streambed dominate these regions. By plants, we are mainly referring to forms of algae, often microscopic and filamentous, that can cover rocks and debris that have settled into the streambed during the summer months.

Note: If you have ever stepped into a stream, the green, slippery slime on the rocks in the streambed is representative of this type of algae.

Although the filamentous algae seem well anchored, strong currents can easily lift them from the streambed and carry them downstream, where they become a food resource for low-level consumers. One factor that influences the productivity of a stream is the width of the channel; a direct relationship exists between stream width and the richness of bottom organisms. Bottom-dwelling organisms are essential to the ecosystem, as they provide food for other, larger benthic organisms through consuming detritus.

9.7.13 Benthic Plants and Animals

Vegetation seldom grows in slow streams but may occur along their banks. Green and blue-green algae, along with common water moss, attach to rocks in swift streams. Mosses and liverworts often climb up the sides of the channel onto the banks as well. Some reed-like wetland plants with long, narrow leaves root deeply and withstand strong currents. Aquatic insects and invertebrates dominate slow-moving streams. Most aquatic insects, such as blackflies, caddisflies, and stoneflies, spend most of their life as larvae or nymphs. Adult water beetles and water bugs are also abundant. Insect larvae and nymphs provide the primary food source for many fish species, including American eel and brown bullhead catfish. Representatives of crustaceans, rotifers, and nematodes (flatworms) are sometimes present. The abundance of leeches, worms, and mollusks (particularly freshwater mussels) depends on stream conditions and is typically higher in environments with low phosphate levels. Slow-moving streams and rivers are home to larger animals like newts, tadpoles, and frogs. As mentioned, the important characteristic of all life in streams is adaptability to withstand currents.

9.8 THE BOTTOM LINE

Streamflow, geology, elevation, temperature, DO, seasonal life cycle patterns, substrate, and riparian habitat are factors that influence the abundance and diversity of aquatic life in a waterbody. While natural factors impact aquatic life, human activities usually have a more substantial influence. Consider, for example, human-caused turbidity and sedimentation in a waterbody resulting from reduced riparian vegetation or other causes of erosion can eliminate food sources and habitat for aquatic organisms. In addition, excess nutrients can promote algal blooms. The eventual death and decomposition of the excessive algae depletes DO, reducing the survival of aquatic organisms. Toxic pollutants such as pesticides, metals, and other contaminants can shift the abundances of aquatic life toward favoring more pollution-tolerant species. Changes in waterbody pH can harm aquatic life by weakening shells or decreasing species' tolerance to altered alkalinity.

Water managers are using AI and IoT more often in stream ecology to enhance ecosystem monitoring and management. In the stream environment, IoT techniques are employed to establish ecosystem monitoring systems, allowing for real-time data collection and analysis. IoT monitoring sensors enhance next-generation compliance and streamflow monitoring. IoT sensors provide real-time water quality monitoring to measure parameters such as temperature, oxygen, and pH, allowing for continuous monitoring of water quality.

REFERENCES

ASTM. 1969. *Manual on Water*. Philadelphia, PA: American Society for Testing and Materials.

Barlocher, R., and Kendrick, L. 1975. Leaf conditioning by microorganisms. *Oecologia*, 20: 359–362.

Benfield, E.F. 1996. Leaf breakdown in streams ecosystems. In *Methods in Stream Ecology*, F.R. Hauer, and G.A. Lambertic (Eds.). San Diego: Academic Press, pp. 579–590.

Benfield, E.F., et al. 1977. Leaf pack processing in a pastureland stream. *Oikos*, 29: 99–103.

Carson, R. 1962. *Silent Spring*. Boston: Houghton Mifflin.

Cummins, K.W. 1974. Structure and function of stream ecosystems. *Bioscience*, 24: 631–641.

Cummins, K.W., and Klug, M.J. 1979. Feeding ecology of stream invertebrates. *Annual Review of Ecology and Systematics*, 10: 631–641.

Darwin, C. 1998. *The Origin of Species*, G. Suriano (Ed.). New York: Gramercy.

Dolloff, C.A., and Webster, J.R. 2000. Particulate organic contributions from forests to streams: Debris isn't so bad. In *Riparian Management in Forests of the Continental Eastern United States*, E.S. Verry, J.W. Hornbeck, and C.A. Dolloff (Eds.). Lancaster, PA: Technomic Publishing Company.

Ecosystems Topics. 2000. *Croplands, Forests, Coastal Zones, Freshwater Systems and Grasslands*. New York: World Resources Organization.

Marshall, P. 1950. *Mr. Jones: Meet the Master*. Washington, DC: New York Avenue Presbyterian Church Lecture.

McCafferty, P.W. 1981. *Aquatic Entomology*. Boston: Jones and Bartlett Publishers, Inc.

Odum, E.P. 1971. *Fundamentals of Ecology*, 3rd ed. Philadelphia: Saunders.

Odum, E.P. 1983. *Basic Ecology*. Philadelphia: Saunders College Publishing.

Paul, R.W., Jr., et al. 1978. Effects of thermal discharge on leaf decomposition in a river ecosystem. *Verhandlugen der Internationalen Vereinigung für Thoeretsche and Angewandte Limnologie*, 20: 1759–1766.

Peterson, R.C., and Cummins, K.W. 1974. Leaf processing in woodland streams. *Freshwater Biology*, 4: 345–368.

Price, P.W. 1984. *Insect Ecology*. New York: John Wiley & Sons.

Smith, R.L. 1996. *Ecology and Field Biology*. New York: HarperCollins College Publishers.

Spellman, F.R. 1996. *Stream Ecology and Self-Purification*. Lancaster, PA: Technomic Publishing Company.

Suberkoop, et al. 1976. *Leaf Breakdown in Streams*. Washington, DC: U.S. Geological Survey.

Tomera, A.N. 1989. *Understanding Basic Ecological Concepts*. Portland, ME: J. Weston Walch, Publisher.

USDA. 1999. *Autumn Colors—How Leaves Change Color*. Accessed 02/08/2023 @ http://www.na.fs.fed.us/spfo/pubs/misc/autumn/autumn_colors.htm.

10 Water and AI in Agriculture

10.1 INTRODUCTION

Water Managers in agriculture and their subfield, forestry, are constantly monitoring their water supplies. A reliable supply of clean water is vital to the success of the U.S. agricultural sector. In farming, irrigation allows the cultivation of otherwise non-arable land and increases the productivity of farms, especially in the Great Plains and the West. Likewise, water is an essential element in livestock and aquaculture operations. This section characterizes the role of water in agriculture, focusing on the following topics:

- The economic importance of U.S. agriculture and its place in the global economy
- The use of water in the U.S. agricultural sector
- Supply issues affecting agricultural water use and ways in which competition for water is influencing its use
- Water quality requirements associated with agricultural water use
- Available estimates of the value of water used in agriculture
- Precision agriculture using artificial intelligence (AI)

Agriculture is a significant component of the U.S. economy. In 2010, agriculture accounted for approximately 1.1% of the U.S. gross domestic product (U.S. CIA, 2020); however, in 2023, agriculture represented about 5.5% of the total share. While the relative significance of the agricultural sector has decreased over the last century, U.S. output of crops and livestock has grown steadily, the result of significant productivity advances associated with new machinery, soil science, digital applications with drones, and other agricultural technology.

10.2 U.S. AGRICULTURAL SECTOR

The National Agricultural Statistics Service (NASS) within the U.S. Department of Agriculture conducts a periodic census of U.S. agriculture. Table 10.1 draws on the census data (2023) to provide a basic economic profile of the U.S. agricultural sector. In 2007, U.S. agricultural products were valued at $297 billion, with crops and livestock making similar contributions. This production occurred on about 2.2 million farms. Most farms are small family-run operations and contribute only a minor portion to overall production. In contrast, about 64% of all market value is generated by a small number of large farming operations having annual sales of $500,000 or more (USDA/NASS, 2023).

DOI: 10.1201/9781003655848-15

TABLE 10.1
Overview of the U.S. Agricultural Sector (2007)

Sector Characteristic	Estimate
Total number of farms	1.8 million
Employment	1.17 million
Land used in farming (acres)	876,000,000
Irrigated land (acres)	*56,599,305*

Source: USDA/NASS (2023), BLS (2023).

U.S. farm employment is decreasing, even though about 114,000 farming positions are expected to open in 2025. Some people point to advances in technology as the reason for the reduction in farm employment. However, the fact is that there are still over 100,000 farm jobs available right now, and this is a conservative estimate. In a study conducted by my upper-class and graduate environmental engineering students, who were assigned this study when probed about the decrease in farming, acreage, and employees, the standard reply I received was the following question: "Who wants to be a farmer today?" I was surprised by this answer.

Neither the Bureau of Labor Statistics (BLS) nor the USDA maintains comprehensive data on employment in the agriculture sector. A rough estimate can be obtained by adding the number of farm proprietors and the number of paid employees in crop and animal production. This computation suggests that the agriculture sector employs approximately 1.17 million individuals.

The market value of U.S. farm products has grown over time and saw particularly marked growth from 2002 to 2025. This growth results from increased productivity and strong domestic and export markets. A specific driver has been the growth in U.S. corn production and increases in corn prices. Corn demand has increased due to its use in new foods like sweeteners, as livestock feed, and in subsidized biofuels such as ethanol (USDA/ERS, 2008). Be advised that President Trump has requested the soft drink industry to use cane sugar instead of corn syrup.

DID YOU KNOW?

Agricultural operations—crop irrigation, livestock watering, and aquaculture—withdraw approximately 140 billion gallons of water per day. Irrigation accounts for over 90% of this use and is the largest single consumptive water use in the United States.

10.3 U.S. AGRICULTURE IN A GLOBAL CONTEXT

The U.S. agricultural sector is part of a complex global system of agriculture, food processing, and food marketing. Agricultural exports have outpaced imports, creating a consistent trade surplus.

Exports have undergone a significant transformation in recent years. While bulk grains (e.g., corn, wheat) historically accounted for growth in U.S. exports, high-value products now play a larger role. These products include meats, poultry, fruits, and vegetables. This change is the direct result of growth in population and incomes worldwide; demand from this growth has driven U.S. exports. Trade agreements have helped expand U.S. export markets. Since NAFTA was fully implemented, Canada and Mexico have surpassed Japan and others as the top markets for U.S. agricultural exports. Exports account for about 30% of the total U.S. agricultural production value.

Numerous factors influence the competitiveness of the U.S. agricultural sector in the global economy. These factors include the prices of key inputs such as labor, seed, fertilizer, and machinery; the status of trade agreements with various nations; changes in subsidies and other agricultural policies; technological improvements; and access to capital. Summarizing competitiveness is challenging since U.S. performance differs by trading partner and agricultural product, but productivity reliably indicates the resilience and competitiveness of U.S. agriculture. The USDA maintains an index measuring agricultural productivity, defined as the difference between the growth in farm output and the growth of all agricultural inputs; this measure of agricultural productivity has increased steadily over time, rising by 152% in the last 60 years. Through productivity gains and efficient use of inputs, the U.S. agricultural sector continues to improve its competitive position and maintain a positive balance of trade.

DID YOU KNOW?

As mentioned, agricultural use is a major consideration in the competition for water, especially in western states. Increasingly, agricultural interests that hold water rights are participating in water transfer initiatives through which they sell water to municipalities and other users.

In the context of this text, a key question is whether a reliable supply of water includes the productivity and competitiveness of the nation's agricultural sector. U.S. producers have a comparative advantage due to several water-related factors, such as the following:

- A temperate climate that allows for non-irrigated production in many parts of the country
- An efficient and technologically advanced system of irrigation in areas where it is required
- A reliable supply of high-quality groundwater and surface water for irrigation and livestock watering
- The incorporation of AI and drones in advancing productivity in agriculture at warp speed

A detailed comparison of the United States to other nations concerning these factors is beyond the scope of this book. Still, personal experience and research, along with the literature, suggest that few nations enjoy the advantages enjoyed by U.S. producers. First, most countries are more reliant upon irrigation for food production than is the United States. Approximately 42% of total freshwater withdrawals are used for irrigating over 59 million acres of cropland. Globally, irrigated agriculture contributes to 40% of the food supply, despite only being 29% of the world's cultivated land. Studies suggest that over 850 million people worldwide do not have a reliable source of food; most of these people live in the arid nations of southern Asia and sub-Saharan Africa, where water access is poor. Finally, recommendations for increasing food security emphasize the need for expanding and improving the efficiency of irrigation systems worldwide (Molden, 2007).

Studies suggest that emerging climate trends could reinforce the competitive advantage of U.S. agriculture. In general, climate change is anticipated to yield increased temperatures and altered precipitation patterns, intensifying competition for limited water resources in many areas—an analysis by Lobell et al. Between 1980 and 2008, climate change reduced global maize production by 3.8% and wheat production by 5.5% (2011). The warming pattern that caused the production decrease, however, was absent in U.S. farming areas. Strzepek and Boehlert (2010) analyzed how future climate change patterns could increase competition for irrigation water. The study projected changes in municipal, industrial, environmental, and agricultural water demand under different climate scenarios. The authors found that threats to agricultural water supplies are likely to intensify in Africa, Latin America, and the Caribbean. In contrast, the danger to agricultural water supplies in North America is expected to decline on net, primarily due to wetter climate conditions and reduced need for water to maintain in-stream flows (USEPA, 2012b).

DID YOU KNOW?

The absence of formal markets complicates efforts to value water applied in agriculture, and studies yield vastly different estimates depending upon the methodology used. Estimates based on delivery cost and factor input methods yield relatively low estimates (generally less than $100 per acre-foot). Data from water transfers suggest much higher values. Transfers between farms show values averaging about $1,800 per acre foot. Transfers from farms to municipalities show even higher average values (over $4,000 per acre-foot).

10.4 WATER USE

Note: Some of the data presented below is dated but representative; this is the case because some of the current data is not always available.

As mentioned, agricultural applications account for approximately 34% of all water withdrawals. Irrigation uses account for over 90% of these withdrawals, while

TABLE 10.2
Irrigation Water Use

Source	Year	Acre-Feet Applied	Acres Irrigated	Average Acre-Feet Per Acre
USGS	2020	>140,000,000	66,100,000	2.35
USDA Farm and Ranch Irrigation Survey	2015	>91,000,000	55,900,000	1.7

Source: USGS, 2015; USDA/NASS, 2023.

watering livestock and supplying to aquaculture make up the balance. The discussion below provides a more detailed review of how water is used in these three agricultural applications.

10.4.1 Irrigation

Estimates of the total amount of water used for irrigation on U.S. farms vary by year and data source. Table 10.2 shows that while the U.S. Geological Survey (USGS) estimates that 150 million acre-feet were used in irrigation in 2015, the USDA estimates that only 91.2 million acre-feet were used for irrigation in 2008. The USDA survey is more recent and provides more detail on irrigation methods and patterns; therefore, the discussion below focuses on the USDA figures when possible.

Note: 1 acre-foot of water equals 320,000 gallons, or enough water to cover an acre of land 1 foot deep.

USDA/NASS (2008) points out that just over half of all irrigation water is pumped from on-farm groundwater sources. Off-farm sources, such as irrigation districts, commonly supply surface water. The source of irrigation water can significantly influence incentives to use agricultural water more efficiently.

In all forms of irrigation, the preferred outcome is to deliver water directly to the crop. In practice, however, the ultimate disposition of irrigation water is more complex. Water can be absorbed by crops, seep into uncultivated land, evaporate during or after application, or flow back to surface water via runoff or drainage for potential reuse. Studies by USGS and USDA typically consider the "consumptive" component of irrigation water use to include the fraction that is "evaporated, transported, incorporated into products or crops . . . or otherwise removed from the immediate water environment" (USGS, 2009). Based on data from 1995, the USDA estimates that 61% of irrigation water is consumed, making irrigation the most dominant consumptive water use (Wiebe and Gollehon, 2006).

U.S. farmers typically use one of three irrigation methods to distribute water.

- **Gravity Systems**—These systems use ditches, furrows, or pipes to deliver water by gravity. Although low-cost, they are typically less efficient than other distribution methods.

- **Sprinkler Systems**—Variations of sprinkler systems include rolling systems, which can be moved across fields, and rotating or pivot systems that direct water over a wide area by moving the water stream. Sprinkler systems were applied on over half the irrigated acreage in the United States in 2008.
- **Drip/Trickle or Micro Sprinkler Systems**—Drip/trickle systems use small-diameter tubes placed on or below the soil surface, applying water frequently and slowly. Water is applied directly to the root zone of the plants. Micro sprinkler systems use low-volume sprinkler heads positioned just above the soil surface. Though relatively expensive, these low-flow methods are highly efficient and typically used for valuable, perennial crops (Wiebe and Gollehon, 2006). Low-flow drip irrigation systems are used on only about 7% of total acreage.

DID YOU KNOW?

While water-use efficiency has improved through increased use of new irrigation technologies, there is still room for improvement. Historically, efficiency incentives were often reduced by abundant water supplies made available through large public infrastructure projects. These projects created interdependencies between agriculture and other water users, such as municipalities and power generators.

Multiple factors have promoted more efficient irrigation practices. A major driver has been the increasing scarcity of water in key agricultural areas and increases in the explicit or implicit price paid for water. In addition, technological shifts and advances have allowed for greater irrigation efficiency. One technological change is simply the method of water delivery. Data show an increase in the use of sprinkler systems and a reduction in the use of flood or gravity systems (which lose water in conveyance). Between 1985 and 2005, the amount of land irrigated by sprinkler systems increased from 22 million to more than 30 million acres. In that same period, the amount of land irrigated by flood systems fell from 35.0 million to 26.6 million acres (USGS, 2009). Although land use for farming has dropped today, in 2025, irrigation continues to expand in the United States.

Irrigation water application rates further demonstrate trends in water-use efficiency. The USGS estimates that in 1950, the average application rate was 3.55 acre-feet per acre; by 2005, this figure had fallen to 2.35 acre-feet per acre. The 2008 USDA irrigation survey estimates per-acre application rates to be as low as 1.7 acre-feet per acre. It is important to note that increases in irrigation efficiency do not necessarily translate into a decrease in water consumption. Because gravity systems deliver more water to fields than is required by crops, a significant portion of that water runs off. It may be available to downstream users through irrigation return flows. By contrast, drip irrigation systems are much more precise in the amount of water delivered to crops, meaning that plants consume a much higher percentage of the water used for irrigation. To the extent that efficiency gains reduce losses to evaporation, they can decrease total consumptive use of water in this sector.

Despite recent increases in irrigation efficiency, significant room for improvement remains. As discussed below, distortions in the prices that some irrigators pay for water can limit the economic incentive to improve efficiency. Comparative studies suggest that the United States is not among the most efficient users of irrigation water. Many operations that could employ precision technologies are still using inefficient gravity systems (Gleick, 2006). Likewise, few farmers are taking advantage of scheduling and measurement technologies in irrigation. Most irrigators still rely heavily on the feel of the soil or the condition of the crop. The 2008 USDA irrigation survey suggests that only about 9% use soil moisture sensors; likewise, only 2% use plant moisture sensors.

Notwithstanding the continuing tendency of many farmers to use personal judgment and old habits, like personal feel of the soil or condition of the crop, in determining proper irrigation treatments, one thing is sure. There is little doubt that irrigation has and does contribute significantly to the productivity of U.S. agriculture.

The importance of irrigation is also reflected in its effect on crop yields. [*Note*: Several irrigated products for which yield information is not available, including forage, orchard crops, rice, and vegetables, are excluded herein]. In all cases, irrigation improves the recorded yield per acre. Alfalfa and wheat yields increase twofold with full irrigation.

The value of irrigated crops reflects two key trends: (1) the importance of irrigation for farm productivity and (2) increased efficiency in the use of irrigation water. Despite stable water use, the market value of irrigated crops has consistently risen. Moreover, it is important to note that irrigation water use is concentrated in a limited number of states. This is the result of some obvious factors, such as an arid climate, as well as complex factors that include demographic patterns and institutional decisions that have enhanced the availability of water for irrigation. Seven states account for about two-thirds of all irrigation water use: California, Arkansas, Texas, Nebraska, Idaho, Arizona, and Colorado. California is by far the largest user, accounting for about one-quarter of all irrigation water.

10.4.2 Livestock

U.S. farmers and ranchers withdraw more than two billion gallons of water per day to maintain livestock. Approximately 60% of this supply is drawn from groundwater sources. Most of the water is used and not returned to surface sources or aquifers. USGS studies of water consumption estimate consumptive-use coefficients ranging from 84% to 100%, depending on the predominant animal type, geographic region, and other factors (Shaffer and Runkle, 2007).

Few states require that livestock operations report water usage. As a result, the USGS method relies on water-use coefficients to estimate livestock water withdrawals. These coefficients reflect total daily water usage for each significant category of livestock; the coefficients are combined with USDA data on animal inventories to estimate water use. Where available, USGS applies water-use coefficients specific

TABLE 10.3
USGS Livestock Water Use Coefficients

Animal Type	Water-Use Coefficient (Gal/Day)
Dairy cows	35
Beef and other cattle	12
Hogs and pigs	3.5
Laying hens	0.06
Broilers and other chickens	0.06
Turkeys	0.1
Sheep and lambs	2
Goats	2
Horses	12

Source: USEPA (2012b), Lovelace (2009a).

to a given state. When state-specific coefficients are not available, USGS applies the median values shown in Table 10.3.

Water needs differ widely among animal types. This is partly due to the water consumption requirements of larger versus smaller animals; it also reflects differences in the use of water for purposes other than consumption. These purposes include waste disposal, sanitation, cooling, and other needs. Water usage for waste disposal differs widely based on the manure management method used. Dairy operations are especially water-intensive; they require water for cleaning cow udders before milking, sanitation of equipment, and cooling of storage tanks (Lovelace, 2009a). This intensive water use is reflected in the high water-use coefficient for dairy cows.

As stated earlier, livestock operations such as dairy procedures require water for activities like cleaning cow udders and sanitizing equipment. Obviously, in cleaning cow udders and in sanitizing dairy operation equipment, the quality of the water used is important (or it certainly should be!). To illustrate, we will provide a concise discussion of water quality issues associated with range cattle.

Water is the most essential nutrient for range cattle. It is required for all life processes. A loss of 20% of the body's water will be fatal. The total body water of cattle is 56%–81% of body weight. Physiological state and body composition affect the body's water content. Lactating cows possess the most significant amount of body water (from 62 to 69%). Fat cows contain less water than thin lactating cows, and younger animals have higher water content than older animals (USDA, 2013).

Loss of body water occurs through milk production, fecal excretion, urine excretion, sweat, and vapor loss. Water losses through milk production are equal to approximately 33% of the cow's daily consumption. Fecal losses are usually comparable to milk losses, and urinary losses are approximately 50% of fecal losses. Fecal water losses are affected by intake, dry matter content of the diet, and digestibility. Urinary losses range from one to nine gallons per day. Urinary water excretion is related to water availability, water absorbed, nitrogen and potassium content of urine, and dry matter content of the diet.

Five criteria can be used to define livestock water quality; a detailed discussion of water quality appears later in the text. They are as follows:

1. Odor and taste
2. Chemical properties (pH, dissolved solids, total dissolved oxygen, and hardness)
3. Toxic compounds (heavy metals, toxic minerals, organophosphates, and hydrocarbons)
4. Presence of excess minerals (nitrates, sodium sulfate, and iron)
5. Presence of living organisms (bacteria)

Cattle can meet their water requirement from three sources:

- Drinking free water
- Ingestion of water contained in feed
- Water produced by the body's metabolism

The following has been estimated:

- Cows requires 2.6 to 3.0 lb of water for every lb of milk produced.
- In moist pastures, cattle may consume only 40% by drinking.
- Salt availability influences consumption. Cattle require an average intake of 11 to 15 grams of salt per day to maintain adequate nutrition. And proper intake of salt also contributes to better productivity and overall herd health.
- High salt or protein may stimulate water consumption. For every gram of sodium consumed (28 grams = 1 ounce), water intake generally will increase by 0.1 lb.

Water is essential during periods of heat stress.

- Properties for cooling include heat conductivity and vaporization. Effective temperatures rise from 65 to 86°F, and consumption increases by 30%.
- Losses from urine, sweating, and respiration increase 15%, 59%, and 50% with temperature increases.
- No shade during the summer will increase water consumption by 18%.
- A critical water quality parameter is salinity; it is expressed as total dissolved solids (TDS) expressed as mg/l or parts per million (ppm):
- sodium chloride, carbonates, nitrates, sulfates, calcium, magnesium, and potassium.

The primary symptom of high salinity water is diarrhea. If the TDS is high, cattle will be reluctant to drink, so they will drink a large amount at once. This can cause the animal to become very sick and potentially die. The following are the guidelines:

- Less than 1,000 ppm. Safe.
- Greater than 1,000 ppm. May cause diarrhea. It may reduce the availability of other minerals, leading to decreased performance.

- 3,000 to 5,000 ppm. Marginally safe. May be reluctant to drink. May reduce performance and affect health. Testing for sulfates is also necessary.
- 5,000 to 7,000 ppm. Poor water. Performance and health slump, especially when temperatures are high. Test for sulfates. Use with low-value stock.
- 7,000 ppm. Unsuitable. Performance and health effects are expected. Limit use with lactating and pregnant stock. Sulfates are most likely high.

Another important livestock water quality parameter is sulfates, which are commonly high in groundwater in South Dakota and Montana. Adult cattle are more resistant to sulfates than calves. Sulfates can cause secondary deficiencies of copper, zinc, iron, and manganese. The form of sulfates can vary. All forms are laxative.

- They will drink less and have diarrhea at a lower sulfate concentration with sodium.
- 2,000+ ppm diarrhea may start, but cattle will adapt.
- The most common forms are sodium sulfate and magnesium sulfate.
- Iron sulfate may cause the most severe rejection of drinking water.

Guidelines for interpretation of sulfate results:

- Less than 500 ppm. Safe.
- 500 to 1,500 ppm. Safe. Trace mineral availability may be reduced. Decreased performance of confined cattle.
- 1,500 to 3,000 ppm. Marginal. Poor for confined cattle during hot weather. Sporadic cases of polio may be seen in confined cattle. Performance reduced.
- 3,000 to 4,000 ppm. Poor water. Sporadic cases of polio are prevalent in confined cattle. The performance of grazing cattle is affected.
- 4,000 plus ppm. Dangerous. Health problems are expected. Reduction in performance.

The results of two South Dakota studies evaluating the performance of three groups of yearling cattle in pens fed a roughage diet and different qualities of water showed the following:

1. Good—400 ppm sulfate/1,000 ppm TDS
2. Poor—3,100 ppm sulfates/4,800 ppm TDS
3. Poorer—3,900 ppm/6,200 ppm TDS

The average daily gain of steers declined from 1.4 to 1 lb per day.

The steers drinking the 3,100 and 3,900 ppm sulfate water had:

- reduced water intake, dry matter intake, and feed efficiency
- 15% of the cattle on the high sulfate water exhibited polio.
- daily sulfur content of the diet was 0.27%, 0.74%, and 0.93%. The requirement is nearly 0.2%.

In another set of studies, steers had access to four sources of water with either

1. 400 ppm sulfates/1,000 ppm TDS
2. 1,700 ppm sulfates/3,000 ppm TDS
3. 2,900 ppm sulfates/5,000 ppm TDS
4. 4,600 ppm sulfates/7,000 TDS.

Steer gains were 1.8, 1.6, 1.5, and 0.6 lbs per day for each sulfate treatment (400, 1,700, 2,900, and 4,600 ppm). Forty-seven percent of steers on the two highest sulfate levels had symptoms of polio, and 33% died (USDA, 2013).

While some technological advances have enhanced water quality and livestock efficiency, improvements in the latter are limited. The drinking component of water use dominates, and animals' water requirements are constant. As a result, livestock water use has varied little over the last 60 years. The growth in water use largely reflects growth in the overall size of the U.S. livestock inventory.

The geographic distribution of withdrawals corresponds to where livestock are raised. Water use is highest in Texas, California, and Oklahoma due to their large dairy and beef cattle operations.

10.4.3 Aquaculture

USGS reports that aquaculture operations account for 2% of all U.S. water withdrawals (Lovelace, 2009b); this percentage of water withdrawals was still 2% in 2015, with about 79% from surface water and the remainder from groundwater (USGS, 2015). USGS considers all the withdrawals to be self-supplied. Available data are limited but suggest that aquaculture water use is skyrocketing, as indicated by an estimated 52% increase in withdrawals for this purpose from 2000 to 2015. Aquaculture water use is primarily non-consumptive and varies by method. Primary methods include the following:

- **Raceways**—Some production (e.g., trout, salmon) occurs in flow-through raceways. In raceways, water is temporarily diverted from a spring or stream to maintain a flowing environment for the fish. This method uses a lot of water but returns almost all of it to the source. USGS reports that about 10% of all aquaculture operations use raceways (Lovelace, 2009b).
- **Ponds**—Ponds are another standard aquaculture method and are used to grow species such as catfish. The amount of water added to ponds varies, depending on precipitation, evaporation, leakage, and the species grown. Therefore, ponds can represent a more consumptive water use in comparison to raceways but generally do not require large water withdrawals.
- **Tanks**—Tanks are used to grow a diverse array of fish species, including trout, salmon, bass, tilapia, perch, and others. Water loss in waste management occurs through evaporation, but most systems use recirculating filtration, so consumption remains low. USGS reports that about 17% of all aquaculture operations use tanks (Lovelace, 2009b).

- **Other Methods**—Other types of aquaculture operations include pens and cages (used for both fish and shellfish), as well as egg incubators.

Five states account for two-thirds of aquaculture water use: Idaho, North Carolina, Alaska, Oregon, and California. Idaho, the top producer of rainbow trout in the United States, relies on raceways for daily operations.

Aquaculture is growing rapidly in the United States, although comprehensive data are not readily available to characterize this growth. USDA only began reporting complete aquaculture information in its 2007 Census. Catfish, trout, and mollusks were and continue to be the most significant products, in terms of both market value and the number of operations. The listed species vary significantly in aquaculture methods and water use. For example, while catfish are growing in human-made ponds, trout are frequently raised in in-stream raceways that effectively consume little or no water.

SIDEBAR 10.2 RECLAMATION ACT AND APPROPRIATION DOCTRINE[1]

Less than 1% of the water on Planet Earth is suitable for people to use, and that small percentage must be shared with nature. The other 99%+ is either too salty or brackish, or it is unavailable in icecaps, glaciers, the atmosphere, and groundwater. Due to the American West's arid climate, both Native Americans and settlers depended on its limited water supply for farming, living, and industry.

As settlers moved into the West, they had to watch the rush of spring and early summer runoff flow away from their towns and crops. They knew they had lost water that would not be available in the dry days of late summer when water shortages are common in the West. Settlers built basic water projects and established complex Western water laws. Varied in detail, western water law systems permanently allocated the right to use water based on the concept of prior appropriation (Appropriation Doctrine) (first in time, first in right) for beneficial use.

At first, water development projects were simple. Settlers diverted water from a stream or river and used it nearby, but in many areas, the demand for water outstripped supply. As water demands increased, settlers wanted to store runoff, which they considered "wasted," for later use. Storage projects would help maximize water use and make more water available for use when needed. Private and state-sponsored irrigation projects were frequently unsuccessful due to insufficient funding or engineering expertise. This resulted in mounting pressure for the Federal Government to develop water resources.

In the jargon of the day, advocates called irrigation projects *"reclamation projects."* The concept was that irrigation would "reclaim" or "subjugate" western arid lands for human use. Many pressures contributed to the discussions that influenced American public opinion, Congress, and the executive branch to support "reclamation." Among them were:

- John Wesley Powell's scientific explorations in the West, including in 1869 the first exploration of the dangerous and almost uncharted canyons of the Green and Colorado Rivers, and his published articles and reports
- Private pressures through publications, irrigation organizations, and irrigation "congresses"
- Nonpartisan political pressures in the West
- Federal Government studies conducted by the U.S. Army Corps of Engineers and USGS

Before 1900, the United States Congress had already invested heavily in America's infrastructure. Significant subsidies were allocated to roads, river navigation, harbors, canals, and railroads. A tradition of government subsidization of settlement in the "West" was long-standing when Congress passed "An Act Granting the Right-of-Way to Ditch and Canal Owners over the Public Lands, and for other Purposes" in 1866. A sampling of subsequent congressional actions promoting irrigation includes the passage of the Desert Land Act in 1877 and the Carey Act in 1894 (also known as the Federal Desert Land Act)—both intended to encourage irrigation projects in the West. In addition, beginning in 1888, Congress appropriated money to the USGS to study irrigation potential in the West. Then, in 1890 and 1891, while the irrigation study continued, Congress passed legislation reserving rights-of-way for reservoirs, canals, and ditches on lands that were then in the public domain. However, westerners wanted more; they wanted the Federal Government to invest directly in irrigation projects. The "reclamation" movement demonstrated its strength when pro-irrigation planks found their way into both Democratic and Republican platforms in 1900. In 1901, "reclamation" gained an assertive and aggressive supporter in Theodore Roosevelt when he became president after the assassination of William McKinley.

President Roosevelt supported the "reclamation" movement because of his personal experience in the West and because of his "conservation" ethic. At that time, "conservation" meant a movement for sustained exploitation of natural resources through careful management for the good of the many. Roosevelt also believed that "reclamation" would permit "homemaking" and support the agrarian Jeffersonian ideal. Reclamation supporters believed that the program would make homes for Americans on family farms. Passed in both houses of Congress by wide margins, President Roosevelt signed the Reclamation Act on June 17, 1902.

In July 1902, Secretary of the Interior Ethan Allen Hitchcock established the United States Reclamation Service (USRS) within the USGS. There is Charles D. Walcott, director of the USGS and the first "director" of the USRS, who placed the original activity in the USRS' Division of Hydrography. Frederick Newell became the first "Chief Engineer" while continuing his responsibilities as chief of the Division of Hydrography.

The Reclamation Act required that:

Nothing in this act shall be construed as affecting or intended to affect or in any way interfere with the laws of any state or territory relating to the control, appropriation, use, or distribution of water . . . or any vested right acquired thereunder, and the Secretary of the Interior . . . shall proceed in conformity with such laws.

That meant implementation of the act required Reclamation to comply with numerous and often widely varying state and territorial legal codes. The development and ratification over the years of multiple interstate compacts governing the sharing of stream flows between states, as well as several international treaties governing the sharing of streams by the United States with Mexico or Canada, have made Reclamation's efforts to comply with United States, state, and territorial water law even more complex.

In its early years, the Reclamation Service relied heavily on the USGS Division of Hydrography's previous studies of potential projects in each western state or territory. As a result, between 1903 and 1906, about 25 projects were authorized throughout the West. These projects were funded by the "Reclamation Fund," into which Congress ordered the deposit of revenues from public land sales in the West. Later, other revenue sources were also directed into the "Reclamation Fund." Because Texas had no federal lands, it was not one of the original "reclamation" states. It became a reclamation state only in 1906.

During its early years, Reclamation, under the direction of the various administrations and Congress, developed several basic principles for the reclamation program. The details have changed over the years, but the general principles remain:

- Federal monies on reclamation water development projects would be repaid by the water users who benefited.
- Projects remain Federal property even when the water users repay their share of Federal construction costs.
- Reclamation contracts with the private sector for construction work.
- Reclamation employees administer contracts and inspect construction to assure that contractors' work meets government specifications.
- In the absence of acceptable bids on a contract, Reclamation, especially in its early years, would complete a project by "force account" (i.e., would use Reclamation employees to do the construction work).
- Hydroelectric power revenues could be used to repay project construction charges.

In 1907, the USRS separated from the USGS to become an independent bureau within the Department of the Interior. The Congress and the Executive Branch, including USRS, were just beginning a learning period during which the economic and technical needs of Reclamation projects became clearer. Initially, overly optimistic about the ability of water users to repay construction costs, Congress established a ten-year repayment period. Subsequently, Congress increased the repayment period to 20 years, then to 40 years, and ultimately to an indefinite period based on the "ability to pay." Other issues that arose included soil science problems related both to construction and to the ability of soils to grow crops; economic viability of projects (repayment potential), including climatic limitations on the value of crops; waterlogging of irrigated lands on projects, resulting in the need for expensive drainage projects; and the need for practical farming experience for people to successfully take up project farms. Many projects were far behind their repayment schedules, and settlers were vocally discontented.

The learning period for Reclamation and Congress resulted in substantial changes when the USRS was renamed the Bureau of Reclamation in 1923. Then, in 1924, the Fact Finder's Act began major adjustments to the basic Reclamation program. Those adjustments were suggested by the Fact Finder's Report, which resulted from an in-depth study of the economic problems and settled unrest on the reclamation's 20-plus projects. Elwood Mead, one of the members of the Fact Finder's Commission, was appointed Commissioner of Reclamation in 1924 as the reshaping of Reclamation continued. A signal of the changes came in 1928, for instance, when Congress authorized the Boulder Canyon Project (Hoover Dam), and for the first time, large appropriations began to flow to Reclamation from the general funds of the United States instead of from the Reclamation Fund or loans to the Fund.

In 1928, the Boulder Canyon Act ratified the Colorado River Compact and authorized construction of Hoover Dam and the All-American Canal—key elements in the implementation of the Compact. Subsequently, during the Great Depression, Congress authorized almost 40 projects for the dual purposes of promoting infrastructure development and providing public works jobs. Among these projects were the beginnings of the Central Valley Project in California, the Colorado-Big Thompson Project in Colorado, and the Columbia Basin Project in Washington.

Ultimately, of Reclamation's more than 180 projects, about 70 were authorized before World War II. The remainder were authorized during and after World War II in minor and major authorizations, such as the Pick-Sloan Missouri Basin Program (1944), the Colorado River Storage Project (1956), and the Third Power Plant at Grand Coulee Dam (1966). Reclamation's most recent major project authorization was in 1968, when Congress passed the Colorado River Basin Project Act, which included the Central Arizona Project, Dolores Project, Animas-La Plata Project, and portions of the Central Utah Project.

During construction, one problem confronted by Reclamation was laboratory testing of special problems. Reclamation performed model testing at several sites, including Montrose, Estes Park, Colorado State University, and downtown Denver. In 1946, Reclamation located its primary laboratory at the Denver Federal Center. These research labs conducted studies on hydraulic structures, concrete technology, electrical issues, design innovations, groundwater, canal and reservoir weed control, environmental concerns, water quality, ecology, drainage, evaporation, and other technical topics.

Passage and implementation of the various Reclamation Projects and the associated laboratory testing described earlier all contribute significantly to or are a result of and in line with Reclamation's mission statement: "The mission of the Bureau of Reclamation is to manage, develop, and protect water and related resources in an environmentally and economically sound manner in the interest of the American public." Under this mission statement, Reclamation's number one priority is always to deliver water, but that priority is often affected by the available water supply and the constraints imposed by various laws, regulations, and court rulings. Each year, over 180 Reclamation projects supply water for agriculture, irrigating roughly ten million acres—about one-third of the West's irrigated land. Reclamation also delivers water used by about one-third of the people in the West.

The second major product delivered to the American public by Reclamation is hydroelectricity. Although the earliest hydroelectric plants on Reclamation projects went into operation in 1908 and 1909, it was only during the 1930s that the generation of hydroelectric power became a principal benefit of Reclamation projects. Reclamation built the large hydroelectric power plants at Hoover, Grand Coulee, and Shasta dams only after a complex public debate about whether the Federal Government should become involved in public power production or whether private power production should be the rule. The Hoover Dam precedent enabled Reclamation to become a leading hydroelectric producer. Once the issues were publicly aired during discussions on the approval of the Hoover Dam, hydroelectric projects became a feature of many additional Reclamation dams. Hydroelectric revenues have subsequently proven to be an essential source for funding the repayment of Reclamation project costs. While the water available in a system determines how much power it is possible to generate, in 1993, for example, Reclamation had 56 power plants online and generated 34.7 billion kilowatt hours of electricity. In 1999, Grand Coulee's hydroelectric generation revenues accounted for approximately two-thirds of Reclamation's total appropriated budget.

DID YOU KNOW?

About 195 Bgal/d, or 8% of all freshwater and saline-water withdrawals for 2000, were used for thermoelectric power. Most of this water was derived from surface water and used for once-through cooling at power plants. About 52% of fresh surface-water withdrawals and about 96% of saline-water withdrawals were for thermoelectric-power use. Withdrawals for thermoelectric power have been relatively stable since 1985.

Regarding reclamation and interstate waters, commissioners representing Arizona, California, Colorado, Nevada, New Mexico, Utah, and Wyoming met in Santa Fe in 1922, with Secretary of Commerce Herbert Hoover moderating, to address the allocation of the water of the Colorado River. The commissioners developed and signed the Colorado River Compact (Compact) to divide and allocate the waters of the Colorado River. The Compact then had to be ratified by each state legislature before going to Congress for ratification by the United States. For Reclamation, this is the most complex interstate compact, ratified by Congress in 1928 without Arizona's agreement. California and Arizona argued for years over how to calculate Arizona's share of the waters of the lower Colorado River. The Arizona legislature ratified the Compact only in 1944 and later sued California over how to interpret the Compact. The lawsuit lasted until 1964. Concerns over the Compact have increased as the Colorado River system contains less water than anticipated. In addition, the Compact required the upper and lower basins to share the burden of delivering 1.5 million acre-feet of water promised to Mexico in a later 1944 treaty. Reclamation is deeply involved in these complicated Colorado River issues because Reclamation reservoirs largely store and regulate the flow of the Colorado River. Reclamation dams in the Upper Colorado River Basin deliver water to Glen Canyon Dam, which then stores the water in Lake Powell. From Lake Powell, the water is delivered by the terms of the Colorado River Compact and operating agreements among the Colorado River Basin states to the Lower Colorado River Basin states. Once delivered to the Lower Colorado River Basin, the water is held until needed in Lake Mead behind Hoover Dam.

It is essential to point out that all over the West, Reclamation is affected and guided by other compacts under the terms of which states share the waters of interstate streams. It is also important to note that even though Reclamation's traditional area of operation is the 17 arid states of the American West, it has at times been assigned work outside that traditional operational area. For instance, during the late 1920s, Reclamation studied "planned group settlement" in the South in cut-over areas and swamps. The project aimed to establish new farms but was abandoned due to the Depression's effects on agriculture. Other projects in the eastern United States were also undertaken, and Reclamation's photograph collection includes hundreds of photographs from areas outside the arid West. Reclamation was also involved in the technical issues and design of many of the Tennessee Valley Authority's dams.

Beginning in the 1930s, Reclamation studied possible projects in Hawaii, and in 1954, Congress authorized investigations on Oahu, Hawaii, and Molokai among the Hawaiian Islands. During the 1940s and 1950s, Reclamation evaluated Alaskan water projects and constructed the Eklutna hydroelectric project near Anchorage. The Department of the Interior transferred the Eklutna Project out of Reclamation in 1967.

Since 2000, Reclamation has supervised additional studies in Hawaii, and Commissioner Bob Johnson was tasked to work with the states of Georgia, Alabama, and Florida over Atlanta's water supply, the flows of the Chattahoochee River, and the affected endangered species.

Early on, Reclamation played a key role in Indian Service irrigation projects for tribes such as the San Carlos, Blackfeet, and Yuma. However, the majority of the reclamation project water went to non-Indians. In the early years, Reclamation's mission to develop water supplies appeared to carry the potential for injuring the rights of tribes. If non-Indians began using Reclamation-provided water, it was feared they would establish a senior right under the Appropriation Doctrine, leaving little or no water for the tribes when they were ready to develop their reservation lands. In a landmark 1908 decision, *Winters v. United States*, the Supreme Court attempted to reconcile this potential conflict and laid out the principles that came to be known as the "Winters Doctrine." This case concerned the Milk River in Montana and delayed the development of Reclamation's Milk River Project for several years. The Winters

Doctrine established the principle of reserved rights. Indian tribes with reservations have reserved water rights in sufficient quantities to fulfill the purposes for which the reservation was established, and the date of the reserved right is the date of the treaty or executive order setting aside the reservation. The dates of reserved rights are very early in relation to non-Indian settlement and, thus, establish a very high priority for Indian water rights. Further, unlike appropriated water rights, a reserved water right does not have to have been used to remain in effect, regardless of how many years have passed. A congressionally authorized and funded Reclamation project could not take precedence over senior or reserved water rights. However, for assorted reasons, some tribes have encountered difficulties in attempting to develop their senior reserved water rights.

In recent years, Reclamation has designed projects that include provisions for delivering water to tribes. Among notable examples are the Central Arizona Project, the Dolores Project, and the Animas-La Plata Project, as well as rural water distribution systems such as the Mid-Dakota in the Dakotas, which provides rural culinary water supply in a large area that includes several reservations. Reclamation staff members also often serve on negotiating teams or provide technical expertise to negotiating teams working for the Secretary of the Interior to develop water solutions for Native Americans.

Reclamation has also been involved with conservation projects that are environmentally sensitive and important. However, the nature of conservation and environmental issues and how they have affected Reclamation has changed considerably. For example, very early in Reclamation's history, between 1908 and 1912, there was a public outcry about the conservation of Lake Tahoe's natural lake level and scenic beauty. Reclamation had proposed to build a dam both to increase storage capacity and to occasionally lower the existing lake level to benefit the Newlands Project. Due to public opposition, the project was not built. In a distinctly different direction, Reclamation's Belle Fourche Project in South Dakota was explicitly designed to avoid mixing hazardous industrial mining wastes in Whitewood Creek with canal irrigation water.

Subsequently, proposals for reclamation projects raised public consciousness about major dams and their impacts on various resources. By the mid-1930s, Reclamation was looking at fishery issues as it addressed the construction of Grand Coulee and other dams. On another front, in the mid- to late-1930s, Coloradoans and their congressional representatives pushed Reclamation to build the Colorado-Big Thompson Project, which would require construction on the fringe of and under Rocky Mountain National Park. The project was built because the law establishing Rocky Mountain National Park allowed a water development project within the park. In the 1950s, the controversy over the construction of Echo Park Dam in Dinosaur National Monument heightened public awareness of issues surrounding the construction of a dam in a National Park Service-managed area. Before public awareness, the importance of preserving scenic beauty didn't make sense to many state residents, who saw the monument as a barren wasteland, or to Mormons, who believed that creating an oasis in the desert was their mission and God's will. At the same time, as many Westerners demanded equal water rights, members of the growing nation's "wilderness movement" saw the Echo Park dam development as an opportunity to prevent a loss equivalent to that of Yosemite's Hetch Hetchy

Valley. The Wilderness Society, the National Parks Association, the Sierra Club, the National Wildlife Federation, the Izaak Walton League, the American Planning and Civic Association, the Wildlife Management Institute, the Audubon Society, and the National Council of State Garden Clubs were among the approximately 30 conservation groups involved in the Echo Park issue. These groups and public opinion forced the cancellation of plans for Echo Park Dam, resulting in the construction of an alternative, Glen Canyon Dam. By the 1960s, public opinion forced the cancellation of plans for the Echo Park dam and resulted in the construction of an alternative, Glen Canyon Dam. By the 1960s, Marble Canyon and Bride Canyon dams were proposed. Still, Secretary of the Interior Stewart Udall canceled those dams because of public pressure to preserve parts of the Grand Canyon. Opposition arose partly because the public and environmentalists viewed nuclear power as a practical alternative for meeting the West's increasing electricity demands (USBOR, 2013).

During the 1960s, Reclamation's work began to change substantially as public awareness changed. Americans became increasingly concerned and proprietary about the use and protection of natural resources. This change resulted, in part, from improved communication and, in part, from improved science, resulting in a clearer understanding of the complex interactions of the communities of nature with Western water issues. Americans were beginning to understand the problems of the West better and to consider the West "mine" or "ours"—even though they lived elsewhere.

Rachel Carson's *Silent Spring*, published in 1962, helped build public support for more environmentally sensitive project development. While even popular music expressed growing environmental concerns, increased public consciousness and support manifested itself in political action when the Congress passed the Wilderness Act in 1964, amendments to the Fish and Wildlife Coordination Act in 1965, the National Historic Preservation Act in 1966, the Wild and Scenic Rivers Act of 1968, the National Environmental Policy Act (NEPA) of 1969, the Endangered Species Act in 1973, and many other laws. Accompanying and buttressing these federal laws were presidential executive orders, federal regulations, and state and local laws, orders, and regulations.

The specific effects of Reclamation projects were also better understood in this period. Dam construction affected fish populations and often altered the flow characteristics and ecology of rivers and streams. Land "reclamation" and construction projects affected plant, animal, fish, and bird populations through habitat displacement or destruction. In addition, land development for agriculture or subdivisions often destroys historic or archeological resources. Destruction of non-arable wetlands was a special environmental problem. Hydroelectric production, usually considered "pollution-free," was recognized as causing environmental effects by altering water temperatures, flow regimes, and natural fluctuation patterns, thus affecting native fish populations' environment, migration, and spawning. Environmental issues that conflicted with the traditional bureau's mission were not unique to Reclamation. Americans identified and looked toward the resolution of many ecological effects caused by construction and natural resource exploitation programs in both the government and private sectors.

Because of the new laws and regulations and increasing public/political pressure, Reclamation hired fresh staff to deal with environmental and historic preservation issues. Reclamation now invests a great deal of time and money in topics such as endangered species; in-stream flows; preservation and enhancement of quality freshwater fisheries below dams; preservation of wetlands; conservation and enhancement of fish and wildlife habitat; dealing with Endangered Species Act issues; controlling water salinity and sources of pollution; groundwater contamination; and the recovery of salmon populations, particularly on the Trinity/Klamath, Columbia/Snake, and San Joaquin/Sacramento River systems. Reclamation implemented a "reoperation" (revision of the way hydroelectric power generation is scheduled and conducted) of hydroelectric facilities at Glen Canyon Dam on the Colorado River to better achieve environmental objectives. Reclamation has made costly modifications to dams such as Shasta and Flaming Gorge to achieve ecological goals. There is also a significant effort underway among federal and state agencies and other interest groups to improve environmental and water quality in the delta at the mouth of the Central Valley of California, where the San Joaquin and Sacramento rivers flow into San Francisco Bay.

Ironically, Reclamation's attempts, in partnership with the Fish and Wildlife Service, to use drainage water to support environmental objectives at the Kesterson National Wildlife Refuge in the Central Valley of California resulted in unexpected and complex ecological problems. The drainage water mobilized and concentrated selenium in the water of the refuge, which caused death and deformity among affected animal populations. The selenium issue was a problem, now resolved, that neither Reclamation nor the Fish and Wildlife Service foresaw (USBOR, 2013).

In addition to public awareness of the potential negative impact of reclamation projects on the surrounding environment, the public also recognized that reclamation reservoirs provide a positive impact all over the West (i.e., regarding flatwater recreation opportunities). While Westerners quickly identified and began to enjoy recreation opportunities on and in the water captured behind Reclamation dams, recreation was not legally recognized as a project use until 1937. Reclamation transferred Lake Mead, behind Hoover Dam, to the National Park Service for recreation management in 1936 and initiated the still-existing pattern of seeking other agencies to manage recreation at Reclamation facilities. That pattern means that today, Reclamation directly manages only about one-sixth of the recreation areas on its projects. From the 1930s to the early 1960s, recreation for specific projects was authorized; however, in the mid-1960s, Congress began to give Reclamation more generalized authorities for funding recreation on all projects. Fishing, hunting, boating, picnicking, swimming, and other recreational opportunities have developed over the years. In addition to flatwater recreation opportunities, many stretches of "blue ribbon" fishing developed all over the West in favorable conditions below Reclamation dams.

In 2010, Reclamation had 289 recreation areas located on about 6.5 million acres of land, most of which were open for public outdoor recreation. In recent years, Reclamation has "reoperated" some facilities seeking to improve fishing and whitewater recreational opportunities. Three recreation areas managed by the National Park Service—Lake Roosevelt behind Grand Coulee Dam, Lake Mead behind Hoover Dam, and Lake Powell behind Glen Canyon Dam—as well as the U.S. Forest Service's Lake Shasta behind Shasta Dam, are among the most prominent National Recreation Areas on Reclamation projects. Other managing partners for recreation areas include other federal agencies, state agencies, counties, and cities. These partnerships result in millions of recreation days of use on reclamation projects annually and raise numerous issues in terms of interagency coordination, water quality, public safety, public access, cost-sharing, law enforcement, etc. As water is converted from rural to urban uses in the West, and the metropolitan population increases, recreation visits to Reclamation projects are expected to continue to increase.

Other benefits gained from reclamation projects are flood control and drought relief. Reclamation operated its facilities to prevent millions of dollars of flood damage between 1950 and 1992. Reclamation projects have averted over $8.3 billion in flood damage. During periods of drought, Reclamation becomes involved in drought management activities. Reclamation projects have carryover storage, which often can provide water during a few consecutive years of drought. In some areas, however, growing demand stresses the water supply even in normal water years. Water shortages, often drought-influenced, will increase in the West, thus forcing more effective and efficient use of water supplies. Reclamation drought activities are pretty varied; for example, assisting water users with planning for use and allocation of limited water supplies, participating in cooperative contingency planning for future droughts, water conservation, loans, cooperation in water banking, deepening wells, and water purchases are among the many possible activities.

Reclamation currently has more than 180 projects in the 17 western states, which are managed out of over 20 area offices. The area offices are within five regions, which are organized around western watersheds. Many projects are operated and maintained by the water users. Reclamation's projects supply agricultural, municipal, and industrial water to approximately one-third of the western population. Farmers on Reclamation projects produce a significant percentage of the value of all crops in the United States, including about 60% of vegetables and 25% of fruit and nut crops. Because of continuing initiatives begun under the presidency of Bill Clinton, Reclamation's staffing levels have trended downward in recent years. In 2010, Reclamation staff was about 29% smaller than in 1993 (USBOR, 2013).

10.5 RESPONSES TO INCREASED COMPETITION FOR WATER

Agricultural interests are experiencing heightened competition for water resources from various sectors, especially in the western regions. The sources of this competition include:

- **Municipal Water Use**—Growth in many arid and semi-arid cities in the western United States has required increasing water deliveries for municipal water systems.
- **Energy Sector**—Hydropower projects may store water and make it available for agricultural use, creating interdependencies between farms and power generators. If water shortages arise, however, power generation may compete with agriculture for access to adequate supplies of water.
- **Conservation and Recreation**—Increasingly, environmental regulations require the maintenance of in-stream flows for the benefit of key wildlife species, as well as for recreational and other uses (CBO, 1997).

Data and modeling suggest that competition among these interests has increased the potential for irrigation water shortages. USDA survey data indicate that, in 2008, 33,000 farms reported diminished crop yields due to irrigation problems. Of these, over 17,000 farms reported reduced crop yields due to irrigation problems. Of these, over 17,000 highlighted a shortage of surface water as the specific problem encountered, which was the single most significant cause reported. Another 3,400 farms reported a shortage of groundwater (USDA/NASS, 2008).

In recent years, stakeholders have responded to the increased competition for water in a variety of ways. Some of these efforts are regulatory. For instance, Congress passed the Central Valley Project Improvement Act (CVPIA) in 1992. Established in 1935, the Central Valley Project (CVP) in California is the largest of Reclamation's water resource projects. It delivers water to the Sacramento Valley in Northern California and the San Joaquin Valley in Central California. About 85% of the CVP's water supply is used for irrigation, while 15% is delivered to municipal and industrial users. The CVPIA sought to address problems associated with the dewatering of wetlands and rivers in the region. It mandates the release of more water to supplement rivers and wetlands, habitat restoration, water temperature control, water conservation, and other steps.

DID YOU KNOW?

Within 20 years, the population of the South Platte River basin outside Denver, Colorado, is projected to grow by 1.9 million. This growth is expected to strain available water supplies, a significant share of which is currently used to irrigate alfalfa and other forage crops. Through the Lower South Platte River Irrigation and Research Demonstration Project, agriculture experts from Colorado State University are collaborating with the Parker Water and Sanitation District to address this issue. Initiated in 2007, the project is identifying and implementing irrigation efficiency techniques designed to generate surplus water that can be traded to the municipal water authority. The project also involves proceedings in Colorado's Water Court to demonstrate that agriculture-to-municipal trades will have no net impact on in-stream flows in the South Platte River (Lytle et al., 2008).

Other initiatives seek to harness the power of market pricing signals to achieve more efficient water use. Many of these efforts fall under the general umbrella of "water transfers" or "voluntary water marketing." In general, these arrangements introduce flexibility into traditional water rights systems, bringing regional water users together in a collaborative trading setting. Specifically, water transfers involve transactions where one party sells or leases water rights to another party. These transactions frequently entail the transfer of water rights from agricultural to municipal interests. They may involve the retirement of irrigated cropland or may focus on improved irrigation efficiency and the transfer of surplus water generated by these efficiency gains.

Research has demonstrated the potential for economic welfare gains from water transfers. An analysis of 1987–2005 water transfers found that irrigator-to-municipality sales consistently resulted in higher prices. The sample of over 1,000 agriculture-to-municipal sales showed a median sales price of over $2,600 per acre-foot (Brewer et al., 2007). Economists have concluded that because water can command higher prices for municipal and other non-irrigation uses, there is significant economic gain to be found in agriculture-to-urban transfers (Eden et al., 2008).

But there are many obstacles to water transfer and marketing. Markets tend to be active only where water law, institutional flexibility, and physical infrastructure combine to allow trading. Likewise, trading may generate unintended externalities for third parties (e.g., reduced groundwater supplies for farms neighboring the irrigation involved in the trading arrangement). Studies suggest that current water markets involve only 1%–2% of all irrigation withdrawals (Wiebe and Gollehon, 2006).

10.5.1 Long-Term Challenges

Long-term global changes are likely to intensify water competition and the need for an innovative response worldwide. Foremost among these challenges is climate change. Climate change is expected to lead to rising temperatures, shifting patterns of precipitation, and more extreme weather events. Most studies predict that agriculture in lower-latitude countries will suffer the most significant harm (Molden, 2007). These studies show that areas such as sub-Saharan Africa and parts of India and China have high vulnerability (in terms of climate and hydrological conditions) and low adaptability (e.g., in terms of water supply and storage options) (UN, 2009). As noted earlier, modeling of future climate change scenarios also supports the idea that threats to agricultural water could intensify in parts of Africa, Latin America, and the Caribbean (Strzepek and Boehlert, 2010).

Climate change poses a threat to agricultural water supplies in the United States, but to a lesser degree relative to some other nations. As mentioned, increased temperature and drought risk in portions of the Southwest and Great Plains may render current irrigation withdrawals unsustainable (USEPA, 2012a). Studies have forecast that increased drought risk in the southwestern and Rocky Mountain regions may necessitate the development of storage capacity to manage variable water supplies better (Strzepek, 2010).

In addition to climate change, other global trends may increase water competition and affect agricultural water supplies. Population growth worldwide has the combined impact of increasing urban water demand while adding to the demand for food. As economic globalization proceeds, increased incomes in developing countries will continue to enhance the demand for meat and other foods that are resource-intensive. A legitimate comparison of the water resource demands of meat versus crop production is the subject of considerable debate; however, available literature consistently demonstrates that much more water is required to produce a given quantity of meat than a calorie-equivalent quantity of any crop (Peden et al., 2007). Competition for water in U.S. agriculture could increase as producers strive to meet food demand in export markets. Finally, the surge in production of certain crops for biofuels may place additional stress on agricultural water supplies. Worldwide, the production of bioethanol from sugarcane, corn, sugar beets, wheat, and sorghum tripled between 2000 and 2007. Along with Brazil, the United States is the major producer meeting this demand, primarily through corn cultivation (UN, 2009).

In the United States, to meet increasing biofuel demands, agriculture will require greater land and water resources. This will need (1) converting current cropland to biofuel crops; (2) shifting other land uses, such as forests and pasture, to biofuel crops; and (3) increasing fertilizer and agrochemical use (Uhlenbrook, 2002). All these actions will heighten potential agricultural impacts on natural resources. If local agriculture shifts to biofuel/bioenergy crops that require more than current agricultural water supplies, there is a likelihood of deleterious effects on limited water resources. To be sustainable, bioenergy production must conserve and protect natural resources, including fresh water. The bottom line: Future trends in biofuels markets may have important consequences for U.S. agricultural water supplies.

Modeling efforts have attempted to integrate all these long-term economic and demographic factors. A study of worldwide trends in water demand determined that competition among municipal/industrial users, environmental flow requirements, and the agricultural sector would result in an 18% reduction in the availability of irrigation water by 2050 (Strzepek and Boehlert, 2010).

10.6 WATER QUALITY

The adequacy of water resources for agricultural purposes depends not only on the quantity of water supplied but also on its quality. Data suggest that water quality issues do not currently pose a significant problem for U.S. producers. Of the 33,000 farms that reported interruptions in the irrigation operations in 2008, only about 2% attributed the problem to poor water quality (USDA/NASS, 2008). As competition for water grows, however, producers may seek out new water supplies of lower quality, increasing the potential for quality-related problems. The following discussion briefly reviews water quality requirements for irrigation, livestock, and aquaculture applications.

10.6.1 Irrigation

The United Nations Food and Agriculture Organization (FAO) has compiled a comprehensive reference document on water quality issues affecting agriculture (Ayers and Westcot, 1994). The FAO study highlights four major categories of water quality problems affecting irrigation:

- **Salinity**—Salts contained in irrigation water can accumulate in the root zone of crops. Yield reduction may result when salt levels grow so great that plants are unable to extract sufficient water from the soil. The salts that contribute to salinity problems are readily transported in irrigation water. Salinity problems may be exacerbated in the presence of a shallow water table, where salts can accumulate and remain in contact with the crop root zone.
- **Water Infiltration Rate**—Irrigation water may infiltrate soil slowly in some conditions, leaving water on the surface to potentially evaporate.

Although soil factors affect infiltration rates, water quality plays a role as well. High sodium relative to calcium and magnesium content in the water will decrease infiltration (as will high salinity).

- **Toxicity**—Plants can absorb certain toxic constituents in irrigation water, and these contaminants can concentrate to levels that damage the plants or reduce yields. Permanent, perennial crops such as trees are most sensitive to these effects. The constituents of most significant concern include chloride, sodium, and boron.
- **Other**—The FAO study cites additional water quality problems that may affect irrigation but which are less common. Elevated nitrogen concentrations are associated with increased vegetative growth and later crop maturity. Likewise, sprinkling water containing excessive concentrations of bicarbonate, gypsum, or iron can create unsightly deposits on fruit or leaves. Finally, excessive pH can lead to various plant abnormalities.

10.6.2 Livestock

Guidance documents for livestock operations emphasize the importance of safe, healthy water supplies. Ensuring the quality of the water supply can be especially challenging, given the impacts that livestock operations themselves can have on nearby source waters. Several categories of water quality concerns include the following:

- **Salinity and Associated Minerals**—Salinity is the problem most encountered in livestock water. Salinity is correlated with the presence of many specific compounds, including sodium, chloride, calcium, magnesium, and bicarbonate. Available guidance suggests that impacts on animals' digestive systems may begin at total salt levels of 3,000 mg/l; some animals may also refuse water with this level of salt. Concentrations above 10,000 mg/l are considered highly saline and unacceptable for use (Faries et al., no date).
- **Nitrogen**—Nitrogen as nitrate is digested and converted to nitrate in animals. Nitrite reduces the blood's ability to carry oxygen and can therefore poison animals. Guidance recommends against using livestock water with greater than 300 ppm nitrate (Pfost and Fulhage, no date).
- **Blue-Green Algae**—High nutrient loadings in farm runoff can result in excessive growth of blue-green algae in livestock water. When ingested by animals, the algae can be toxic and may result in muscle tremors, liver damage, and death (Spellman, 2013). Guidance recommends chlorination of livestock water to reduce algae growth.
- **Suspended Solids**—Water with elevated levels of suspended solids may discourage animals from drinking adequately. The desired range for suspended solids is below 500 mg/l, while concentrations above 3,000 mg/l are considered problematic (Spellman, 2013).

10.6.3 Aquaculture

Because water constitutes the environment in which fish live, water quality is of utmost importance in aquaculture. Aquaculture operations are often carefully calibrated to create the optimum setting for fish and shellfish to thrive. Although water quality requirements are highly specific to the species being cultured, key considerations include the following:

- **Temperature**—Every species has a water temperature at which it grows most readily. For instance, catfish thrive in temperatures between 70°F and 85°F.
- **Dissolved Oxygen**—To prevent stress or death, dissolved oxygen (DO) in aquaculture systems must be maintained at levels conducive to the species in question. In general, warmwater species (e.g., tilapia, carp, catfish) require DO levels above 3.0 ppm, while coldwater species (e.g., trout) require DO levels above 5.0 ppm (Buttner et al., 1993).
- **Ammonia, Nitrates, and Nitrites**—Most fish and shellfish excrete ammonia as their primary nitrogenous waste. Fish exposed to total ammonia-nitrogen levels above 0.02 ppm may grow more slowly and be more susceptible to disease (Buttner et al., 1993).
- **pH**—Fish survive and grow best in water with a pH between 6 and 9.

Table 10.4 summarizes the primary water quality concerns associated with irrigation, livestock, and aquaculture.

10.7 AGRICULTURAL TECHNOLOGIES

Thanks to advancements in several different technological fields, including the digital field, we have made gigantic strides in agriculture, hopefully helping to meet the challenges facing us in the future. Machines and equipment have dominated agriculture for over a century. And this is still the case today with the addition of digital technology, making the equipment and machines not only more efficient but also more capable. Machine learning and other AI methods are increasingly being used

TABLE 10.4
Summary of Agriculture Water Quality Issues

Agriculture Sector	Primary Water Quality Concerns
Irrigation	Salinity
	Toxicity
Livestock	Salinity and minerals
	Nitrogen
	Algae
	Suspended solids
Aquaculture	Temperature
	Dissolved oxygen
	Ammonia, nitrates, nitrites
	pH

to transform agriculture. These technological methods are helping agriculture keep up with the global increase in population.

Currently, Water Managers in agriculture are catalyzing efforts that harness the power of AI in applications throughout agriculture and the food supply chain. Regarding ML, it is a type of software capable of helping us to understand complex situations that involve many factors simultaneously. Examining data and extracting patterns is made more accurate with ML algorithms. This allows humans to perform tasks that have traditionally required human intelligence, including ML, data imaging, innate language processing, smart decision support, self-directed systems, and innovative applications of these techniques to agriculture and food production.

Water Managers are using ML in irrigation in agriculture because it enhances decision-making and efficiency. Some of the key applications of ML in irrigation include:

- **Predictive Analytics**: Whether using classification models, regression models, clustering models, time series models, or neural network models, this application is used in water management because this branch of data science also leverages ML algorithms, statistical methods, and historical data to make data-driven predictions about future outcomes. Why is this important? In the first place, it allows the Water Manager to make informed decisions. It helps optimize crop yields, predict disease, manage resources efficiently, and boost productivity. When water management can anticipate future trends and outcomes, it can make more strategic decisions. Another benefit of using predictive analytics is in risk management, whereby the management personnel can identify and mitigate potential risks. Predictive analytics also enables optimization processes and resource allocation. The key to sustainable water supply, treatment, and distribution can be summed up in one word: INNOVATION, INNOVATION, INNOVATION! Finally, ML can predict crop water needs based on past historical applications. The bottom line: Predictive analytics allows more informed irrigation; that is, when irrigation should be scheduled.
- **Real-Time Data Analysis**: Water management uses this platform to analyze large volumes of soil and weather data instantly. This allows for immediate adjustments to irrigation operations.
- **Automation and Control Systems**: ML algorithms can automate and control irrigation systems, allowing for real-time adjustments.
- **IoT Platforms**: Water Managers incorporate IoT into irrigation systems because they play a vital role in improving water-use efficiency. IoT-based systems use sensors, wireless communication, cloud computing, and automation technologies to monitor and manage real-time irrigation practices. The types of IoT sensors in use today in irrigation include:
  - **Soil Moisture Sensors**: Measure soil water content to help farmers time irrigation.
  - **Temperature and Humidity Sensors:** These sensors track weather conditions, enabling adjustments in optimal growth conditions.

- **Light Sensors**: Measure light intensity essential for crops.
- **Weather Sensors:** Measure rainfall, wind, and other factors affecting crop growth.

IoT sensors are increasingly being used in agriculture to monitor soil moisture and to control irrigation systems based on soil moisture and weather conditions. These sensors are being used in other applications such as gauge light, soil tension, pH levels, and optimal fertilization patterns. In short, adding IoT sensors enables precision farming.

10.8 DRONE TECHNOLOGY IN AGRICULTURE

Agricultural technology continues to evolve, enabling Water Managers to boost farming efficiency and productivity. In the 1980s, farmers started using remote-controlled planes to survey crops from above. Later, drones were introduced and became widely adopted in agriculture due to technological advancements. This ongoing blitz of technological advances has also included AI-assisted drones. Initially, the drones used were not complicated and were easy to operate with a camera for catching an eagle's-eye view of the farmland. With ongoing technological advances, drones have become more sophisticated due to the integration of GPS and advanced imaging sensors. Today, when we think about drones, if we think about them at all, we may envision companies like Amazon, FedEx, and UPS experimenting with delivering packages to our homes or offices using drones. In agriculture, these sophisticated drones deliver or enable precise mapping and data analysis, enabling informed decisions by farmers.

The integration of AI represents a transformative advancement in modern farming practices. Using drones improves the Water Manager's management of water and other resources in agriculture. Drones help farmers increase efficiency, boost yields, and adopt sustainable practices. Drones enable farmers to conduct crop monitoring in real time for healthier assessments and early disease detection. Drones enable pest and weed identification and treatment with minimal chemical use. Drones enable accurate yield forecasting, harvest time, and efficiency. Drones can provide detailed soil mapping and soil health monitoring. Finally, agricultural drones reduce resource use and promote biodiversity.

If you were to visit one of the U.S. farms that use drones, you might be amazed at how sophisticated farming has become. The old horse and plow and early tractor farming are not today's farming. Tractors now feature electronic systems and AI, including machine learning and generative AI, to assist farmers with crop monitoring and management. The name of the game in farming today is based on the transformative capabilities of AI-driven drones.

Advancements in drone technology, like many other technologies, have advanced because of technological developments in various applications such as thermal cameras, multispectral and hyperspectral sensors, LIDAR (Light Detection and Ranging), and AI-driven data analytics for use in agriculture.

Here is an example of transformative change in modern farming. Remember crop dusting operations using aircraft? Today, traditional crop dusting is being replaced by precision crop spraying with drones. The significant advantage, one of them anyway, is that drone crop dusting allows for targeted application. It's all about precision, allowing for targeted application of pesticides, herbicides, and fertilizers. The agricultural drone is capable of accurately delivering its payload to specific areas that require treatment. This action significantly reduces waste and helps to ensure that environmental contamination is reduced. Drones are capable of transporting payloads to challenging hilly regions where conventional delivery methods may not be effective. Drone agriculture can reduce labor and time requirements, as spraying tasks are completed more quickly and efficiently compared to alternative methods. Operation costs are always a consideration that farmers must make; even though initial financial outlays can be expensive in the long run, they allow for cost savings. Farming is hazardous, so farmers should handle chemicals with caution. Drone spraying will enable operators to keep their distance from the substances.

Along with spraying agricultural fields more efficiently, drones are also used for seeding. To maximize crop yield, plant seeds at the correct depth and spacing. Drones that are precise in planting are available, such as Real-Time Kinematic (RTK) and Post-Processed Kinematic (PPK) GPS.

AI-driven drones are also being used in livestock farming, primarily for herd monitoring and inventory management.

10.9 THE BOTTOM LINE

Not understanding and the practice of misusing water can have profound effects on human and animal life. Fortunately, with the rise of AI solutions for water, farmers can transform how they preserve and manage our most precious resource. AI and its subfields are building blocks that enable farmers to make the right decisions at the right time.

NOTE

1 From Bureau of Reclamation (BOR) (2013). *Brief History of Bureau of Reclamation*. Washington, DC.

REFERENCES

Ayers, R.S., and Westcot, D.W. 1994. *Water Quality for Agriculture*. New York: United Nations Food and Agriculture Organization.

Brewer, J., et al. 2007. *Water Markets in the West: Prices, Trading, and Contractual Forms*. February 8, Tucson, AZ.

Bureau of Labor Statistics. 2023. *Quarterly Census of Employment and Wages*. Washington, DC.

Buttner, J.K., et al. 1993. An introduction to water chemistry in freshwater aquaculture. *NRAC Fact Sheet No. 170–1993*. College Park, MD: Northeast Regional Aquaculture Center.

Congressional Budget Office (CBO). 1997. *Water Use Conflicts in the West: Implications of Reforming the Bureau of Reclamation's Water Supply Policies*. Washington, DC.

Eden, S., et al. 2008. Agriculture water use to municipal use, the legal and institutional context for voluntary transactions in Arizona. *The Water Report*.

Faries, F.C., et al. no date. *Water Quality: Its Relationship to Livestock*. Odessa, TX: Texas Agricultural Extension Service.

Gleick, P.H. 2006. Critical issues on water and agriculture in the United States: New approaches for the 21st century. *Prepared for the NAREEE Advisory Board.* Washington, DC: USDA.

Lobell, D.B., et al. 2011. Climate trends and global crop production since 1980. *Science*, 333, July.

Lovelace, J.K. 2009a. *Method for Estimating Water Withdrawals for Livestock in the United States, 2005.* Washington, DC: USGS.

Lovelace, J.K. 2009b. *Methods for Estimating Water Withdrawals for Aquaculture in the United States, 2005.* Washington, DC: USGS.

Lytle, B.A., et al. 2008. A win-win scenario for urban/rural water supplies. *The Water Report*, February.

Molden, D., ed. 2007. *Water for Food, Water for Life—Summary.* Colombo, Sri Lanka: Prepared for Comprehensive Assessment of Water Management in Agriculture.

Peden, D., et al. 2007. Water and livestock for human development. In *Water for Food—Water for Life.* New York: Routledge.

Pfost, D.L., and Fulhage, C.D. no date. *Water Quality for Livestock Drinking.* Columbia, MO: University of Missouri-Colombia.

Shaffer, K.H., and Runkle, D.L. 2007. *Consumptive Water—Use Coefficients for the Great Lakes Basin and Climatically Similar Areas.* Washington, DC: U.S. Geological Survey.

Spellman, F.R. 2013. *Handbook of Water and Wastewater Treatment Plant Operations*, 3rd ed. Boca Raton, FL: CRC Press.

Strzepek, K. 2010. Characterizing changes in drought risk for the United States from climate change. *Environmental Research Letters*, 5.

Strzepek, K., and Boehlert, B. 2010. Competition for water for the food system. *Philosophical Transactions of the Royal Society*, 365: 2927–2940.

U.S. Bureau of Reclamation (USBOR). 2013. *Brief History of Bureau of Reclamation.* Washington, DC: U.S. Department of the Interior.

U.S. Central Intelligence Agency (CIA). 2020. *The World Factbook.* Accessed 01/12/2019 @ https://www.cia.gov/library/publictions/the-world-factbook/fields/2012.html.

U.S. Department of Agriculture, and Economic Research Service. 2008. Corn prices near record high, but what about food costs? *Amber Waves*, February. Accessed 12/12/2019 @ http://www.ers.usda.gov/AmberWaves/February08/Features/CornPrices.htm.

U.S. Department of Agriculture. 2013. *Livestock Water Quality.* Fort Keogh, MT.

U.S. Department of Agriculture, and National Agricultural Statistics Service. 2008. *2007 Census of Agriculture.* Washington, DC.

U.S. Department of Agriculture, and National Agricultural Statistics Service. 2023. *Farm and Ranch Irrigation Survey (2023).* Washington, DC.

U.S. Environmental Protection Agency. 2012a. National water program 2012 strategy: Response to climate change. *Public Comment Draft*, March.

U.S. Environmental Protection Agency. 2012b. *The Importance of Water to the U.S. Economy Part 1: Background Report*, September. Washington, DC: Office of Water.

Uhlenbrook, S. 2002. Biofuel and water cycle dynamics: What are the related challenges for hydrological processes research? *Hydrological Processes*, 21: 3647–3650.

United Nations. 2009. *Water Assessment Programme, Water in a Changing World.* New York: United Nations World Water Development Report 3.

USGS. 2009. *Estimated Water Use in the United States in 2005.* Washington, DC: U.S. Geological Survey.

USGS. 2015. Water use. *Survey Circular 1441*, 65 p. Washington, DC: United States Geological Survey.

Wiebe, K., and Gollehon, N., eds. 2006. *Agricultural Resources and Environmental Indicators*, 2006 ed. USDA Economic Research Service. Access 01/20/2020 @ http://www.ers.usda.gov/publications/arei/eib16/.

11 Manufacturing, Water and AI

11.1 INDUSTRIAL MANUFACTURING

Water use in manufacturing varies significantly across industries. In industries such as chemical, paper, petroleum, primary metal, and food product manufacturing, water is vital to the production process. In some instances, these industries use water to fabricate, process, wash, dilute, cool, or transport a product; in others, they incorporate water directly into the product. Water is also used for sanitation needs within manufacturing facilities. This chapter characterizes the role of water in manufacturing, focusing on the following topics:

- The economic importance of U.S. manufacturing and its place in the global economy
- The use of water in the U.S. manufacturing sector
- Water quality requirements associated with the use of water in manufacturing
- Available estimates of the value of water used in manufacturing
- The AI and water mix

DID YOU KNOW?

Manufacturers withdraw approximately 18.2 billion gallons of water per day, which represents 4% of total water withdrawals in the United States. In addition to these direct withdrawals, the most recent data available suggest that approximately 12% of publicly supplied water withdrawals (4.74 billion gallons of water per day in 1995) were used for manufacturing.

11.2 U.S. MANUFACTURING SECTOR

Manufacturing is a major component of the secondary mega-sector of the U.S. economy. The United States produces a wide range of durable and non-durable goods. Chemical manufacturing contributes the largest share of value added to the U.S. economy, followed by transportation equipment and food products.

Water use in manufacturing varies significantly across industries.

DOI: 10.1201/9781003655848-16

DID YOU KNOW?

Manufacturers withdraw approximately 18.2 billion gallons of water per day, which represents 4% of total water withdrawals in the United States. In addition to these direct withdrawals, the most recent data available suggest that approximately 12% of publicly supplied water withdrawals (4.74 billion gallons of water per day in 1995) were used for manufacturing.

Manufacturing is a major component of the secondary mega-sector of the U.S. economy.

The United States produces a wide range of durable and non-durable goods. Chemical manufacturing contributes the largest share of value added to the U.S. economy, followed by transportation equipment and food products. Overall, these three industries account for approximately 38% of the value added by manufacturing.

11.3 WATER USE

Well-supplied manufacturing water withdrawals account for approximately 4% of total water withdrawals in the United States (18,200 MGD). In addition, some industries use publicly supplied water. The most recent published estimate of the use of publicly supplied water in manufacturing dates from 1995. At that time, 12% of publicly supplied water withdrawals (4,750 MGD) were used for manufacturing. This amount represented 18% of water use in manufacturing in 1995.

The withdrawal of water for manufacturing purposes is heavily concentrated in a limited number of states; 11 states account for approximately 70% of self-supplied manufacturing water use. Louisiana, Indiana, and Texas are the top three users; combined, they account for 38% of self-supplied manufacturing withdrawals. The withdrawal of water for manufacturing purposes is the greatest east of the Mississippi and in the Gulf Coast states.

The withdrawal of water for manufacturing purposes does not necessarily correspond to state-level manufacturing output. California has the highest level of manufacturing output, but manufacturers in this state withdraw relatively little water. On the other hand, states such as Texas and Pennsylvania host a large manufacturing sector and report significant withdrawals of water for manufacturing purposes. The inconsistency in manufacturing output and water withdrawals across states can be explained in part by the varying levels of water usage across industries.

More recent data have been collected on water pollution abatement operating costs, which give a sense of the levels of water use across industries. The data on pollution abatement costs are like earlier water use data; the five sectors representing the highest water pollution abatement costs are chemical, food, paper, petroleum and coal, and primary metal manufacturing in that order.

DID YOU KNOW?

The pollution control requirements introduced under the Clean Water Act are one factor that has contributed to the decline in water withdrawals by the manufacturing sector. The cost of complying with these requirements provides a strong economic incentive to reduce effluent discharges, which in turn encourages greater efficiency in water use.

Overall withdrawals of water by the manufacturing sector declined for several years, but with the rise of AI, water use is increasing exponentially and continues to grow. Note that many uses of water in manufacturing (e.g., cooling and processing) do not require water of exceptionally high quality, allowing manufacturers to reuse water multiple times before treating and discharging it as wastewater. In addition, substitutes, such as using air for cooling instead of water, allow manufacturers to decrease their water use. Sector-specific gains in water-use efficiency are described in greater detail below.

While most industries in the manufacturing sector are primarily concerned with access to adequate supplies of water, some are also concerned with the water's quality. These industries include the food and beverage sector and the electronics sector. Water used in the processing of food and beverages must meet health and safety standards. For example, we know that water is essential for living. If the quality of water is not satisfactory, it may result in ill health. Many diarrheal illnesses are caused by the consumption of contaminated water. Diarrheal illnesses constitute one of the leading infectious causes of death and disability worldwide, especially in areas with poor sanitation. Therefore, access to clean water is an important consideration when locating bottling and food processing facilities. Ironically, as compared to water quality or purity requirements for food, in the electronics sector, the water used to manufacture (usually used in cleaning processes) high-tech products such as semiconductors and microchips must be ultrapure water. In these sectors, water quality is so important that manufacturers are likely to rely on sophisticated systems to purify their source water, regardless of its initial quality (Strzepek and Boehlert, 2010). Nonetheless, high-quality source water can, at least to some extent, reduce the costs associated with water purification and treatment.

11.3.1 Chemical Manufacturing

Chemical and allied products manufacturing is likely the leading user of water in the U.S. manufacturing sector. It accounted for 29% of industrial water use in 1983 (U.S. Census Bureau, 2008). The primary uses of water in chemical manufacturing are for non-contact cooling, steam applications, and product processing. Water use varies by product—silicone-based chemicals require larger quantities of water to produce. In contrast, many of the top manufactured chemicals by volume (including nitrogen, ethylene, ammonia, phosphoric acid, propylene, and polyethylene) require less water during production. In fact, water use per unit of production decreased steadily from 1954 to 1973 in part due to water recycling (David, 1990). In addition, improvements in overall efficiency and the substitution of air in place of water during certain cooling processes have contributed to the reduction in water use per unit of production (David, 1990). Again, before AI data centers, the increase in water reuse was the major reason for the decrease in water use.

The importance of water quality varies across uses within the chemical manufacturing industry. Cooling accounts for approximately 88% of the gross water used in chemical manufacturing; 67% of this water is recycled (Ellis et al., 2003). An excellent example of the use of recycled water for cooling in manufacturing is the trend toward using treated wastewater effluent for machinery cooling (Spellman, 2020). In general, the quality of water used for cooling is not of great importance. On the other hand, certain processes, such as rinsing parts, may require high-quality water (Michigan Chamber of Commerce, 2008).

11.3.2 Paper Manufacturing

Paper and allied products manufacturing accounted for almost a quarter of all industrial water use. Paper is produced from raw materials containing cellulose fibers, such as wood, recycled paper, and agricultural residues. The main steps in paper manufacturing are raw material preparation, for example, wood debarking and chip making, pulp manufacturing, paper manufacturing, and fiber recycling (World Bank Group, 1998). Water is used at various points in this process. In the initial step of raw material preparation, much water is needed to clean the wood, transport wood from one place to another in the facility, and cool machinery used for conveying, debarking, and chipping. To produce pulp, more water is used to steam-cook wood chips, which are then washed and screened. In the paper manufacturing step, the pulp is further diluted before being drained, heat-dried, and pressed. The water that drains off can be reused in the paper manufacturing step (David, 1990).

Water use in paper manufacturing has decreased due to recycling water at various points in the production process and improved technology, such as high-pressure, low-volume showers. Although positive from the perspective of total resource use, water use reductions can increase the concentration of contaminants in process water, leading to high rates of scale deposition and other unwanted effects. This buildup of contaminants may increase production costs and decrease product quality (Ellis et al., 2003).

11.3.3 Petroleum and Coal Products Manufacturing

Petroleum and coal products manufacturing accounted for approximately 20% of industrial water use in 1983. Within this industry, petroleum refining accounts for between 1,000 million and 2,000 million gallons of water used daily, compared to 400 million gallons per day for natural gas processing and pipeline operations. Refineries use about one to 2.5 gallons of water for processing and cooling per gallon of product (U.S. DOE, 2006). This figure has decreased by more than 95% since 1975 (Ellis et al., 2003).

Cooling is the primary use of water in petroleum refining; a typical refinery may use ten times as much cooling water as process water. However, petroleum refineries have the highest rate of water recycling of any primary industry (Ellis et al., 2003). Water is used approximately 7.5 times before being discharged (David, 1990). As with chemical manufacturing, water quality is not a critical consideration in the use of water for cooling. The use of water for other purposes, however, may be more sensitive to water quality considerations.

11.3.4 Primary Metal Manufacturing

Primary metal products manufacturing accounts for more than 15% of industrial water use. Information on water use in primary metal manufacturing focuses primarily on iron and steel manufacturing. Water is used in the steel industry for three purposes: Material conditioning, air pollution control (i.e., use in wet scrubbers), and cooling (CH2M Hill, 2003). Cooling represents over 70% of the water used in most processes. The exceptions are sinter plants and pickling, where air pollution controls the largest water use.

The intensity of water use within the steel industry has declined in recent years, primarily due to the recycling and reuse of water in production facilities. In addition, process changes in steel production, such as the replacement of basic oxygen furnaces with electric arc furnaces, have decreased water demand. Water use within the industry is expected to continue to decline, and facilities may reduce their use of surface and groundwater by moving toward the reuse of treated municipal effluent. In addition, internal treatment and recycling of water are expected to increase (CH2M Hill, 2003).

11.3.5 Food and Beverage Manufacturing

Food and kindred products manufacturing account for approximately 4% of industrial water use. Although this industry accounts for a relatively small portion of overall industrial water use, water quality is paramount in food and beverage manufacturing. Within the food and beverage industry, water is used as an ingredient in products, a medium for mixing and seeping in food processing, and a medium for cleaning and sanitizing operations.

Water use varies by food product for the processing of various water-intensive foods. Sugar refining is a large user of water within the food processing industry. Unlike most food products, which require water primarily for processing, sugar refineries use about half of their intake water for cooling. Beverage manufacturers also use large quantities of water for cooling (David, 1990).

Unlike the other industries discussed earlier, water use in the food products industry has not declined dramatically over the last several decades. Food processing techniques have changed little, and only minimal water recycling and reuse occur within the industry. Water recycling and reusing are limited by safety concerns. The water used in food processing must meet human health and safety standards.

The price elasticity of demand for water in manufacturing captures the relationship between the quantity of water demanded and its price—specifically, it measures the percentage change in quantity demanded for a given percentage change in price. Both economic theory and empirical evidence suggest that the demand for water in manufacturing is inelastic at current prices, meaning that when its price increases, the quantity demanded falls but at a correspondingly lower rate. The available evidence suggests that industrial firms' water demands may be relatively more price sensitive than agricultural or domestic water demands (Renzetti, 2005).

11.4 MINING AND ENERGY RESOURCE EXTRACTION

Compared to other sectors, the mining and energy resource extraction sector uses a small amount of water: approximately 1% of the nation's total use. This figure has remained relatively constant since 1985. Nonetheless, water plays a crucial role in many production processes, including hydraulic fracturing (fracking), secondary oil recovery, and the extraction and processing of oil shale. This sector examines the economic importance of water in mining and energy resource extraction. It includes the following:

- Background information on the mining and energy resource extraction industry
- A discussion of the role of water in the production of minerals, crude oil, and natural gas
- Available estimates of the value of water to this sector

The mining and energy resource extraction sector of the U.S. economy is part of the primary mega-sector described earlier. Most of its output flows to the manufacturing or utility sectors, with relatively little output flowing directly to other sectors of the economy.

DID YOU KNOW?

The distinction between meaning and energy resource extraction is a simple one: Mining refers to the extraction of any solid mineral other than coal and uranium, while energy resource extraction includes coal and uranium, as well as raw petroleum and natural gas.

11.4.1 MINING

Mining includes the extraction of two general categories of minerals: metals (iron, gold, copper, etc.) and industrial materials (clays, feldspar, salt, sand and gravel, etc.). In 2009, the United States removed approximately 5.0 billion metric tons of material in mining operations, including approximately 3.6 billion metric tons (73%) of crude ore and 1.4 billion metric tons (27%) of waste material. Of this total, 51% was associated with the production of metals, and 49% was associated with the production of industrial materials (USGS, 2011).

11.4.2 Energy Resource Extraction

The United States is one of the leading producers of energy resources in the world; as of 2010, the United States ranked third worldwide in crude oil production, second in natural gas production, and second in coal production (world statistics for uranium producers were unavailable). The United States is the only nation to rank in the top three in each of these categories.

DID YOU KNOW?

The use of water in mining and energy resource extraction is relatively insensitive to source water quality. For example, much of the water used in oil and gas extraction is water that is withdrawn during the drilling process. Reuse of this water, which is often unsuitable for other purposes, helps to offset the demands of water-intensive processes like secondary oil recovery.

Water is used in mining and energy resource extraction for a variety of purposes. In mineral extraction, it is used for processes such as milling, wet-screening, and hydraulic mining, while in petroleum and natural gas extraction, it is used for procedures like hydraulic fracturing, secondary oil recovery, and the extraction and processing of oil shale.

DID YOU KNOW?

Scarcity of water in the west may constrain the exploitation of the region's oil shale deposits, a potentially significant source of petroleum. This issue has encouraged research into less water-intensive processes.

SIDEBAR 11.1 FRACKING (FRACING) WATER SUPPLY

The drilling and hydraulic fracturing of a horizontal shale gas well requires water—two million to four million gallons of water (Satterfield et al., 2008), with about three million gallons being most common. Note that the volume of water needed may vary substantially between wells. In addition, the volume of water needed per foot of wellbore appears to be decreasing as technology and methods improve over time.

Water for drilling and hydraulic fracturing of these wells frequently comes from surface water bodies such as rivers and lakes but can also come from groundwater, private water sources, municipal water, and reused produced water. Most of the production shale produced by gas shale basins contain large amounts of local water sources.

Even though the water volumes needed to drill and stimulate shale gas wells are large, they generally represent a small percentage of the total water resource used in the shale gas basins. This volume is small in terms of the overall surface water budget for an area; however, operators need this water when drilling activity is occurring (on demand), requiring that the water be procured over a relatively short period of time. Water withdrawals during periods of low stream flow could affect fish and other aquatic life, fishing and other recreational activities, municipal water supplies, and other industries such as power plants.

One alternative that states and operators are pursuing is to make use of seasonal changes in river flow to capture water when surface water flows are greatest. Utilizing seasonal flow differences allows planning of withdrawals to avoid potential impacts to municipal drinking water supplies or to aquatic or riparian communities. In the Fayetteville Shale play of Arkansas, one operator is constructing a 500-acre-ft impoundment to store water withdrawals from the Little Red River obtained during periods of high flow (storm events or hydroelectric power generation releases from Greer's Ferry dam upstream of the intake) when excess water is available (Chesapeake Energy Corporation, 2008). The project is limited to 1,550 acre-feet of water annually. As additional mitigation, the company has constructed extra pipelines and hydrants to provide portions of this rural area with water for fire protection. Also included is monitoring of in-stream water quality, as well as game and non-game fish species in the reach of the river surrounding the intake. This design provides a water recovery system similar in concept to what some municipal water facilities use. It will minimize the impact on local water supplies because surface water withdrawals will still be limited to times of excess flow in the Little Red River. This project was developed with input from a local chapter of Trout Unlimited, an active conservation organization in the area, and represents an innovative environmental solution that serves both the community and the gas developer.

These water needs may challenge supplies and infrastructure in new areas of shale gas development; that is, in areas where the impact of shale gas operations is new and the potential impact is unknown to local inhabitants and governing officials. As operators look to develop new shale gas plays, failure to communicate with local officials is not a viable option. Communication with local water planning agencies can help operators and communities to coexist and effectively manage local water resources. Understanding local water needs can help operators develop a water storage or management plan that will be accepted in neighboring communities. Although the water needed for drilling an individual well may represent a small volume over a large area, the withdrawals may have a cumulative impact on watersheds over the short term. This potential impact can be avoided by working with the local water resource manager to develop a plan outlining when and where withdrawals will occur (i.e., avoiding headwaters, tributaries, small surface water bodies, or other sensitive sources).

Before a gas shale play hydraulic fracking operation is developed, not only is it a good idea to communicate with state and local government officials (and obtain proper permits, if required), but it is also prudent to get information and data related to the urban water cycle (i.e., if the gas shale operation is near or has an impact on the surrounding area). Moreover, a study of gas shale hydraulic fracking operations on the indirect water reuse process is called for. In the planning stage, scrutiny related to water sources, outfalls, annual precipitation levels, drought histories, indirect water reuse, and (if located in or near an urban area) the urban water cycle is prudent and necessary.

Extracting minerals such as coal, hard rock, sand and gravel, and metal ores from the earth can involve several procedures that are water-intensive, particularly those that include reducing the size of the extracted material. Wet screening, for instance, in which the mined material is filtered by water through a series of screens, can use anywhere from 30 to 250 gallons of water per ton of mined material. Milling (or grinding), in which the mined material is broken down into smaller particles, also uses a large amount of water—anywhere from 125 to 300 gallons per ton of mined material. Water is also used in drilling for minerals; however, usage is highly variable, depending on the diameter of the hole, depth, orientation, etc.

The use of water in mineral extraction is greatest when processing softer minerals, such as kaolin (a type of clay) and silica sand. Kaolin clay, a material used primarily in papermaking, goes through several processing procedures, such as suspension and dispersion, screening, grit removal, brightening, and flocculation, all of which are water-intensive. While the exact amount varies from facility to facility, roughly 2,000 gallons of water are used to process a ton of kaolin. Kaolin is also shipped via slurry pipelines in a mixture of 70% kaolin and 30% water (Mavis, 2003).

In coal mining, water is used for several purposes, mainly for the cooling of machinery, dust suppression, and safety (i.e., dousing fires). Most of the water used in mining is for dust control, which utilizes approximately 5.2 gallons per ton of coal produced. Water use statistics for other coal mining processes are not readily available.

Water quality is of little concern in most mining operations; much of the water that is utilized is later reused for the same process. Some procedures, particularly those that increase mineral concentration, are sensitive to water quality, yet these are so rare as to be inconsequential (Mavis, 2003).

11.4.3 Crude Oil

In petroleum extraction, water is used in many of the same ways that it is for mineral extraction: dust suppression, cooling of machinery, etc. Much of the water used in oil extraction (and in gas extraction) is produced water, or water that is generated during the drilling process. This water is generally saline but can range from fresh to hypersaline.

Water is a primary input to the process known as secondary oil recovery, which is used to maintain production at wells that would otherwise be abandoned. Water (or steam) is injected into the wells to extract additional oil. The amount of water used in secondary oil recovery varies greatly, as anywhere from two to 350 gallons of water can be used per gallon of oil extracted. While this process can be extremely water-intensive, much of the water used is produced water and is otherwise unusable.

In addition to conventional crude oil reserves, the United States holds one of the largest deposits of oil shale in the world. These reserves are not heavily utilized at present but could have a significant impact on the nation's future oil output. DOE's Energy Information Administration (EIA) estimates that, by 2035, shale oil could account for 2% of total U.S. oil production (USDOE, 2011). A potential impediment to that production, however, is the resource-intensive process for turning mined oil shale into usable crude oil. The process, retorting (i.e., the heating process that separates the oil fractions of oil shale from the mineral fractions), requires two to five gallons of water per gallon of refinery-ready oil. Moreover, most oil shale deposits are in areas of the West in which water is scarce. The scarcity of water increases the costs of retorting and may, in some cases, render it economically infeasible. To address this issue, the industry is attempting to develop in situ retorting processes that are less water-intensive. These processes could dramatically reduce reliance on water for oil shale production.

11.4.4 Natural Gas

As with mineral extraction and oil extraction, natural gas extraction utilizes water in a variety of ways, such as for cooling machinery and dust suppression. Water is also used extensively in the mining of unconventional natural gas sources, such as tight gas (gas that is trapped beneath sandstone formations), coal bed gas (gas that is generated by coal, then stored within its seams), and shale gas (gas stored within low-porosity shale, which also acts as a source). The extraction of gas from these unconventional sources often (almost always in the case of shale gas) requires water-intensive hydraulic fracturing.

Hydraulic fracturing involves high-pressure injection of a solution—typically made up of 98% water and sand and 2% chemical additives—into the shale where the gas is trapped. This creates cracks or fractures in the shale, allowing for easier extraction of the gas. This process allows for access to gas that is otherwise unreachable, greatly increasing natural gas production. While shale gas accounted for only 16% of U.S. natural gas production in 2009, the EIA estimates that by 2035 its share of the total output may increase to roughly 43%.

Much of the projected increase in shale-gas production is contingent on the use of hydraulic fracturing. This process is under scrutiny due to environmental concerns, mainly the potential contamination of groundwater resources. The greatest concern is that fracturing may allow gas and other contaminants, such as those in the fracturing fluid, to seep into underground sources of drinking water. The EPA is now investigating these issues, conducting a study that is not expected to be completed until 2014; however, the agency recently released a draft report detailing tests done at hydraulic fracturing sites in Wyoming's Pavillion gas field. In this report, the EPA noted that domestic groundwater sources located near hydraulic fracturing sites had a high number of organic and inorganic contaminants associated with hydraulic fracturing (DiGulio et al., 2011). While these findings are not definitive and the agency has reached no conclusions about the safety of hydraulic fracturing, the results suggest that environmental externalities associated with the production of shale gas may, at least in some cases, affect the use of groundwater for other purposes.

11.5 WATER RESOURCE CONSTRAINTS

Because once-through cooling systems require large quantities of water, thermoelectric power plants using such systems are particularly vulnerable to drought conditions and other water shortages. In recent years, water shortages have curtailed power generation at several facilities. For example:

- In 1999, drought in the Susquehanna River basin in New York and Pennsylvania prevented power plants in the region from obtaining sufficient water supplies to meet operational needs.
- In 2006, drought along the Mississippi River caused power plants in Illinois and Minnesota to restrict operations.
- In 2007, drought in the southeastern U.S. caused several nuclear power plants to reduce output by up to 50%, due to low river levels.

The move away from once-through cooling systems has somewhat mitigated this vulnerability. However, a substantial portion of the country's electricity-generating capacity still relies on regular access to large quantities of water.

The proposed areas of research include improving dry cooling technology, reducing water loss from cooling towers, using impaired water, and developing decision support tools to anticipate and respond to water shortages and climate change. Dry cooling systems, or hybrid dry-wet cooling systems, could drastically reduce water withdrawals and consumption, but they currently have much higher costs than wet cooling systems and can negatively affect plant operating efficiency. Reduced operating efficiency in turn leads to higher fuel consumption per unit of electricity produced, with associated environmental consequences of fossil fuel extraction and combustion. Use of impaired water for thermoelectric cooling—effluent from wastewater treatment plants or low-quality groundwater, for example—could also reduce the sector's use of freshwater. Still, such water might require pretreatment to prevent damage to cooling equipment (Carter, 2010). Future research into both dry cooling and the use of impaired water could help reduce the dependence of the electric power generation sector on reliable access to large quantities of water.

11.5.1 Water Quality Constraints

Thermoelectric cooling typically has low water quality requirements, with around 30% of its water withdrawals using saline water. Water discharged from thermoelectric plants with once-through cooling systems, however, can have a detrimental impact on the quality of the receiving water. In addition to the temperature difference between discharged cooling water and receiving water (which can disrupt aquatic habitats), chemicals used to protect cooling equipment can also affect downstream water quality and use (Carter, 2010). Thermoelectric cooling typically has low water quality requirements, with around 30% of its water withdrawals using saline water.

Section 316 of the Clean Water Act (CWA) gives EPA the authority to regulate the use of water for industrial cooling, with Section 316(a) regulating the temperature of discharged water and Section 316(b) regulating cooling water intake structures. These regulations have played a significant role in driving the shift from once-through cooling systems to recirculating cooling systems. New plants are already required to install recirculating cooling systems, and the EPA is currently developing regulations to update requirements for existing plants with once-through systems.

11.5.2 Interaction With Other Uses of Water

As a primary off-stream user of water, the thermoelectric power generation sector competes for water with several other industries, particularly agriculture and domestic supply. In Western states, prior appropriation water rights laws give precedence to those who first obtained legal rights to use the water, which typically include agricultural and municipal users. Under drought conditions, users who obtained legal rights to use water later—typically including thermoelectric power generators—are the first to suffer restrictions in water supply. In contrast, the development of hydroelectric power has often served as a complement to other water use sectors, as large reservoirs created by dams are frequently used to provide water for domestic and agricultural users, in addition to serving as a setting for recreational activities such as boating, swimming, and fishing. As noted previously, however, these regulations have limited hydropower development in some regions, and providing fish passage at large dams can raise operating costs significantly. Negative impacts on wildlife, such as salmon populations in the Pacific Northwest. Regulations have limited hydropower development in some regions, and providing fish passage at large dams can raise costs. Regulations have limited hydropower development in some areas, and providing fish passage at large dams can raise operating costs significantly. Regulations have limited hydropower development in some areas, and providing fish passage at large dams can raise costs.

11.5.3 Future Considerations

The factors that in the future are likely to have the greatest effect on the use of water in the electric power generation sector are projected limitations on the availability of water, the potential impacts of climate change on both power demand and water supply, and changing water demands due to increased reliance on renewable energy sources.

A 2011 EPRI study attempted to identify the regions of the United States most likely to face future constraints on thermoelectric power generation because of constraints on water supplies (EPRI, 2011). The authors first projected water uses through 2030 and developed a water sustainability risk index that evaluated water supply constraints according to several dimensions, including the extent of development of available renewable water and groundwater, susceptibility to drought, expected growth in water demand, and the likelihood of increased need for storage (to ensure water availability during seasonal dry periods).

The study found that about 250,000 MW of thermoelectric generation capacity, or 22% of total generating capacity in 2010, is in counties with eight high or extreme levels of water supply sustainability risk.

Climate change and responses to climate change could affect the relationship between water and electric power generation in several ways. First, the increased frequency of floods and droughts predicted by many climate models could significantly compromise the reliability of water access for both thermoelectric cooling and hydropower. Increased temperatures could also increase demand for electricity (e.g., for air conditioning during summer months). Second, regulations to reduce greenhouse gas emissions through carbon capture and storage could potentially increase demand for water in electricity generation. Carbon capture and storage increases water demands at fossil fuel-burning power plants, because operating carbon capture equipment requires both energy (thereby reducing a plant's generating efficiency) and additional cooling.

Federal and state regulations could help mitigate the effects of future water constraints on electric power generation. California and Arizona, for example, have actively worked to minimize the use of fresh water in thermoelectric power generation. At the same time, other states have no official policies or permitting requirements for power plant water use.

11.6 THE AI CONNECTION

AI in water management in industrial manufacturing is being revolutionized by providing real-time data and analytics to optimize water use and reduce waste. A few key examples of applications of AI in water management that are transforming the industry by enabling more sustainable practices and improving overall efficiency include the following:

- **Optimized Water Management**: Water Managers are beginning to buy into and use AI systems to analyze water availability, usage patterns, and population growth to optimize distribution networks and water allocation. This enables Water Managers to minimize water waste and enhance network performance.
- **Condition-based Maintenance (CBM), also known as Predictive Maintenance)**: AI can predict equipment failures and improve maintenance schedules by analyzing data from IoT sensors, equipment logs, and historical maintenance records, reducing the chance of disruptive downtimes.
- **Energy Optimization**: Water treatment, including wastewater treatment, typically requires significant energy input. AI algorithms can optimize energy consumption in treatment by analyzing historical data and environmental factors. This not only improves treatment operations but also reduces costs.
- **Predicting Water Needs**: AI can predict water needs based on historical data and environmental factors, allowing industry to adjust water usage based on supply conditions.
- **Reduction of Water Waste**: This is one of the Water Manager's most valuable AI applications. AI can assist in reducing water waste by about a quarter of a percent and thus reduce operational costs at the same time.

11.7 THE BOTTOM LINE

As industrial manufacturing continues to evolve in the digital age, the integration of AI and manufacturing is an essential ingredient in the Water Manager's mix to secure and manage water needs and usage.

REFERENCES

Carter, N.T. 2010. Energy's Water demands: Trends, vulnerabilities and management. *Congressional Research Service*, November. Washington, DC.

CH2M Hill. 2003. Water use in industries of the future. In *Industrial Water Management: A Systems Approach*, 2nd ed. Corvallis, OR.

Chesapeake Energy. 2008. *Little Red River Project*. Presentation at Trout Unlimited.

David, E.L. 1990. Trends and associated factors in off-stream water use: Manufacturing and mining water use in the United States, 1954–83. *USGS National Water Summary 1987—Hydrologic Events and Water Supply and Use, Water-Supply Paper 2350*. Washington, DC.

DiGulio, D.C., et al. 2011. Investigation of groundwater contamination *Near Pavillion, Wyoming*. *Draft*, December. Washington, DC: USEPA.

Electric Power Research Institute (EPRI). 2011. *Water Use for Electricity Generation and Other Sectors: Recent Changes (1985–2005) and Future Projections (2005–2030)*. Palo Alto, CA.

Ellis, M., et al. 2003. *Industrial Water Use and Its Energy Implications*, December. Washington, DC: U.S. Department of Energy.

Mavis, J. 2003. Water use in industries of the future: Mining industry. *Individual Water Management: A Systems Approach*, July.

Michigan Chamber of Commerce. 2008. *Water Conservation Plan: A Member of the Chemical Manufacturing Sector*, February. Lansing, Michigan.

Renzetti, S. 2005. Economic instruments and Canadian industrial water use. *Canadian Water Resources Journal*, 30(1): 21–30.

Satterfield, J.M., et al. 2008. *Chesapeake Energy Corp. Managing Water Resource's Challenges in Select Natural Gas Shale Plays*. Presented at the GWPC Annual Meeting.

Spellman, F.R. 2020. *Handbook of Water and Wastewater Treatment Plant Operations*, 3rd ed. Boca Raton, FL: CRC Press.

Strzepek, K., and Boehlert, B. 2010. Competition for water for the food system. *Philosophical Transactions of the Royal Society*, 365: 2927–2940.

U.S. Census Bureau. 2008. *Pollution Abatement Costs and Expenditures: 2005*. Washington, DC.

U.S. Department of Energy. 2006. *Energy Demands on Water Resources: Report to Congress on the Interdependency of Energy and Water*, December. Washington, DC.

U.S. Department of Energy. 2011. Energy information administration. *Annual Energy Outlook 2011*, April. Washington, DC.

U.S. Environmental Protection Agency. 2008. *2008 Sector Performance Report*. Washington, DC.

USGS. 2011. *2009 Minerals Yearbook*, June. Washington, DC.

World Bank Group. 1998. Pulp and Paper Mills. *Pollution Prevention and Abatement Handbook*, July. Washington, DC.

12 Water Treatment and AI

On December 3, 1998, U.S. President Bill Clinton visited a water treatment plant in Newport, R.I., to announce new measures that would strengthen drinking water protection for at least 140 million Americans.

> The new standards are a critical milestone on the path toward a new model for solving complex, high-stakes environmental problems. Where a comprehensive solution could be extremely costly and require essential information that is currently unavailable, the best approach may be to make decisions in stages. The initial decision may be a commitment to a process that will result in a comprehensive answer, starting with information-gathering, analysis, and research, and accompanied by a broad set of consensuses, "no regrets" actions. Under such a model, however, taking all the initial actions, as we have now done, is vital to sustaining the good-faith commitment to complete the process. This is such a promising model that it is being applied to address the highly controversial problem and could have even broader applications. This one useful step for safe drinking water could become a giant leap for the protection of public health and the environment.
>
> —J. Charles Fox, 1999[1]

> Many developing cities with private water-management companies face service failures, high costs, and corruption. In Manila, where the water system is controlled by Suez, San Francisco-based Bechtel, and the prominent Ayala family, water is only reliably available for a few hours a day, and rate increases have been so severe that the poorest families must choose each month between paying for water and two days' worth of food. In 2001, the government of Ghana agreed to privatize local water systems as a condition for an IMF loan. To attract investors, the government doubled water rates, setting off protests in a country where the average annual income is less than $400 a year and the water bill (for those fortunate enough to have running water) can run upwards of $110. In Cochabamba, the third-largest city in Bolivia, water rates shot up by 35 percent after a consortium led by Bechtel took over the city's water system in 1999; some residents found themselves paying 20 percent of their income on water. An initial round of peaceful street protests led to riots in which six people were killed. Eventually, the Bolivian government voided Bechtel's contract and told the company's officials it could not guarantee their safety if they stayed in town. Privatization has also spawned protests (and, in some cases, even dominated elections) in Paraguay, where police turned water cannons on anti-privatization protesters, Panama, Brazil, Peru, Colombia, India, Pakistan, Hungary, and South Africa.
>
> —John Louma, 2004[2]

12.1 CASH COWS OR CASH DOGS?

Maintaining the sustainable operations of water and wastewater treatment facilities is expensive. If funding is not available from federal, state, or local governmental

DOI: 10.1201/9781003655848-17

entities, then the facilities must be funded by ratepayers. Water and wastewater treatment plants are usually owned, operated, and managed by the community (the municipality) where they are located. While many of these facilities are privately owned, the majority of Water Treatment Plants (WTPs) and Wastewater Treatment Plants (WWTPs) are Publicly Owned Treatment Works (POTWs) (i.e., owned by local government agencies).

These publicly owned facilities are managed and operated on-site by professionals in the field. On-site management, however, is usually controlled by a board of elected, appointed, or hired directors/commissioners, who set policy, determine budget, plan for expansion or upgrading, hold decision-making power for large purchases, set rates for ratepayers, and, in general, control the overall direction of the operation.

When final decisions on matters that affect plant performance are in the hands of, for example, a board of directors comprised of elected and appointed city officials, their knowledge of the science, the engineering, and the hands-on problems that those who are on-site must solve can range from nothing to everything. Matters that are of critical importance to those in on-site management may mean little to those on the board. The board of directors may indeed also be responsible for other city services and have an agenda that encompasses more than just the water or wastewater facility. Thus, decisions that affect on-site management can be affected by political and financial concerns that have little to do with the successful operation of a WTP or POTW.

Finances and funding are always of concern, no matter how small or large, well supported or underfunded, in the municipality. Publicly owned treatment works are generally funded from a combination of sources. These include local taxes, state and federal monies (including grants and matching funds for upgrades), as well as usage fees for water and wastewater customers. In smaller communities, the water/wastewater (W/WW) plants may be the only city services that actually generate income. This is especially true in water treatment and delivery, which is commonly seen as the cash cow of city services. As a cash cow, the water treatment plant generates cash in excess of the amount of money needed to maintain the treatment works. These treatment works are "milked" continuously with as little investment as possible. Funds generated by the facility do not always stay with the facility. Funds can be reassigned to support other city services—and when facility upgrades are needed, funding for renovations can be problematic. On the other end of the spectrum, spent water (wastewater), treated in a POTW, is often looked upon as one of the cash dogs of city services. Typically, these units make only enough money to sustain operations. This is the case, of course, because managers and oversight boards or commissions are fearful, for political reasons, of charging ratepayers too much for treatment services. Some progress has been made, however, in marketing and selling treated wastewater for reuse in industrial cooling applications and some irrigation projects. Moreover, wastewater solids have been reused as soil amendments; also, ash from incinerated biosolids has been used as a significant ingredient in forming cement revetment blocks used in areas susceptible to heavy erosion from river and sea inlets and outlets (Spellman and Drinan, 2012).

DID YOU KNOW?

Again more than 50% of Americans drink bottled water occasionally or as their major source of drinking water—an astounding fact given the high quality and low cost of U.S. tap water.

Planning is essential for funding, controlling expenses, and ensuring water and wastewater infrastructure sustainability. The infrastructure we build today will be with us for a long time and must be efficient to operate, offer the best solution in meeting the needs of a community, and be coordinated with infrastructure investments in other sectors such as transportation and housing. It is both critical and challenging to ensure that a plan is in place to renew and replace it at the right time, which may be years away. Replacing an infrastructure asset too soon means not benefiting from the remaining useful life of that asset. Replacing an asset too late can lead to expensive, emergency repairs that are significantly more expensive than those that are planned. Additionally, making retrofits to newly constructed infrastructure that was not designed or built correctly is expensive. Doing the job correctly the first time requires planning and a certain amount of competence.

We live on a planet with a surface that is three-fourths covered with water, so we recognize the irony inherent in the fact that many areas of the world face critical shortages of drinking water. Most of Earth's water is salty seawater, making it unsuitable for drinking. Of the little "fresh" water that remains, most, of course, is trapped in polar ice caps, where harnessing it for use by the world's population is difficult.

Much of the accessible natural supply of potable water faces stresses from a growing world population, which increases the basic demand for this natural resource while reducing the supply further through contamination.

Major population centers in developing nations (those without established waste treatment or water treatment infrastructures) often suffer from epidemics of waterborne disease. In these areas, raw sewage usually directly contaminates the rivers and streams used for drinking, washing, and cooking. In other cases, unchecked industrialization leads to water contamination through improperly disposed of chemical and nuclear wastes. The drinking water provider must ensure the water is safe to drink. The primary reason for the development and installation of a public water system is the protection of public health. Basically, a properly operated water system serves as a line of defense between disease and the public.

Properly operated water treatment and supply systems are defined as those that

- remove or inactivate pathogenic microorganisms, including bacteria, viruses, and protozoans
- reduce and/or remove chemicals that can be detrimental to health
- provide quality water to prevent customers from seeking alternative sources that may be unsafe.

This last point is critical but one often overlooked in the operation and management of public water systems. When the water produced by a system is objectionable because of odor, taste, or appearance, the customer will often seek another source for their drinking water. Ironically, these sources, while appearing, tasting, and smelling fine (better than "city water"), could contain microorganisms and/or chemicals that are harmful.

In this chapter, we discuss the Drinking Water Manager's most crucial function: Ensuring that water delivered to the public is properly treated and arrives as the clean, wholesome, safe product that it must be. More of the Water Manager's responsibilities include the following:

- **Product Quality**—Works to ensure that regulatory and reliability requirements are met, meaning that the waterworks operate with customer, public health, economic, and ecological needs.
- **Customer Satisfaction**—Actively solicits timely feedback from rate payers and is committed to delivering reliable, responsive, and cost-effective services.
- **Employee Development**: New employees are required to participate in a training program that includes on-the-job instruction. In water and wastewater treatment operations, operators must be licensed. In addition to on-the-job training, new operator trainees require pre-licensure training to prepare for the licensure examination.
- **Operational Optimization**—Water Managers must be aware of and adopt operational technology improvements, including AI advances, and ensure operators are properly trained in digital plant operations and oversight of equipment parameter operations.
- **Financial Viability**—Water Managers must be capable of making financial decisions that are based on sound financial planning, outlays, and accountability.
- **Infrastructure Viability**—Water and wastewater management personnel need to ensure not only optimum operation of their waterworks or wastewater treatment plants but also distribution and interceptor line sustainability. Exterior pipelines and pumping stations should be regularly monitored for operation and condition, with prompt repairs or replacements as needed.
- **Problem Avoidance**: Water managers should anticipate potential scenarios that could impact operations. Risk management is essential in management, addressing legal issues, organizational integrity, system reliability, personnel and visitor safety, and proactive decision-making.
- **Community/Ratepayer Interface:** Water Managers should engage in community affairs to understand the needs of customers and make informed decisions that support sustainability. Communicate to residents that their water and wastewater utilities are maintaining or improving environmental standards. To ensure community viability, water management personnel must conduct operations in a manner that is both efficient and cost-effective.
- **Watershed Viability**—Water Managers need to be able to see the whole picture, so to speak. The operation of water and wastewater treatment

involves protecting the source of a community's water supply and ensuring that treated wastewater is suitable for disposal or reuse.

- **Customer Satisfaction**—In addition to establishing a community/ratepayer interface, Water Managers must listen to customers to ensure that they are receiving the needed water supply and that customers understand the need for water sustainability by cutting down waste.

12.2 BASIC WATERWORKS OPERATION

Waterworks managers should be licensed engineers, experienced managers, or Class 1 operators with strong backgrounds in water treatment, distribution, and leadership. Relying on seniority rather than merit for promotions is a significant weakness in water management. On the basis of my experience, choosing the right leader is often more challenging than it seems.

Waterworks managers should be licensed engineers, experienced managers, or Class 1 operators with strong backgrounds in water treatment, distribution, and leadership. Relying on seniority rather than merit for promotions is a significant weakness in some water utilities.

On the basis of my experience, choosing the right leader is often more challenging than it seems.

A typical water treatment plant (waterworks) for turbid surface water uses these processes: screening, coagulation, flocculation, sedimentation or settling, filtration, hardness treatment, disinfection, and fluoridation (see Figure 12.1). This is a summary of the main unit processes used in standard surface water drinking treatment systems.

12.2.1 Screening

Water intakes have screening devices. *Screening* (the first important step in treating water containing large solids) is defined as the process whereby relatively large and suspended debris is removed from the water before it enters the plant. River water (the source of water used in this discussion) frequently contains suspended and floating debris varying in size from small rags to logs. Removing these solids is essential,

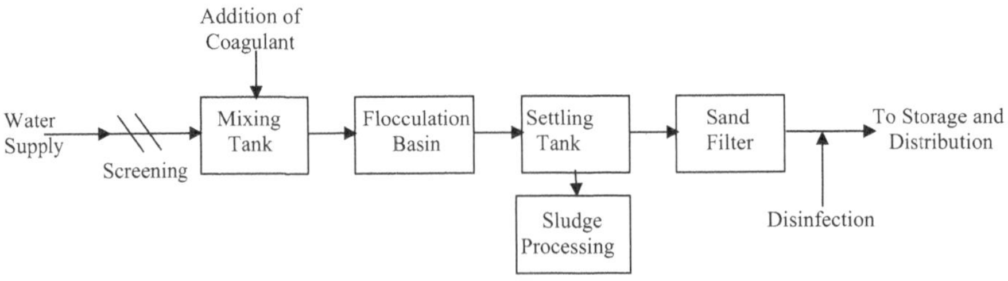

FIGURE 12.1 Typical water treatment unit processes used in treating surface waters.

not only because these items have no place in potable water but also because this river trash may cause damage to downstream equipment (clog and damage pumps, etc.), may increase chemical requirements, may impede hydraulic flow in open channels or pipes, or may hinder the treatment process (Pankratz, 1995). The most important criteria used in the selection of a particular screening system for water treatment technology are the screen opening size and flow rate. Other essential criteria include costs related to operation and equipment, plant hydraulics, debris handling requirements, operator qualifications, and availability.

Large surface water treatment plants may employ a variety of screening devices, including trash screens (or trash rakes), traveling water screens, drum screens, bar screens, or passive screens. We briefly discuss each of these screening devices in the following sections.

12.2.1.1 Trash Screens (Rakes)

Trash screens or *trash rakes* are used to remove rough or large debris retained on a trash rack. They protect pumping equipment and may be used as a preliminary screening device to protect finer screens—drum or traveling water screens, for example.

A trash screen consists of one or more stationary trash rakes and a screen raking device. Trash rack bar spacings range from 1.5 to 4 inches and are mostly constructed of steel bars. Screens made from high-density polyethylene polymers are increasingly being used instead of traditional steel bar models. These synthetic screens are lighter and less prone to microbes, corrosion, and ice.

Raking mechanisms are available for use in a variety of intake configurations, including installation on vertical building and dam walls. Rakes are usually fixed to clean one trash rack, mounted on wheels to move across wide racks, or suspended from an overhead gantry.

12.2.1.2 Traveling Water Screens

Traveling water screens (sometimes called band screens) are placed in a channel of flowing water to remove floating or suspended debris. These automatically cleaned screening devices protect pumping or other downstream equipment from debris in surface water intakes.

Consisting of a continuous series of wire mesh panels bolted to basket frames or trays and attached to two matched strands of roller chain, the traveling water screen operates in a vertical path over a sprocket assembly through the flow. As raw water passes through the revolving baskets, debris is collected and retained on the upstream face of the wire mesh panels. The debris-laden baskets are lifted out of the flow and above the operating flow, where a high-pressure water spray directed outward removes the impinged debris. This process can be continuous or intermittent. For intermittent operation, the screen activates when a specified head loss or time elapsed has occurred.

When located on a river, traveling water screens may be subject to significant fluctuations in flow conditions, debris loading, water depths, and salinity. Depending on the application, the traveling water screen's size is determined by considering such factors as maximum-average flow, maximum-minimum-average water levels, wire mesh size, velocity through mesh, basket or channel width, the number of screens, type of service, and/or starting/operating headloss requirements.

12.2.1.3 Drum Screens

A *drum screen* (or cylinder screen) has very few moving parts and is mounted on a horizontal axis with a series of wire mesh panels attached or mounted on the periphery of its cylinder. The cylinder slowly rotates on its axis. Because of its simplicity of construction, the maintenance and operating costs of a drum screen installation are usually less than those of a traveling water screen.

12.2.1.4 Bar Screens

Though primarily used in wastewater treatment applications, bar screens are also employed in some water treatment facilities. A *bar screen* is made of straight steel bars joined to two horizontal members and cleaned automatically by power-operated rakes. As a rake is operated up the face of the bar rack, it removes accumulated debris (usually large solid objects and rags) and elevates in and out of the flow. At the top of the rake's operating cycle, the debris is swept from the rake into a debris receptacle by a wiper mechanism. When installed in a waterway, the bar screen assembly is normally placed at an angle of 60–80° from the horizontal.

12.2.1.5 Passive Screens

Passive intake screens (stationary screening cylinders) have no moving parts and require no debris handling or debris removal equipment. Passive intake screens are positioned in surface water to use natural currents and low through-screen velocities, reducing debris accumulation.

Usually mounted on a horizontal axis and oriented parallel to the natural current flow within the water body, the current flow action works to keep the screen clean. A maximum intake velocity of about 0.5 fps is typical and works to minimize debris impingement on the screen surface.

12.2.2 Coagulation

Coagulation, the second step in water purification, is a unit process that has been used for several years in the treatment of raw water. Basically, coagulation works to settle very fine material of suspended solids.

Note: Chemicals employed for coagulation are expected to be safe for drinking water when used according to standards.[3]

12.2.2.1 The Coagulants

Typically, after screening, raw water is pumped into large *settling basins* (aka clarifiers or sedimentation tanks). Within the confines of the settling basin, the screened raw water is allowed to sit for some predetermined time. Although screened, the raw water still contains impurities that may be either dissolved or suspended. The settling basin provides the most convenient way to remove suspended matter: It lets the force of gravity to do the work. In the basin, denser particles settle at the bottom of the tank during calm, low-turbulence conditions. This process is called *sedimentation*, and the layer of accumulated solids at the bottom of the tank is called *sludge* (or *biosolids* in some wastewater treatment processes).

The size and density of the suspended particles have a direct bearing on the speed at which they will settle toward the bottom of the basin. The larger and heavier particles will, of course, settle faster than smaller or lighter particles. The forces opposing the downward force of gravity include buoyancy and drag (friction). Particle settling rate is also affected by the temperature and viscosity of the water.

Note: The nature of the sedimentation process also varies with the concentration of suspended solids and their tendency to interact with one another.

Note that in the sedimentation process just described, not all suspended solids or particles can be completely removed from water, even when given very long detention times. For example, very small particles called *colloids* (and other particles such as bacteria, particles of color, and turbidity, which may be colloidal) will not settle out of suspension by gravity without some help. This is where *coagulants* come into play. If we rapidly mix chemical coagulants in the water and then slowly stir the mixture before allowing sedimentation to occur, the colloidal particles will settle.

Colloids, or finer particles, must be chemically coagulated to produce larger flocs that are removable in subsequent settling and filtration.

The coagulation process (along with flocculation) works to neutralize or reduce the natural repelling electrical force of particles in water, keeping them apart and in suspension. Particles in water usually carry a negative electrical charge. Since all these particles carry the same negative electrical charge, they repel each other in the same way as like poles of a magnet do.

The object of coagulation (and subsequently flocculation) is to turn the small particles into larger flocs, as either precipitates or suspended particles. These flocs are then conditioned for ready removal in subsequent processes. Stated another way, in this text, we define *coagulation* as a method to alter the colloids so that they will be able to approach and adhere to each other to form larger floc particles.

DID YOU KNOW?

Coagulation and flocculation convert nonsettleable turbidity particles into a settable form for their effective removal by gravity.

Two types of coagulants are used in the coagulation process: coagulants and coagulant aids. Generally, the type of coagulants and aids available is defined by the plant process scheme. To determine optimum chemical dosages for treatment, jar tests are normally used.

We discuss jar tests in the following section:

DID YOU KNOW?

Selection of the proper coagulant and coagulant aid is critical for a water supply. A jar test for water should be run to determine which coagulant and flocculant aids are economical and most effective.

12.2.2.2 Jar Test

Jar tests are widely used to simulate a full-scale coagulation (and flocculation) process to determine optimum chemical dosages—the cost-effective dose of a coagulant for the time and intensity of agitation selected. Used for many years by the water treatment industry, the test conditions are intended to reflect the normal operation of a chemical treatment facility. In this way, the type and quantity of sludge and the physical properties of the floc can be evaluated.

The test can be used to:

a. choose the appropriate chemical
b. select the optimum dosage
c. determine the value of a flocculant aid and the proper dose.

The testing procedure requires a series of samples to be placed in testing jars and mixed at 100 rpm. Varying amounts of the process chemical or specified amounts of several flocculants are added (one volume/sample container). Mix for one minute. The mixing slows to 30 rpm to provide gentle agitation, and then the floc is allowed to settle. The flocculation period and settling process are observed carefully to determine the floc strength, settleability, and clarity of the supernatant liquor (defined as the water that remains above the settled floc). The supernatant can then be tested to determine the efficiency of the chemical addition for removal of TSS, BOD_5, and phosphorus.

The equipment required for the jar test includes a 6-position variable speed paddle mixer, six two-quart wide-mouth jars, an interval timer, and assorted glassware, pipettes, and graduates.

DID YOU KNOW?

Rapid mixing is the fast and thorough mixing—flash mixing—of the various chemicals, such as coagulants and coagulant aids, with water for their proper chemical reactions.

12.2.2.3 Types of Coagulants

Several different chemicals can be used for coagulation. Commonly used metal coagulants are those based on aluminum (aluminum sulfate) and those based on iron (ferric sulfate).

TABLE 12.1
Common Coagulant Chemicals

Common Name	Comments
Aluminum sulfate	Most common coagulant in the United States; often used with cationic polymers
Ferric chloride	May be more effective than alum in some applications
Ferric sulfate	Often used with lime softening
Ferrous sulfate	Less pH-dependent than alum
Aluminum polymers	Synthetic polyelectrolytes: large molecules
Sodium aluminate	Used with alum to improve coagulation
Sodium silicate	The ingredient of activated silica coagulant aids

Source: Adapted from *Water Treatment Plant Design* (1990).

The most common coagulant is aluminum sulfate [aka alum, $Al_2(SO_4)^3$]. We list other common coagulation chemicals in Table 12.1.

12.2.2.4 Types of Coagulant Aids

Difficulties with coagulation often occur due to slow-settling precipitates or fragile flocs that can easily be fragmented under hydraulic forces in basins and filters (Hammer and Hammer, 1996). AWWA (1995) describes a coagulant aid as a chemical added during coagulation to improve coagulation, build stronger, more settleable floc, overcome the effect of temperature drops that slow coagulation, reduce the amount of coagulant needed, and reduce the amount of sludge produced. Coagulant aids enhance flocculation by improving the settling qualities and toughness of flocs. Polymers are the most widely used materials. Synthetic polymers are water-soluble, high-molecular-weight organic compounds with multiple electrical charges along a molecular chain of carbon atoms. In drinking water treatment, polymers are extensively used as coagulant aids to build large flocs before sedimentation and filtration. Other coagulant aids are activated silica, adsorbent-weighting agents, and oxidants.

12.2.2.5 Coagulation: Process Operation

The standard coagulation unit process operation involves the addition of coagulant chemicals by rapid mixing. The detention time in the rapid mix tank is typically on the order of minutes (Masters, 1991). During this mixing process, polymer (or some other coagulant aid) is added and blended into the destabilized water prior to or during flocculation.

The removal of impurities by coagulation depends on their nature and concentration, the use of both coagulants and aids, and other characteristics of the water, including pH, temperature, and ionic strength. Because of the complex nature of coagulation reactions, chemical treatment is based on empirical data derived from jar testing or other laboratory tests and field studies (Viessman and Hammer, 1998).

12.2.3 Flocculation

The destabilized particles and chemical precipitates resulting from coagulation are designed to enhance their settling qualities and thus their removal from water. However, even after coagulation has taken place, these particles and chemical precipitates may still settle very slowly (too slowly). To speed up the settling process, the next step in the water treatment process, flocculation, is employed.

Note: Flocculation is the "clumping" of coagulated particles as the result of coagulation. Although often used interchangeably, the two terms have different meanings.

Flocculation is the most critical factor affecting particle-removal efficiency. In operation, flocculation is a slow mixing process in which these particles are brought into contact so that they will collide, stick together, and grow (agglomerate) to a size that will readily settle. Enough mixing must be provided (gentle agitation for approximately one-half hour) to bring the floc particles into contact with each other and to keep the floc from settling in the flocculation basin (Note: The heavier the floc and the higher the suspended solids concentration, the more mixing is required to keep the floc in suspension). The most common type, "mixer" or flocculator, is paddle type, which uses redwood slats mounted horizontally on motor-driven shafts. Rotating slowly at about one revolution per minute, the paddles provide gentle agitation that promotes floc growth. The rate of agglomeration or flocculation depends on the number of particles present, the relative volume that they occupy, and the velocity gradient in the basin.

Note: When we state that the rate of agglomeration or flocculation depends on *velocity gradient*, we refer to the fact that too much mixing shears the floc particles, tearing them apart again, so that the floc becomes small and finely dispersed (a situation we are obviously trying to avoid). Thus, the velocity gradient must be controlled within a relatively narrow range.

The theory of flocculation is complex and beyond the needs of this text. Still, on an elemental level, we can say that flocculation is generally affected by slowly rotating large-diameter mixers. Recent practice incorporates dispersion of the coagulant (flash mixing), flocculation, and sedimentation in a single unit called a *contact clarifier.*

Note: Flocculation is a principal mechanism in removing turbidity from water.

12.2.4 Sedimentation

Note: In a conventional water treatment plant, coagulation/flocculation precedes the sedimentation process for better results and improved utilization of the settling basins and is then followed by the filtration process. In the past, but not necessarily confined to the past, filtration may be preceded only by coagulation. Here, filtration is provided only after a few minutes of contact, with consequent additional stress to the filters. Without sedimentation, filters become less reliable during sudden changes in water quality (De Zuane, 1997).

Sedimentation (also known as *clarification*) is the gravity-induced removal of particulate matter, chemical floc, and precipitates from suspension, as well as other settleable solids. Sedimentation separates the liquid from the solids. The process takes place in a rectangular, square, or round tank called a settling or sedimentation tank or basin. Flow patterns within such basins may be rectilinear flow in rectangular basins, radial flow in center-feed settling tanks, radial flow in peripheral-feed settling tanks, spiral flow in peripheral-feed settling, or radial flow in square settling tanks.

DID YOU KNOW?

An adequate detention time is important for complete sedimentation. The shorter the detention time, the less the settling and the higher the turbidity.

Sedimentation, in the conventional water treatment process, is typically the step between flocculation and filtration. Design criteria are based on empirical data from the performance of full-scale sedimentation tanks. The common criteria for sizing settling basins are detention time (typically from one to ten hours), overflow rate, weir loading, and, with rectangular tanks, horizontal velocity.

In water treatment, the majority of settling basins are essentially upflow clarifiers where the water rises vertically for discharge through effluent channels. More specifically, in the idealized sedimentation tank, water flows horizontally through the basin, then rises vertically, overflowing the weir of a discharge channel at the tank surface. Floc settles downward, opposite the upflow of water, and is removed from the bottom by a continuous mechanical sludge removal apparatus. The particles with a settling velocity greater than the overflow rate are removed (settled), while lighter flocs are carried out in the effluent. The effluent is then filtered.

Note: Sedimentation basins, either circular or rectangular, are designed for slow, uniform water movement with a minimum of short-circuiting.

DID YOU KNOW?

Short-circuiting causes a short detention time, resulting in inefficient sedimentation.

12.2.5 Filtration

Even after chemical coagulation and sedimentation by gravity, not all of the suspended solids or impurities are removed from water. About 5% of suspended solids remain as non-settleable floc particles. Even with only this small percentage left, the water is not sufficiently clean. This remaining floc would cause problems (including noticeable turbidity), and particles shield microorganisms from the subsequent disinfection processes. The goal of water treatment is to produce potable water that is perceptually crystal clear and meets the Safe Drinking Water Act requirement of 0.5 NTU (for turbidity). To accomplish this, water needs an additional treatment step that follows coagulation, flocculation, and sedimentation.

This next step is the physical process of *filtration*. Filtration (sometimes called a polishing process) involves the removal of suspended particles from water by passing it through a layer or *bed* of a porous granular material—sand, for example. As water flows through the filter bed, the suspended particles become trapped within the pore spaces of the filter material (or *filter media*). In purifying a surface water source (as we are discussing in this case), filtration is a crucial step, even though it is only one part of the overall treatment process.

Note: Filtration is the process that occurs naturally as surface waters migrate (percolate) through the porous layers of soil to recharge groundwater. This natural filtration removes most suspended matter and microorganisms and is the reason many wells produce water that doesn't require any further treatment.

The Surface Water Treatment Rule (SWTR) specifies certain filtration technologies. The most common treatment filter systems include rapid gravity filters, either built on-site or packaged plants, and pressure filters. Another type is direct filtration. Two other types are slow sand filters and the diatomaceous earth (DE) filters. The SWTR allows the use of "alternate" filtration technologies. Cartridge filters are an alternative technology, for example.

DID YOU KNOW?

The deeper the filtration bed, the better the filtration.

Standard water purification filtration methods include slow or rapid sand filtration, DE filtration, and package systems. Slow and rapid filter systems refer to the rate of flow per unit of surface area. Filters are also classified by the type of granular material used in them. Sand, anthracite coal, coal-sand, multi-layered, mixed bed, or DE are examples of different filtering media. Filtration systems may also be classified by the direction the water flows through the medium. Some examples are downflow, upflow, fine-to-coarse, or coarse-to-fine. Lastly, filters are commonly distinguished by whether they are gravity or pressure filters. Gravity filters rely only on the force of gravity to move the water down through the grains and typically use upflow for washing (backwashing) the filter media to remove the collected foreign material. Gravity filters are free-surface filters that are much more commonly used for municipal applications. Pressure filters are entirely enclosed in a shell so that most of the water pressure in the lines leading to the filter is not lost and is used to push the water through the filter.

12.2.5.1 Rapid Filter Systems

Slow fine sand filtration has been used in the United States since 1872, but although still used in many older plants, it is not commonly used today in most modern water treatment plants, due to various associated problems. One of the problems is related to the tiny size of the pore spaces in the fine sand, which slows down the water, hindering its progress through the filter bed. These filter types also have problems with suspended particles clogging the surface, requiring the filter to be manually scraped clean. These units also take up a considerable amount of land area because slow filtration rates mean higher filter surface area to produce the required water quality.

In modern water treatment plants, the *rapid filter* has largely replaced the slow sand filter. The rapid filter consists of a layer of carefully sieved silica sand ranging from 0.6 to 0.75 m in depth on top of a bed of graded gravels. The pore openings between the grains of sand are often greater than the size of the floc particles that are to be removed, so much of the filtration is accomplished by means other than simple straining.

Note: The ideal filter media possesses the following characteristics: coarse enough for large pore openings to retain large quantities of floc, yet sufficiently fine to prevent passage of suspended solids, adequate depth to allow relatively long filter runs, and graded to permit effective cleaning during backwash.

Adsorption, ongoing flocculation, and sedimentation within pore spaces represent key mechanisms for contaminant removal. When the filter becomes clogged with particles (which occurs approximately once a day, depending on the turbidity of the water), the filter is shut down for a short period of time and cleaned by forcing water backward through the sand for 10–15 minutes. After washing, the sand settles back in place, and the operation resumes.

DID YOU KNOW?

In granular filtration, the higher the temperature, the better the filtration.

12.2.5.2 Other Common Filter Types

Rapid-flow filters are the most common type used for treating water supplies, primarily because they are the most reliable. But other types of filters are sometimes used to clarify water, including pressure filters and DE filters.

A *pressure filter* works like a rapid filter, letting water pass through a granular bed. However, instead of being open to the atmosphere and using the force of gravity, the pressure filter is enclosed in a cylindrical steel tank, and the water is pumped through the bed under pressure. They are not as reliable as rapid filters, because pressure may force solids through the bed in the effluent. Consequently, these systems are rarely utilized in municipal water treatment facilities; rather, they are primarily implemented for industrial water filtration or within swimming pools.

DE filters use a thin layer of natural powder material made from diatom shells. These filters are mainly used for industrial or swimming pool purposes, as their reliability differs from that of rapid sand filters.

DID YOU KNOW?

Reverse osmosis is suitable when water has a very high concentration of dissolved substances such as chlorides, nitrates, and fluorides, in addition to other contaminants (e.g., salty water).

12.3 HARDNESS TREATMENT

Two commonly used methods to reduce hardness are the lime-soda process and ion exchange. The lime-soda process is applicable for extensive facilities, while ion exchange is usually employed in smaller waterworks. The *lime-soda process* typically leaves about 100 mg/l of residual hardness as $CaCO_3$. Greater reductions are not economical and may have adverse health consequences as well (McGhee, 1991). In this text, our discussion focuses on ion exchange. The lime-soda process typically leaves about 100 mg/l of residual hardness as $CaCO_3$.

Ion exchange is accomplished by charging a resin with sodium ions and allowing the resin to exchange the sodium ions for calcium and/or magnesium ions.

Standard resins, such as zeolites (both natural and synthetic), extract specific ions like sodium or $KMnO_4$ from solutions. They either exchange these ions, as in water softening, or use them for oxidation, such as in iron or manganese removal.

The negative side of using ion exchange is that even though the process softens water by removing all (or nearly all) of the hardness and adds sodium ions to the water, the water may be more corrosive than before. The addition of sodium ions to the water may also increase the health risk of those with high blood pressure.

12.3.1 Disinfection

At the turn of the century, 35,000 out of 1,000,000 people did not reach 20 years of age. Today, according to USA TODAY (1998), while 101 deaths occur every minute, 261 births occur simultaneously. Curbing waterborne disease through disinfection has significantly contributed to birth rates outpacing death rates worldwide.

The Safe Drinking Water Act requires that public water supplies be disinfected and that the USEPA set standards and establish processes for treatment and distribution of disinfected water to ensure that no significant risks to human health occur. The USEPA Science Advisory Board ranks pollutants in drinking water as one of the highest health risks meriting USEPA's attention because of large-scale population exposure to contaminants, including lead, disinfectant by-products (DBPs), and disease-causing organisms. Disinfectants are used by virtually all surface water systems in the United States and by an unknown percentage of systems that rely on groundwater. For nearly a century, chlorine has been the most widely used and most cost-effective disinfectant. However, disinfection treatments can produce a wide variety of by-products, many of which have been shown to cause cancer or produce other toxic effects. Recently, concern has risen over water quality deterioration, which can increase dramatically during distribution—unless systems are properly designed and operated. While disinfection is an integral part of water treatment, filtration before disinfection is necessary to reduce pathogen levels, making disinfection more reliable by removing turbidity and other interfering constituents.

To solve the disinfection by-products problem, we need innovation to upgrade existing techniques as well as develop new approaches to address these problems. Areas of interest include the following:

- Alternatives to chlorine disinfection for removing pathogenic microorganisms, including innovative applications of ultraviolet (UV) radiation and processes that improve overall effectiveness while using reduced amounts of disinfectant.
- Development of innovative unit processes, particularly for small systems, for the removal of organic and inorganic contaminants (such as arsenic), particulates, and pathogens [e.g., cyst-like organisms and emerging pathogens like caliciviruses, microsporidia (septate and enterotoxin), hepatitis A virus, Mycobacterium avium intracellular, Helicobacter pylori, Legionella pneumophilar, adenovirus 40/41/1-39, and Toxoplasma gondii].
- Development of efficient, cost-effective treatment processes for removing disinfection by-product precursors (e.g., trihalomethanes, haloacetic acids) and for ozonation (bromate, aldehydes), chlorination (chloropicrin, haloacetonitriles) and chloramination (organic chloramines, cyanogen chloride).
- Enhanced pathogen control using coagulation, settling, filtration, or other cost-effective methods.
- Drinking water contamination control between the treatment plant and the user, especially considering potential chemical leaching from distribution system materials and surfaces (e.g., lead, copper, iron, and other pipe materials, and protective coatings) because of instability, interaction with microorganisms, disinfection agents, and water treatment chemicals.

To this point, the unit processes described—screening, coagulation, flocculation, sedimentation, and filtration—together comprise a type of treatment called *clarification*. Along with removing turbidity and suspended solids, clarification also removes many microorganisms from the water. However, clarification by itself is not sufficient to ensure the complete removal of pathogenic bacteria and viruses.

Earlier, we stated that one of the primary goals of water treatment is to treat raw water to the point where we can deliver to the consumer a water product that is perceptually crystal clear. The consumer does not want to drink a glass full of mud, a glass full of slime, a glass full of metal-colored, foul-smelling water—or even a glass of water that looks like it was dipped from the creek. Would you?

The point is, once the water is treated to the point of crystal clarity, the treatment process must be taken a step further to ensure that the water is completely free of disease-causing microorganisms. To accomplish this, the final treatment process in water treatment plants occurs—*disinfection*, which destroys or inactivates pathogens.

12.3.1.1 Key Disinfection Terms[4]

Before we move on to a discussion of the major disinfection methods used in treating water for human consumption, we must first define a few pertinent terms related to disinfection in general. To begin with, we need to establish the distinction between primary and secondary disinfection.

1. *Primary Disinfection*: the initial killing of *Giardia* cysts, bacteria, and viruses.
2. *Secondary Disinfection*: the maintenance of a disinfectant residual, which prevents regrowth of microorganisms in the water distribution system between treatment and consumer.

Disinfect: to inactivate virtually all recognized pathogenic microorganisms, but not necessarily all microbial life (cf. pasteurize or sterilize).

Disinfectant: (1) Any oxidant, including but not limited to chlorine, chlorine dioxide, chloramine, and ozone, added to water in any part of the treatment or distribution process that is intended to kill or inactivate pathogenic microorganisms. (2) A chemical or physical process that kills pathogenic organisms in water. Chlorine is often used to disinfect sewage treatment effluent, water supplies, wells, and swimming pools.

Disinfectant time: the time water takes to move from one point of disinfectant application (or the previous point of residual disinfectant measurement) to a point before or at the point where the residual disinfectant is measured.

Disinfectant contact time (T in CT calculation): the time in minutes that water takes to move from the point of disinfectant application or the previous point of disinfection residual measurement to a point before or at the point where residual disinfectant concentration ("C") is measured. Where only one "C" is calculated, "T" is the time in minutes that water takes to move from the point of disinfectant application to a point before or at which residual disinfectant concentration ("C") is measured. Where more than one "C" is calculated, "T" is:

(1) For the first measurement of "C," the time in minutes that water takes to move from the first or only point of disinfectant application to a point before or at the point where the first "C" is measured and

(2) For subsequent measurements of "C," the time in minutes that water takes to move from the previous "C" measurement point to the "C" measurement point for which the particular "T" is being calculated.

Disinfection: a process that inactivates pathogenic organisms in water using chemical oxidants or equivalent agents.

Disinfection by-product: a compound formed by the reaction of a disinfectant such as chlorine with organic material in the water supply.
Waterborne disease: caused by pathogenic organisms in water.

Note: Disinfection should not be confused with sterilization. *Sterilization* is the destruction of *all* microorganisms. Sterilizing potable water requires the application of a much higher dose of chemical disinfectants, which would greatly increase operating costs and would create taste problems for the consumer. Excessive application of disinfectants also generates excessive levels of unwanted disinfection by-products. Because of these problems, present treatment practices are used for turbidity removal and subsequent disinfection to the extent necessary to eliminate known disease-causing organisms sufficient to protect public health.

Presence/Absence of Coliforms: The presence of coliform bacteria in water is an indication that the water may be contaminated by pathogenic organisms. If coliform bacteria are absent, the source is reliable, chlorine levels are maintained, and the supply has a good track record, pathogens are considered unlikely to be present.

DID YOU KNOW?

Organic matter, including microorganisms, reacts with chlorine to produce chloroorganic compounds, which show as residual chlorine and can cause taste and odor problems.

12.3.1.2 Disinfection Methods

Although chlorination is the best-known and most common disinfection method, other methods are available and can be used in various situations. The three general types of disinfection are as follows:

- *Heat treatment*—probably one of the first methods employed to disinfect water was to boil it. For small quantities of water, boiling water is still a good emergency procedure to use.
- *Radiation treatment*—employs the use of UV radiation to disinfect water.
- *Chemical treatment*—employs the use of chemicals to disinfect water. Chemical disinfectants encompass a variety of substances, including oxidizing agents such as chlorine, ozone, bromine, iodine, and potassium permanganate; metal ions like silver, copper, and mercury; as well as acids and alkalis.

Several different disinfectants are available for use in treating water, and we'll discuss several of these in detail in subsequent sections. For now, you should understand that even though several choices are available, whichever disinfectant is chosen must meet certain criteria—more specifically, to be effective for disinfecting water (and wastewater), the disinfectant chosen must possess certain desirable characteristics. They are as follows:

1. It must act in a reasonable time.
2. It must not be affected by temperature or pH changes.
3. It must be non-toxic.
4. It must not add unpleasant taste or odor.
5. It must be readily available.
6. It must be safe and easy to handle and apply.
7. It must be easy to determine the concentration of the water.
8. It must be able to provide residual protection.
9. Pathogenic organisms must be more sensitive to disinfectants than non-pathogens.
10. It must be capable of being applied continually.
11. The amount applied must be sufficient to produce safe water.

In addition to the desirable characteristics of a disinfectant listed earlier, the disinfectant chosen must be able to kill off or deactivate pathogenic microorganisms by one of several possible methods, including (1) damaging the cell wall; (2) altering the ability to pass food and waste through the cell membrane; (3) altering the cell protoplasm; (4) inhibiting the cell's conversion of food to energy; or (5) inhibiting reproduction.

12.3.1.3 Chlorination

For the past several decades, chlorine dispensed as a solid (calcium hypochlorite), liquid (sodium hypochlorite), or gas (elemental chlorine, Cl_2) has been the disinfectant of choice, particularly in the United States. Chlorine, known for its effectiveness and low cost, is widely used for disinfection. It also leaves a residual in water systems, helping keep water safe from disease.

Gaseous chlorine (Cl_2), 2.5 times as heavy as air, is a greenish-yellow toxic gas. One volume of liquid chlorine confined in a container under pressure yields about 450 volumes of gas. Significant water treatment works usually use chlorine gas, supplied in liquid form, in high-strength, high-pressure steel cylinders. The liquid immediately vaporizes in the form of gas when released from these pressurized containers. Chlorine gas is lethal at concentrations as low as 0.1% air by volume. In non-lethal concentrations, it irritates the eyes, nasal membranes, and respiratory tract.

Sodium hypochlorite is most used in smaller systems because it is simpler to use and has less extensive safety requirements than gaseous chlorine; in the form used, it is less toxic. Recently, many larger water facilities that have used chlorine for disinfection are beginning to shift to sodium hypochlorite because of regulatory pressure.

Note: Two regulations (OSHA's Process Safety Management Standard, 29 CFR 1910.119, and USEPA's Risk Management Program, Clean Air Act Section 112(r)(7)) have come to be known in industry as the "chlorine killers" because of their effect on industrial processes. The USEPA is attempting to steer industry away from the use of chlorine. While they cannot outlaw this substance from use, they are following the path of simply regulating it to death. At present, that is what is occurring. To avoid compliance with strict (in some cases unworkable) regulations, substitution with some other chemical product that is not regulated (at least for the moment), such as sodium hypochlorite, is being affected by many water/wastewater facilities in the United States.

Sodium hypochlorite provides 5% to 15% available chlorine (common laundry bleach is a 5% solution of sodium hypochlorite). Usually diluted with water before application as a disinfectant, it is very corrosive and should be handled and stored with care and kept away from equipment that can be damaged by corrosion. Sodium hypochlorite solution is more costly per pound of available chlorine and does not provide the level of protection (the "killing" power against pathogens) of chlorine gas.

Calcium hypochlorite is a white solid in granular, powdered, or tablet form containing 65% available chlorine. In packaged form, calcium hypochlorite is stable—more stable than solutions of sodium hypochlorite, which deteriorate over time. However, calcium hypochlorite is *hygroscopic*, which means it readily absorbs moisture. It reacts slowly with moisture in the air to form chlorine gas. It is a corrosive material with a strong odor and requires proper handling. Some practical difficulty is involved in dissolving calcium hypochlorite. It must be kept away from organic materials such as wood, cloth, and petroleum products. Reactions between it and organic materials can generate enough heat to cause a fire or explosion.

12.3.1.3.1 Chlorine Use

Whatever form of chlorine is used for disinfection (elemental chlorine, sodium hypochlorite, or calcium hypochlorite), it may be added to the incoming flow (perchlorinating) to assist with the oxidation of inorganics or to arrest biological action that may produce undesirable gases in the bottom of clarifiers. More often, however, chlorine is added just prior to filtration to keep algae from growing on the medium surface and to prevent large populations of bacteria from developing within the filter medium. Safe and effective application of chlorine requires specialized equipment and considerable care and skill on the part of the plant operator. Various means of feeding chlorine have been developed, but probably one of the most widely used and safest types of chlorine feed devices is called an *all-vacuum chlorinator*. Mounted directly on the chlorine cylinder, the gaseous chlorine is always under a partial vacuum in the line that carries it to the point of application. In a typical vacuum chlorine feed system, the vacuum is created by water flowing through the ejector unit at high velocity.

Hypochlorite is usually applied to water in liquid form by means of positive-displacement-type pumps, which deliver a specific amount of liquid on each stroke of a piston or flexible diaphragm. Chlorine, when added to water, reacts with various substances or impurities in the water (e.g., organic materials, sulfides, ferrous iron, and nitrites), which creates a *chlorine demand*. Chlorine demand is a measure of the amount of chlorine that will combine with impurities and is therefore available to act as a disinfectant.

Chlorine combines with ammonia or other nitrogen compounds to form chlorine compounds that have some disinfectant properties. These compounds are called the combined *available chlorine residual.* In the context used here, "available" means available to act as a disinfectant. The uncombined chlorine that remains in the water after combined residual is formed is called the *free available chlorine residual.* Free chlorine is a much more effective disinfectant than combined chlorine.

The factors important to successful chlorination are

- concentration of free chlorine
- contact time
- temperature
- pH
- turbidity.

The effectiveness of chlorination is directly related to the contact time with and concentration of free available chlorine. At lower chlorine concentrations, contact times must be increased. Maintaining a lower pH will also increase the effectiveness of disinfection. The higher the temperature, the faster the disinfection rate. Chlorine, like other disinfectants, works only when it contacts the target organisms. Good contact between chlorine and microorganisms is prevented whenever high turbidity levels exist. For these and aesthetic reasons, turbidity should be reduced where necessary through the coagulation and sedimentation methods previously discussed. Chlorine, like other disinfectants, works only when it contacts the target organisms.

12.3.1.3.2 Chlorination By-Products

A serious disadvantage of chlorination is the potential formation of by-products. Chlorine, for example, can mix with organic compounds in water (such as decaying vegetation) to form trihalomethanes (THMs). One THM (chloroform) is a suspected carcinogen. Other THMs like chloroform may also be carcinogenic.

At present, an estimate about 90% of the U.S. water utilities use chlorine to disinfect water. Although chlorine has virtually eliminated the risks of waterborne diseases such as typhoid fever, cholera, and dysentery, recent studies have shown risks associated with by-products of chlorine, a reason why water utilities have already been looking at alternative methods for disinfecting water.

Several approaches for reducing harmful chlorination by-products have been used. For example, one approach is to remove more of the organics before any chlorination takes place. This can be accomplished (to a degree) by not chlorinating the incoming raw water before coagulation and filtration, thus reducing the formation of THMs. Aeration or adsorption on activated carbon will remove organic materials at higher concentrations or those not removed by other techniques. Another approach is to reevaluate the amount of chlorine used—the same degree of disinfection might be possible with lower chlorine dosages. Changing the point in treatment where chlorine is added is another approach commonly employed (i.e., instead of adding chlorine as a chemical feed, during coagulation, sedimentation, or filtration, it can instead be added after filtration). Another current approach is using alternative disinfection methods.

12.3.1.4 Alternative Disinfection Methods

Currently, several alternative disinfection methods are available for use in treating water. However, in this text (for water treatment), we focus on two alternatives: ozonation and UV radiation. These commonly used alternatives (especially in small water treatment systems) are also increasingly being substituted for existing chlorination systems at larger plants because of regulatory pressure.

Note: Before beginning a brief discussion of the ozonation and UV disinfection alternatives, we must point out that neither one of these two alternatives is an easy solution to problems created by chlorination. While it is true that each has the advantage of not creating THMs and not being covered by the requirements under PSM/RMP, it is also true that each has uncertainties and known disadvantages that have restricted their more widespread use. In addition, ozonation and UV irradiation cannot be used as disinfectants by themselves. Both require a secondary disinfectant (usually chlorine) to maintain a residual in the distribution system.

12.3.1.4.1 Ozonation

Ozone (O3), a gas at ordinary temperature and pressures, is a very powerful disinfectant (it disinfects by breaking up molecules in water) and is even more effective against some viruses and cysts than chlorine. It has the added advantage of leaving no taste or odor and is unaffected by the pH or ammonia content of the water. When ozone reacts with reduced inorganic or organic compounds, it adds an oxygen atom (not a chloride) to the organics, producing environmentally acceptable products. But since ozone is unstable and cannot be stored, it must be produced on-site. Ozonation usually costs more than chlorination as well.

12.3.1.4.2 UV

UV light is electromagnetic radiation just beyond the blue end of the light spectrum, outside the range of visible light. It has a much higher energy level than visible light, and in large doses, it inactivates both bacteria and viruses. UV energy is absorbed by material within microorganisms, disrupting their ability to reproduce and survive, provided the radiation reaches microorganisms without obstruction from turbidity. The significant advantage of UV disinfection over chlorine and ozone is that UV does not involve chemical use. Generally, UV light used for disinfecting water is generated by a series of submerged, low-pressure mercury lamps. Continuing advances in UV germicidal lamp technology are making UV disinfection a more reliable and economical option for disinfection in many plants.

12.3.1.5 Non-Conventional Water Treatment Technologies

Stage 1 of the USEPA's Disinfectant/Disinfection By-product Rule (D/DBP) and the new Interim Enhanced SWTR (designed to significantly lower THMss—by-products of chlorine disinfection in water) has driven (along with the regulatory requirements of PSM/RMP) many water (and wastewater) utilities to find and use alternative disinfection methodologies. Ozonation, UV irradiation, and using chlorine dioxide instead of chlorine are effective disinfection methods that minimize THM production.

Whichever disinfection alternative is ultimately selected, remember that the selection is driven not only by regulatory requirements but also by site-specific requirements.

The disinfection issues covered to this point are essential—the overall ramifications of regulatory pressure and environmental impact cannot be overstated—but other issues besides disinfection must be considered when deciding which water treatment methodology to employ. For example, clarification by coagulation, flocculation, sedimentation, and filtration removes suspended impurities and turbidity from drinking water. Disinfection (the final step in the process) produces potable water, free of harmful pathogens. Simply put, the water treatment processes discussed in the previous sections of this part of the text are sufficient to render most natural surface water (such as the river source we have used for illustration) potable. In some instances, however, the water supply may contain materials that are not removed by conventional water treatment processes. Other treatment processes may be required, particularly to remove many of the dissolved organic and inorganic substances, for example, groundwater with excessive dissolved solids, and for surface waters containing organic compounds from domestic or industrial wastewaters or naturally occurring organics such as humic acid or products of algal blooms. Processes are available for removing these contaminants.

Note: These additional water treatment processes involve sophisticated equipment, require highly skilled operators, and are therefore quite expensive (Peavy et al., 1985).

Additional water treatment processes may be used in addition to clarification or applied separately, depending on the source and quality of the raw water. Let's take a closer look at groundwater. Does standard treatment suffice for most groundwater sources?

Answer: Groundwater does not normally require processing by the unit treatment steps listed earlier, other than disinfection, because groundwater is filtered naturally by the layers of soil from which it is withdrawn. Disinfection is only applied (in many cases) as a precautionary step required by law for public water systems. Groundwater is usually free of bacteria or other microorganisms. Groundwater contact with soil and rock is a concern. With such contact, groundwater may become contaminated by high levels of dissolved minerals that must be removed.

12.4 FLUORIDATION

Fluoride, when added to drinking water supplies in small concentrations (about 1.0 mg/l), can be beneficial. In some locations, standard practice is to mix a 4% solution of sodium fluoride and feed that into the flow of the water system. The amount that is fed depends on the air temperature and the fluoride levels in the raw water.

Experience has shown that drinking water containing a proper amount of fluoride can reduce tooth decay by 65% in children. Fluoride combines chemically with tooth enamel when permanent teeth are forming. The result, of course, is teeth that are harder, stronger, and more resistant to decay. The USEPA sets the upper limits for fluoride in drinking water supplies based on ambient temperatures, because people drink more water in warmer climates. Fluoride concentrations should be lower in these areas (Spellman, 1998).

12.5 WATER TREATMENT OF ORGANIC AND INORGANIC CONTAMINANTS

Man-made compounds that contain carbon (commonly called SOCs or Synthetic Organic Chemicals) are, from time to time, detected in U.S. water supplies. Some of these are volatile organic compounds (VOCs), such as the solvent trichloroethylene. The problem with VOCs in a water supply (i.e., any water supply used by the public) is twofold. They are quickly absorbed through the skin and can evaporate into gases that are inhaled during showers, baths, or dishwashing.

How do water supplies become contaminated by organic compounds? Good question. Basically, sources of organic contaminants are usually provided by improperly disposed wastes, pesticide use, industrial effluents, and leaking fuel oil tanks (gasoline in particular).

Water supplies may also contain inorganic contaminants consisting mainly of substances occurring naturally in the ground, such as sulfate, fluoride, arsenic, barium, radium, selenium, and radon. Metallic substances from industrial sources can contaminate surface waters. Nitrate ions from fertilizers and feedlot runoff often contaminate groundwater. Another source of inorganic chemical contamination in drinking water supplies is corrosion or deterioration of water supply appurtenances such as plumbing systems, which release metal and nonmetal substances into the water, such as lead, cadmium, zinc, copper, iron, and plumbing cement.

Inorganic contaminants can be treated by corrosion controls and removal techniques. Corrosion controls work to reduce the presence of corrosion by-products (e.g., lead) at the consumer's tap. Removal technologies, such as coagulation/filtration, reverse osmosis (RO), and ion exchange, are used to treat source water that is contaminated with metals or radioactive substances.

The following sections discuss processes for removing inorganic and organic dissolved solids from water intended for potable use. Note that, with some changes, these processes can also serve as tertiary wastewater treatment.

12.5.1 Aeration

Aeration (*air stripping*) is a physical treatment process in which air is thoroughly mixed with water—a technique effective for removing dissolved gases and highly volatile odorous compounds. Air exposure and oxygenation can enhance water quality. When aeration is used as an initial step in well water treatment, it can remove hydrogen sulfide, reduce dissolved carbon dioxide, and add oxygen to help oxidize iron and manganese into rust. One of the most common uses of aeration is for taste and odor control. Sedimentation and filtration are then necessary to clarify the water.

Note: Aeration is rarely adequate in processing surface waters, simply because the odor-producing substances are generally non-volatile.

Several methods for aerating the water are available. The method selected primarily depends on the type and concentration of material to be removed from the water and the available pressure. Aeration in water treatment can be accomplished using spray nozzles, cascade systems, multiple-tray aerators, diffused-air aerators, and mechanical aerators.

12.5.2 Oxidation

Simply stated, *oxidation* is a reaction in which a substance loses electrons, thus increasing its charge. A substance that oxidizes another is called an *oxidizing agent* or *oxidizer*. In water treatment, oxidation is used to remove or destroy undesirable tastes or odors, to aid in the removal of iron and manganese, and to help improve clarification and color removal in source water.

Note: Atmospheric oxygen through aeration can be used to oxidize the organic substances responsible for undesirable tastes and odors, but the process is usually too slow to be of value. However, if dissolved gases such as hydrogen sulfide are the cause of taste and odor problems, aeration will effectively remove them through oxidation and stripping.

Chlorine dioxide, potassium permanganate, and ozone are strong oxidants capable of destroying many odorous compounds. Because they do not produce THMs, these chemicals are favored over heavy chlorination.

12.5.3 Adsorption

Adsorption is primarily considered a surface phenomenon, occurring when one substance adheres to the surface of another. The two most common adsorptive media used in water treatment are activated carbon and activated alumina. These adsorptive materials are generally most effective for taste and odor control and for the removal of organic pollutants. However, the most important applications of adsorption in water treatment are the removal of arsenic and organic contaminants.

Adsorption of organic materials using *activated carbon* has been a common practice in water treatment for many years. Activated carbon is produced from carbon-based materials such as wood, coal, or petroleum residues. A char is made by burning the material in the absence of air and then oxidizing it at higher temperatures to create a very porous structure. This "activation" step provides irregular channels and pores in the solid mass, resulting in a very large surface area per mass ratio. This large surface area gives activated carbon its effectiveness as an adsorbing agent. The larger the surface area of an adsorber, the greater its power. Each activated carbon contains a huge number of pores and crevices into which organic molecules can enter and be adsorbed onto the activated carbon surface.

Activated carbon strongly attracts a range of organic molecules, including aromatic solvents (e.g., benzene, toluene, and nitrobenzene), chlorinated aromatics (PCBs, chlorobenzenes, and chloronaphthalene), phenol and chlorophenols, polynuclear aromatics (acenaphthene and benzopyrenes), pesticides, herbicides, chlorinated aliphatics (carbon tetrachloride and chloroalkyl ethers), dyes, gasoline, amines, and humic substances.

Two forms of activated carbon are used in water treatment: powdered and granular. Powdered activated carbon is often used for taste and odor control. Its effectiveness depends on the source of the undesirable flavors and odors. It is also effective in removing the organic precursors that react with chlorine to form harmful THM compounds after disinfection.

Powdered activated carbon is a finely ground, insoluble black powder that can be added at any point in the treatment process ahead of the filters. It is fed to water either as a dry powder or as a wet slurry. Although adsorption is nearly instantaneous, a contact time of 15 minutes or more is desirable before sedimentation or filtration.

Activated carbon media must periodically be replaced with new or regenerated activated carbon. Replacement cycles can vary from one to three years for taste and odor treatment to as little as four or five weeks for removal of organics. The activated carbon regeneration process involves (1) removing the spent carbon as a slurry; (2) dewatering the slurry; (3) feeding the carbon into a special furnace where regeneration occurs (i.e., the organics are driven from the carbon surface by heat); and (4) returning it to use.

Activated alumina (a highly porous and granular form of aluminum oxide) is also an adsorptive medium used in water treatment. Used primarily to remove arsenic and excess fluoride ions, in use, water is percolated through a column of alumina media. A combination of adsorption and ion exchange performs the actual removal of arsenic and fluoride ions.

Like the regeneration process used to restore used activated carbon to full potency, activated alumina also requires periodic regeneration, accomplished by passing a caustic soda solution through the media. Excess caustic soda is neutralized by rinsing the activated alumina with an acid. Disposal of these wash waters, laden with toxic arsenic and fluoride ions, must be done in accordance with applicable laws.

Note: Powdered activated carbon is much more difficult to regenerate than granular activated carbon.

Note: Granular activated carbon is sometimes used in the filter bed itself, combining both filtration and adsorption in one treatment unit. The major problem associated with granular activated carbon systems is the plugging of the bed by suspended solids in the water.

12.5.4 Demineralization

Demineralization refers to the removal of dissolved solids (inorganic mineral substances) from water. Dissolved solids contain both cations and anions and therefore require two types of ion exchange resins. Cation exchange resins used for demineralization purposes have hydrogen exchange sites and are divided into strong acid and weak acid classes. The anion exchange resins commonly used contain hydroxide ions and are divided into strong and weak base classes.

Demineralization is commonly used in industry in waste treatment for removal of arsenic, barium, cadmium, chromium, fluoride, sulfate, and zinc. Some general advantages of using ion exchange to remove these contaminants are the low capital investment required and the mechanical simplicity of the process. In addition, the ion-exchange process can be used to recover valuable chemicals for reuse or harmful ones for disposal. For instance, this process is frequently employed to reclaim chromic acid from metal finishing waste, enabling its reuse in chrome plating operations. It also has some application in the removal of radioactivity. The major disadvantages are the high chemical requirements needed to regenerate the resins and to dispose of chemical wastes from the regeneration process. These factors make ion exchange more suitable for small systems than for large ones.

12.5.5 Membrane Processes

Membrane processes used in water treatment are primarily demineralization processes. Demineralization of water can be accomplished using thin, microporous membranes. Electrodialysis and RO are the most common membrane processes.

Before we briefly discuss these two membrane processes, you need a basic understanding of osmosis. In osmosis, a semipermeable membrane separates two solutions with different mineral concentrations. Water moves through the membrane from the dilute solution to the concentrated one. This is *osmosis*, and it continues until the buildup of hydrostatic pressure on the more concentrated solution is sufficient to stop the net flow.

In *RO*, the flow of water through the semipermeable membrane is reversed by applying external pressure to offset the hydrostatic pressure. This results in a concentration of minerals on one side of the membrane and pure water on the other side. RO can treat a wide variety of health and aesthetic contaminants in water. Reverse osmosis equipment, when effectively designed, can treat aesthetic contaminants that cause unpleasant taste, color, and odor problems, such as a salty or soda taste caused by chlorides or sulfates. RO can also be effective for treating arsenic, asbestos, atrazine, fluoride, lead, mercury, nitrate, and radium. When used with appropriate carbon prefiltering, additional treatment can also be provided for such "volatile" contaminants as benzene, trichloroethylene, THMs, and radon. Some RO equipment is also capable of treating *Cryptosporidium*. RO can be expected to play a major role in water treatment for years to come.

Reverse osmosis (also called ultrafiltration) is the most common process for reducing the salinity of brackish groundwater. In operation, a semipermeable membrane (the most essential element in the RO method of demineralization) separates salty water from two different concentrations. Concentrations have a natural tendency to become equalized by a flow of water from the dilute side to the concentrated side (osmosis). But high pressure applied to the high concentration side of the membrane can reverse this direction of flow. Freshwater diffuses through the membrane, leaving a more concentrated salt solution behind. RO unit performance relies heavily on water quality parameters. Suspended solids, dissolved organics, hydrogen sulfide, iron, and strong oxidizing agents (chlorine, ozone, and permanganate) are harmful to membranes.

Electrodialysis is the demineralization of water using the principles of osmosis, but it uses ion-selective membranes and an electric field to separate anions and cations in solution. In the past, electrodialysis was most often used for purifying brackish water, but it is now finding a role in industrial waste treatment as well. For example, metal salts from plating rinses are sometimes removed in this way.

12.6 THE WATER-AI CONNECTION

The inclusion of artificial intelligence (AI) is revolutionizing the management and operation of water treatment operations. This is especially the case currently in water monitoring. AI rapidly processes vast data in real time, detects patterns, and predicts issues before they escalate.

AI solves problems in a human-like way. Technology has developed to a stage where AI can perform specific tasks. This ability of AI to adapt and learn, and the ability to sort out irrelevant information, enabling the enhancement of basic scenarios, is one of its strongest benefits.

Using machine learning algorithms to analyze water samples, detecting pollutants, algae blooms, and other contaminants quickly and accurately. Analyzing water samples quickly and accurately is what precision in monitoring water quality is all about.

12.7 THE BOTTOM LINE

These processes, from preliminary screening to filtration, disinfection, and the advanced processes needed for specialized water problems, are in place to serve one primary, essential purpose: to supply the consumer with safe potable water. The consumer may use this water for a wide variety of purposes, from drinking to watering the lawn, but those uses are in many ways beside the point: Safe potable water is essential for human life.

NOTES

1 J.C. Fox (1999). Stakeholder Collaboration Key to New Safe Drinking Water Rules. *Water Environment & Technology*, February.
2 J. Louma (2004). Water Thieves. *The Ecologist.*
3 *AWWA Standards (Coagulation—No. 42402 to 42407).* Denver, CO: AWWA.
4 Adapted from F.R. Spellman (1999). *Disinfection Alternatives.* Lancaster, PA: Technomic Publishing Company.

REFERENCES

AWWA. 1995. *Water Treatment: Principles and Practices of Water Supply Operations.* Denver, CO: American Water Works Association.

De Zuane, J. 1997. *Handbook of Drinking Water Quality*, 2nd ed. New York: John Wiley and Sons.

Hammer, M.J., and Hammer, M.J., Jr. 1996. *Water and Wastewater Technology*, 3rd ed. Englewood Cliffs, NJ: Prentice Hall.

Masters, G.M. 1991. *Introduction to Environmental Engineering and Science*. Englewood Cliffs, NJ: Prentice Hall.

McGhee, T.J. 1991. *Water Supply and Sewerage*, 6th ed. New York: McGraw-Hill, Inc.

Pankratz, T.M. 1995. *Screening Equipment Handbook: For Industrial and Municipal Water and Wastewater Treatment*, 2nd ed. Lancaster, PA: Technomic Publishing Company.

Peavy, H. S., et al. 1985. *Environmental Engineering*. New York: McGraw-Hill, Inc., 1985.

Spellman, F.R. 1998. *The Science of Water: Concepts and Applications*, 2nd ed. Boca Raton, FL: CRC Press.

Spellman, F. and Drinan, J. 2012. *The Science of Environmental Pollution*. Boca Raton, FL: CRC Press.

USA TODAY, October 1998.

Viessman, W., and Hammer, M.J. 1998. *Water Supply and Pollution Control*, 6th ed. Menlo Park, CA: Addison-Wesley.

13 Wastewater Treatment and AI

13.1 HIDDEN FUNCTIONS[1]

Water and wastewater operations are hidden functions. Treatment plants are usually isolated; the piping systems are typically buried underground; the customers and ratepayers hardly think about either—that is, until the rude awakening—until the tap runs dry, the toilet backs up, or the bill is received in the mail or online. Although not often thought of as a commodity (or, for that matter, not thought about at all), water is a commodity—a very valuable, absolutely vital commodity. And even though water is a useful and essential commodity, we often consume it, waste it, discard it, pollute it, poison it, and flush it. We relentlessly modify the hydrological cycles (natural and urban cycles) with total disregard for the consequences: too many people, too little water, water in the wrong places, and in the wrong amounts. The human population is burgeoning, but water demand is increasing twice as fast (Spellman, 2007). It is the author's position that with the passage of time, potable water will become even more valuable. Moreover, with the passage of even more time, potable water will be even more helpful than we might ever imagine—possibly (likely) comparable in pricing, gallon for gallon, to what we pay for gasoline, or even more. From urban growth to infectious disease and newly identified contaminants in water (aka "forever" chemicals—perfluoroalkyl and polyfluoroalkyl chemicals, collectively known as PFAS), greater demands are being placed on our planet's water supply (and other natural resources). As the global population continues to grow, people will place greater demands on our water supply. The fact is—simply, profoundly, without a doubt in the author's mind—water is the "new oil."

Earth was originally allotted a finite amount of water; we have no more or no less than that original allotment today—they are not making any more of it. Thus, it logically follows that, to sustain life as we know it, we must do everything we can to preserve and protect our water supply. Moreover, we also must purify and reuse the water we presently waste (i.e., wastewater).

Key Point: The National Primary Drinking Water Regulations are legally enforceable primary standards and treatment techniques for public water systems.

> **DID YOU KNOW?**
>
> More than 50% of Americans drink bottled water occasionally or as their major source of drinking water—an astounding fact given the high quality and low cost of U.S. tap water.

DOI: 10.1201/9781003655848-18

The term "Sick Water" was coined by the United Nations in a 2010 press release addressing the need to recognize that it is time to arrest the global tide of sick water. The gist of the UN's report pointed out that transforming waste from a major health and environmental hazard into a clean, safe, and economically attractive resource is emerging as a key challenge in the twenty-first century. As a practitioner of environmental health, I certainly support the UN's view on this important topic.

However, when I discuss sick water in the context of this text and many others I have authored or co-authored on the topic, I go a few steps further than the UN in describing the real essence and tragic implications of water that makes people or animals sick or worse, or at least can be classified as sick—again, in my opinion.

Water that is sick is a filthy medium, wastewater—a cocktail of fertilizer runoff and sewage disposal alongside animal, industrial, agricultural, and other wastes. In addition to these listed wastes of concern, other wastes are beginning to garner widespread attention; they certainly have garnered our attention in our research on the potential problems related to these so-called other wastes.

What are these other wastes? Any waste or product we dispose of in our waters that we flush down the toilet, pour down the sink or bathtub, or down the drain of a worksite deep sink. Consider the following example of "pollutants" we discharge to our wastewater treatment plants or septic tanks, which we often overlook as waste products.

Note: The following is deliberate repetition for "Emphasis" of a critical point.

As pointed out earlier, each morning, a family of four wakes up and prepares for the workday for the two parents and school for the two teenagers. Fortunately, this family has three upstairs bathrooms to accommodate each other's needs to prepare for the morning, including natural waste disposal, showering, soap usage, cosmetic application, hair treatments, vitamins, sunscreen, fragrances, and prescribed medications. In addition, the overnight deposit of cat and dog waste is routinely picked up and flushed down the toilet. Let's make a short inventory list of what this family of four has disposed of or has applied to themselves as they prepare for their day outside the home.

- Toilet-flushed animal wastes
- Prescription and over-the-counter therapeutic drugs
- Veterinary drugs
- Fragrance
- Soap
- Shampoo, condition, other hair treatment products
- Body lotion, deodorant, body powder
- Cosmetics
- Sunscreen products
- Diagnostic agents
- Nutraceuticals (e.g., vitamins, medical foods, functional foods)

Even though these bioactive substances have been around for decades, today we group all of them (the exception being animal wastes), substances and/or products, under the title of pharmaceuticals and personal care products called "PPCPs."

I pointed to the human activities of the family of four in contributing PPCPs to the environment, but other sources of PPCPs should also be recognized. For example, residues from pharmaceutical manufacturing, residues from hospitals, illicit drug disposal [i.e., police knock on the door and the frightened user flushes the illicit drugs down the toilet (along with $100 bills, weapons, self-aborted fetuses, dealers' phone numbers, etc.) and into the wastewater stream], and veterinary drug use, especially antibiotics and steroids, and agribusiness are all contributors of PPCPs in the environment.

Regarding personal deposits of PPCPs into the environment and the local wastewater supply, let's return to the family of four for now. After having applied or taken in the various substances mentioned earlier, the four individuals involved add these products, PPCPs, to the environment through excretion (the elimination of waste material from the body), and when bathing later after returning home, and then possibly the disposal of any unwanted medications to sewers and trash. How many of us have found old medical prescriptions in the family medicine chest and decided they were no longer needed? How many of us have grabbed such unwanted medications and simply disposed of them with a single toilet flush? Many of these medications, for example, antibiotics, are not typically found in the environment. And then there are those who dump the deep-fryer grease down the toilet—please, do not do that!

Earlier, I stated that wastewater is a cocktail of fertilizer runoff and sewage disposal alongside animal, industrial, agricultural, and other wastes. When we factor in the addition of PPCPs to this cocktail, we can state analogously that we are simply adding a mixer to the mix.

The questions about our mixed waste cocktail are obvious: Does the disposal of antibiotics and/or other medications into the local wastewater treatment system cause problems for anyone or anything else downstream, so to speak? When we ingest locally treated water, are we also ingesting flushed-down-the-toilet or -the-drain antibiotics, other medications, illicit drugs, animal excretions, cosmetics, vitamins, vaginal cleaning products, sunscreen products, diagnostic agents, crankcase oil, grease, oil, fats, and veterinary drugs each time we drink a glass of tap water?

Well, Joni, bar the door—we certainly hope not. But hope is not always a fact.

The jury is still out on these questions; we do not know what we do not know about the fate of PPCPs or their impact on the environment once they enter our wastewater treatment systems, the water cycle, and eventually our drinking water supply systems. This is the case even though some PPCPs are easily broken down and processed by the human body or degraded quickly in the environment. Moreover, we have known for some time, since the time of the mythical hero Hercules, arguably the world's first environmental engineer, when he was ordered to perform his fifth labor by Eurystheus to clean up King Augeas' stables. Hercules, faced literally with a mountain of horse and cattle waste piled high in the stable area, had to devise some method to dispose of the waste; so, he did. He diverted a couple of river streams to the inside of the stable area so that all the animal waste could be deposited into the river: Out of sight, out of mind. The waste simply flowed downstream—someone else's problem. Hercules understood the principal point in pollution control technology that is pertinent to this very day: that is, *dilution is the solution to pollution.*

As applied to today, the fly in the ointment in Hercules' dilution is the solution to the pollution approach, which is today's modern PPCPs. Although he was able to dispose of animal waste into a running water system where eventually the water's self-purification process would clean the stream, he did not have to deal with today's pharmaceuticals and hormones that are given to many types of livestock to enhance their health and growth.

The simple truth is that studies have shown that pharmaceuticals are present in our nation's water bodies. Further research suggests that certain drugs may cause ecological harm. The EPA and other research agencies are committed to investigating this topic and developing strategies to help protect the health of both the environment and the public. To date, scientists have found no evidence of adverse human health effects from PPCPs in the environment. Moreover, others might argue that even if PPCPs were present today or in ancient (and mythical) times, the amount present in local water systems and others would represent only a small fraction (ppt—parts per trillion, 10^{-12}) of the total volume of water. And the critics would be quick to point out that when we are speaking of ppt, we are speaking of a proportion equivalent to one-twentieth of a drop of water diluted into an Olympic-size swimming pool. I remember when I had a student in my environmental health class who stated that he did not think the water should be termed "sick water" because it was evident to him that if the water contained so many medications, for example, how could it be sick? Instead, it might be termed "getting well water—making anyone who drinks it well, cured, no longer sick, etc."

It is important to point out that the term "sick water" can be applied to not only PPCP-contaminated water but also any filthy, dirty, contaminated, vomit-filled, polluted, pathogen-filled drinking water source. The fact is that dirty or sick water means more people now die from contaminated and polluted water than from all forms of violence, including wars. The United Nations also points out that dirty or sick water is a key factor in the rise of deoxygenated dead zones that have been emerging in seas and oceans across the globe.

DID YOU KNOW?

More than 50% of Americans drink bottled water occasionally or as their major source of drinking water—an astounding fact given the high quality and low cost of U.S. tap water.

13.2 THE PARADIGM SHIFT

Historically, the purpose of water supply systems has been to provide pleasant drinking water that is free of disease organisms and toxic substances. In addition, the purpose of wastewater treatment has been to protect the health and well-being of our communities. Water/wastewater treatment operations have accomplished this goal by (1) prevention of disease and nuisance conditions, (2) avoidance of contamination of water supplies and navigable waters, (3) maintenance of clean water for survival of fish, bathing, and recreation, and (4) general conservation of water quality for future use.

The purpose of water supply systems and wastewater treatment processes has not changed. However, primarily because of new regulations that include (1) protecting against protozoan and virus contamination; (2) implementation of the multiple barrier approach to microbial control; (3) new requirements of the Ground Water Disinfection Rule, the Total Coliform Rule and Distribution System (DS), the Lead and Copper (Pd/Cu) Rule; (4) regulations for trihalomethanes and Disinfection By-Products (DBPs); and (5) new requirements to remove even more nutrients (nitrogen and phosphorus) from wastewater effluent, the paradigm has shifted. We discuss this important shift momentarily, but first it is important to abide by Voltaire's advice; that is, "If you wish to converse with me, please define your terms."

For those not familiar with the term "paradigm," it can be defined in the following ways: A *paradigm* is the consensus of the scientific community: "concrete problem solutions that the profession has come to accept" (Holyningen-Huene, 1993). Thomas Kuhn coined the term paradigm. He outlined it in terms of the scientific process. He felt that "one sense of paradigm is global, embracing all the shared commitments of a scientific group; the other isolates a vital sort of commitment and is thus a subset of the first" (Holyningen-Huene, 1993). The concept of paradigm has two general levels. The first is the encompassing whole, the summation of parts. It consists of the theories, laws, rules, models, concepts, and definitions that go into a generally accepted fundamental theory of science. Such a paradigm is "global" in character. The other level of paradigm is that it can also be just one of these laws, theories, models, etc., that combine to formulate a "global" paradigm. These have the property of being "local." For instance, Galileo's theory that the Earth rotated around the Sun became a paradigm, namely a generally accepted law in astronomy. Yet, on the other hand, his theory combined with other "local" paradigms in areas such as religion and politics to transform culture. Paradigm can also be defined as a pattern or point of view that determines what is seen as reality. I use this definition in this text.

A *paradigm shift* is defined as a major change in the way things are thought about, especially scientifically. Once a problem can no longer be solved within the existing paradigm, new laws and theories emerge and form a new paradigm, overthrowing the old if it is accepted. Paradigm shifts are the "occasional, discontinuous, revolutionary changes in tacitly shared points of view and preconceptions" (Daly, 1980). A paradigm shift represents "a profound change in the thoughts, perceptions, and values that form a particular vision of reality" (Capra, 1982). For our purposes, I use the term "paradigm shift" to mean a change in the way things are understood and done.

13.2.1 A Transformational Change

In water supply systems, the historical focus, or traditional approach, has been to control turbidity, iron and manganese, taste and odor, color, and coliforms. New regulations provided a new focus and thus a paradigm shift. Today, the traditional approach is no longer sufficient. Providing acceptable water has become more sophisticated and costly. To meet the requirements of the new paradigm, a systems approach must be employed. In the systems approach, all components are interrelated. What affects one impacts others. The focus has shifted to multiple requirements (i.e., new regulations require the process to be modified or the plant upgraded).

To show how water supply systems have changed, let's review the traditional disinfection method. Disinfection was used in water to destroy harmful organisms. Currently, disinfection is still used in water to destroy harmful organisms but is now only one part of the *multiple barrier approach*. Moreover, disinfection has traditionally been used to treat coliforms only. Currently, because of the paradigm shift, disinfection now (and in the future) is used against coliforms, *Legionella, Giardia, Cryptosporidium*, and others. (Note: To effectively remove the protozoans Giardia and Cryptosporidium, filtration is required; disinfection is not effective against the oocysts of Cryptosporidium.) Another example of the traditional versus current practices is seen in the traditional approach to particulate removal in water to lessen turbidity and improve aesthetics. Current practice still aims to decrease turbidity to improve aesthetics but now microbial removal plus disinfection is also considered practical.

Another significant factor that contributed to the paradigm shift in water supply systems was the introduction of the Surface Water Treatment Rule (SWTR) in 1989. SWTR requires water treatment plants to achieve 99.9% (3 log) removal/activation/inactivation of Giardia and 99.99% (4 log) removal/inactivation of viruses. SWTR applies to all surface waters and ground waters under direct influence.

As mentioned earlier, removal of excess nutrients such as nitrogen and phosphorus in wastewater effluent is now receiving more attention from regulators (USEPA, 2007b) and others. A major concern is that excess nutrients create oxygen-deficient dead zones in water bodies by promoting algae growth that consumes oxygen.

13.3 THE AI CONNECTION

The introduction of AI and its subfields in wastewater treatment is ongoing. At first mention, at first a conservation of those in charge of wastewater treatment when the topic turned toward possibly using AI and a few of its subfields (IoT and ML and others), many administrators, plant managers, chief operators, lead operators, and operators just shrugged at the idea as being rather far-fetched, space cadet-type talk (whatever that is), and thought it was a joke. It was like they were all turning their heads with a grin on their faces, looking for all the stars of those *Star Trek* movies.

I remember one lecture I attended in 1999, along with several water and wastewater professionals, who listened to a presentation about the future of water and wastewater treatment being digitalized—computerized operations. Again, those listening basically laughed off the presentation, including me. However, after a few days of thinking about digitizing these treatment processes, I had a change of perception about computerizing operations.

I came to realize that automatic samplers, sensors that monitor various stages of treatment for water quality and proper operation, and computer operating stations where licensed operators can monitor the entire plant operation, including each unit process, might be worth buying into. I included each unit process, and several were involved at the beginning of the project rather than later stages.

AI is innovating wastewater treatment. This book is concerned with water, and the mix of wastewater treatment and AI is the focus here, but the concern is not only for the wastewater but also for the enormous amount of energy used in the plant. So, the point is, although focused on the treatment of wastewater in AI, it is important to note that AI helps Water Managers at the plant to also incorporate AI into harnessing energy usage.

13.4 WASTEWATER TREATMENT

In the United States, the Code of Federal Regulations (CFR) 40 CFR Part 403 regulations were established in the late 1970s and early 1980s to help Publicly Owned Treatment Works (POTW) control industrial discharges to sewers. These regulations were designed to prevent pass-through as well as interference at the treatment plants and interference in the collection and transmission systems.

Pass-through occurs when pollutants literally "pass through" a POTW without being properly treated and cause the POTW to have an effluent violation or increase the magnitude or duration of a breach.

Interference occurs when a pollutant discharge causes a POTW to violate its permit by inhibiting or disrupting treatment processes, treatment operations, or processes related to sludge use or disposal.

Unit operations (unit processes), which are the components that are linked together to form a process train (as shown in Figure 13.1, keep in mind the caboose attached to

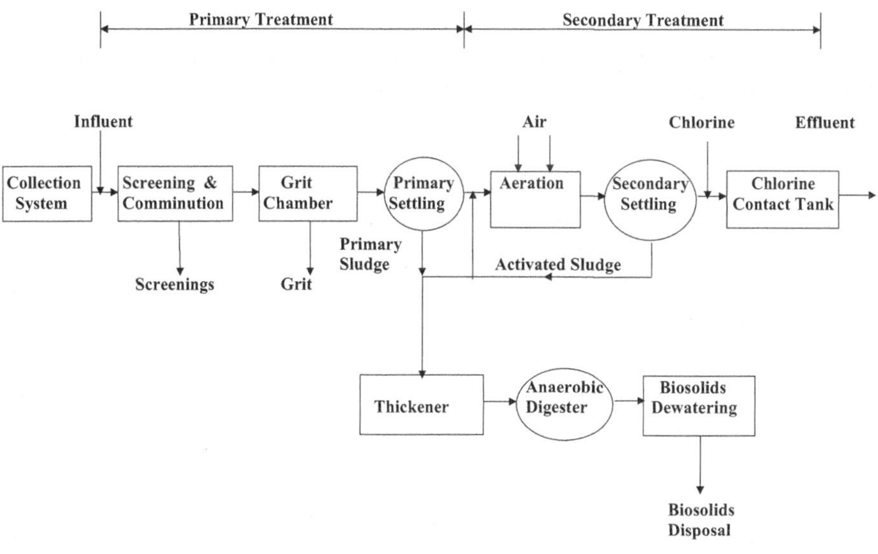

FIGURE 13.1 Schematic of an example wastewater treatment process providing primary and secondary treatment using the activated sludge process.

this train is treated and cleaned wastewater, which when outfalled is usually cleaner than the water in the receiving body), are commonly divided on the basis of the fundamental mechanisms acting with them (i.e., physical, chemical, and biochemical). Physical operations are those, such as sedimentation, that are governed by the laws of physics (gravity). Chemical operations are those in which strict chemical reactions occur, such as precipitation. Biochemical operations are those that use living microorganisms to destroy or transform pollutants through enzymatically catalyzed chemical reactions (Grady et al., 2011).

13.4.1 Wastewater Operators (aka Fluid Mechanics)

Like waterworks operators, wastewater operators are highly trained and artful practitioners and technicians of their trade. Moreover, wastewater operators, again, like waterworks operators, are required by the States to be licensed or certified to operate a wastewater treatment plant. When learning wastewater operator skills, there are several excellent texts available to aid in the training process.

13.4.2 Wastewater Treatment Process: The Model

Figure 13.1 shows a basic schematic of an example of a wastewater treatment process providing primary and secondary treatment using the *activated sludge process*. This is the model, the prototype, and the paradigm used in the book. Though it is true that in secondary treatment (which provides BOD removal beyond what is achievable by simple sedimentation), there are three commonly used approaches—trickling filter, activated sludge, and oxidation ponds—we focus, for instructive and illustrative purposes, on the activated sludge process throughout this handbook. The purpose of Figure 13.1 is to allow the reader to follow the treatment process step-by-step as it is presented (and as it is configured in the real world) and to assist understanding of how all the various unit processes sequentially follow and tie into each other. Therefore, we begin certain sections (which discuss unit processes) with frequent reference to Figure 13.1. It is important to start these sections in this manner because wastewater treatment is a series of individual steps (unit processes) that treat the waste stream as it makes its way through the entire process. Thus, it logically follows that a pictorial presentation along with pertinent written information enhances the learning process. It should also be pointed out, however, that even though the model shown in Figure 13.1 does not include all unit processes currently used in wastewater treatment, we do not ignore the other major processes: trickling filters, rotating biological contactors (RBCs), and oxidation ponds.

13.4.3 Wastewater Terminology and Definitions

Wastewater treatment technology, like many other technical fields, has its own unique terms with their own meaning. Though some of the terms are unique, many are common to other professions. Remember that the science of wastewater treatment is a combination of engineering, biology, mathematics, hydrology, chemistry, physics, and other disciplines. Therefore, many of the terms used in engineering, biology, mathematics, hydrology, chemistry, physics, and other fields are also used in wastewater treatment. Any terms not listed or defined in the following section will be defined as they appear in the text.

Activated sludge—the solids formed when microorganisms are used to treat wastewater using the activated sludge treatment process. It includes organisms, accumulated food materials, and waste products from the aerobic decomposition process.

Advanced waste treatment—treatment technology to produce an extremely high-quality discharge.

Aerobic—conditions in which free, elemental oxygen is present. Also used to describe organisms, biological activity, or treatment processes that require free oxygen.

Anaerobic—conditions in which no oxygen (free or combined) is available. Also used to describe organisms, biological activity, or treatment processes that function in the absence of oxygen.

Anoxic—conditions in which no free, elemental oxygen is present; the only source of oxygen is combined oxygen, such as that found in nitrate compounds. Also used to describe biological activity or treatment processes that function only in the presence of oxygen.

Average monthly discharge limitation—the highest allowable discharge over a calendar month.

Average weekly discharge limitation—the highest allowable discharge over a calendar week.

Biochemical oxygen demand, BOD_5—the amount of organic matter that can be biologically oxidized under controlled conditions (5 days at 20°C in the dark).

Biosolids—from *Merriam-Webster's Collegiate Dictionary, Tenth Edition* (1998): biosolid *n* (1977) Springfield, Massachusetts—solid organic matter recovered from a sewage treatment process and used primarily as fertilizer, a soil amendment—usually used in plural.

NOTE: In this text, biosolids (note: biosolids is in the plural) are used in many places (activated sludge being the exception) to replace the standard term "sludge." The author (along with others in the field) views the term "sludge" as an ugly six-letter word that is inappropriate to use in describing biosolids. Biosolids are a product that can be reused; they have some value. Because biosolids have some value, they should not be classified as a "waste" product, and when biosolids for beneficial reuse are addressed, they are not.

Buffer—a substance or solution that resists changes in pH.

Carbonaceous biochemical oxygen demand, $CBOD_5$—the amount of biochemical oxygen demand that can be attributed to carbonaceous material.

Chemical oxygen demand (COD)—the amount of chemically oxidizable materials present in the wastewater.

Clarifier—a device designed to allow solids to settle or rise and be separated from the flow. It is also known as a settling tank or sedimentation basin.

Coliform—a type of bacteria used to indicate possible human or animal contamination of water.

Combined sewer—a collection system that carries both wastewater and stormwater flows.

Comminution—a process to shred solids into smaller, less harmful particles.

Composite sample—a combination of individual samples taken in proportion to flow.

Daily discharge—the discharge of a pollutant measured during a calendar day or any 24-hour period that reasonably represents a calendar day for sampling. Limitations expressed as weight are the total mass (weight) discharged over the day.

Limitations expressed in other units are average measurements of the day.

Daily maximum discharge—the highest allowable values for a daily discharge.

Detention time—the theoretical time water remains in a tank at a given flow rate.

Dewatering—the removal or separation of a portion of water present in a sludge or slurry.

Discharge Monitoring Report (DMR)—the monthly report required by the treatment plant's NPDES discharge permit.

Dissolved oxygen (DO)—free or elemental oxygen, which is dissolved in water.

Effluent—the flow leaving a tank, channel, or treatment process.

Effluent limitation—any restriction imposed by the regulatory agency on quantities, discharge rates, or concentrations of pollutants that are discharged from point sources into state waters.

Facultative—organisms that can survive and function in the presence or absence of free, elemental oxygen.

Fecal coliform—a type of bacteria found in the bodily discharges of warm-blooded animals and is used as an indicator organism.

Floc—solids that join to form larger particles that will settle better.

Flume—a flow rate measurement device.

Food-to-microorganism ratio (F/M)—an activated sludge process control calculation based on the amount of food (BOD_5 or COD) available per pound of mixed liquor volatile suspended solids.

Grab sample—an individual sample collected at a randomly selected time.

Grit—heavy inorganic solids such as sand, gravel, eggshells, or metal filings.

Industrial wastewater—waste associated with industrial manufacturing processes.

Infiltration/inflow—extraneous flows in sewers, defined by Metcalf and Eddy in *Wastewater Engineering: Treatment, Disposal, Reuse*, 3rd. Ed., New York: McGraw-Hill, Inc., pp. 29–31, 1991, as follows:

- *Infiltration*—water entering the collection system through cracks, joints, or breaks.
- *Steady inflow*—water discharged from cellar and foundation drains, cooling water discharges, and drains from springs and swampy areas. This type of inflow is steady and is identified and measured along with infiltration.

- *Direct flow*—those types of inflow that have a direct stormwater runoff connection to the sanitary sewer, causing an almost immediate increase in wastewater flows. Possible sources are roof leaders, yard and areaway drains, manhole covers, cross connections from storm drains and catch basins, and combined sewers.
- *Total inflow*—the sum of the direct inflow at any point in the system plus any flow discharged from the system upstream through overflows, pumping station bypasses, and the like.
- *Delayed inflow*—stormwater that may require several days or more to drain through the sewer system. This category can include the discharge of sump pumps from cellar drainage as well as the slowed entry of surface water through manholes in ponded areas.

Influent—the wastewater entering a tank, channel, or treatment process.

Inorganic—mineral materials such as salt, ferric chloride, iron, sand, and gravel.

License—a certificate issued by the State Board of Waterworks/Wastewater Works Operators authorizing the holder to perform the duties of a wastewater treatment plant operator.

Mean Cell Residence Time (MRCT)—the average length of time a mixed liquor suspended solids particle remains in the activated sludge process. It may also be known as sludge retention time.

Mixed liquor—the combination of return activated sludge and wastewater in the aeration tank.

Mixed Liquor Suspended Solids (MLSS)—the suspended solids concentration of the mixed liquor.

Mixed Liquor Volatile Suspended Solids (MLVSS)—the concentration of organic matter in the mixed liquor suspended solids.

Milligrams/Liter (mg/l)—a measure of concentration. It is equivalent to parts per million (ppm).

Nitrogenous Oxygen Demand (NOD)—a measure of the amount of oxygen required to biologically oxidize nitrogen compounds under specified conditions of time and temperature.

NPDES Permit—National Pollutant Discharge Elimination System permit that authorizes the discharge of treated wastes and specifies the conditions that must be met for discharge.

Nutrients—substances required to support living organisms. Usually, it refers to nitrogen, phosphorus, iron, and other trace metals.

Organic—materials that consist of carbon, hydrogen, oxygen, sulfur, and nitrogen. Many organics are biologically degradable. All organic compounds can be converted to carbon dioxide and water when subjected to high temperatures.

Pathogenic—disease-causing. A pathogenic organism can cause illness.

Point Source—any discernible, defined, and discrete conveyance from which pollutants are or may be discharged.

Part per million—an alternative (but numerically equivalent) unit used in chemistry is milligrams per liter (mg/l). As an analogy, think of a ppm as being comparable to a full shot glass in a swimming pool.

Return Activated Sludge Solids (RASS)—the concentration of suspended solids in the sludge flow being returned from the settling tank to the head of the aeration tank.

Sanitary wastewater—wastes discharged from residences and from commercial, institutional, and similar facilities, which include both sewage and industrial wastes.

Scum—the mixture of floatable solids and water, which is removed from the surface of the settling tank.

Septic—wastewater that has no dissolved oxygen present. Generally characterized by black color and rotten egg (hydrogen sulfide) odors.

Settleability—a process control test used to evaluate the settling characteristics of the activated sludge. Readings taken at 30 to 60 minutes are used to calculate the settled sludge volume (SSV) and the sludge volume index (SVI).

Settled Sludge Volume—the volume, in percent, occupied by an activated sludge sample after 30 to 60 minutes of settling. Usually written as SSV with a subscript indicating the time of the reading used for calculation, such as SSV_{60} or SSV_{30}.

Sewage—wastewater containing human wastes.

Sludge—the mixture of settleable solids and water, which is removed from the bottom of the settling tank.

Sludge Retention Time (SRT)—see Mean Cell Residence Time.

Sludge Volume Index (SVI)—a process control calculation used to evaluate the settling quality of activated sludge. It requires the results of the SSV_{30} and mixed liquor suspended solids tests to calculate.

Storm sewer—a collection system designed to carry only storm water runoff.

Stormwater—runoff resulting from rainfall and snowmelt.

Supernatant—in a digester, is the amber-colored liquid above the sludge.

Wastewater—the water supply of the community after it has been soiled by use.

Waste Activated Sludge Solids—the concentration of suspended solids in the sludge, which is being removed from the activated sludge process.

Weir—a device used to measure wastewater flow.

Zoogleal Slime—refers to the biological material that develops on fixed film treatment devices. It contains a wide variety of organisms essential to the treatment process.

13.4.4 Measuring Plant Performance

To evaluate how well a plant or treatment unit process is operating, *performance efficiency* (or *percent (%) removal*) is used. The results can be compared with those listed in the plant's operation and maintenance manual (O&M) to determine if the facility is performing as expected.

13.5 WASTEWATER SOURCES AND CHARACTERISTICS

Wastewater treatment is designed to use the natural purification processes (self-purification processes of streams and rivers) to the maximum level possible. It is also designed to complete these processes in a controlled environment rather than over many miles of stream or river. Moreover, the treatment plant is also designed to remove other contaminants, which are not normally subjected to natural processes, as well as treat the solids, which are generated through the treatment unit steps. The typical wastewater treatment plant is designed to achieve many different purposes:

- Protecting public health
- Protecting public water supplies
- Protecting aquatic life
- Preserving the best uses of the waters
- Protecting adjacent lands

Wastewater treatment is a series of steps. Each of the steps can be accomplished using one of the many treatment processes or types of equipment available. The major categories of treatment steps are as follows:

- **Preliminary Treatment**—removes materials that could damage plant equipment or occupy treatment capacity without being treated.
- **Primary Treatment**—removes settleable and floatable solids (may not be present in all treatment plants).
- **Secondary Treatment**—removes BOD_5 and dissolved and colloidal suspended organic matter by biological action; organics are converted to stable solids, carbon dioxide, and more organisms.
- **Advanced Waste Treatment**—uses physical, chemical, and biological processes to remove additional BOD_5, solids, and nutrients (not present in all treatment plants).
- **Disinfection**—removes microorganisms to eliminate or reduce the possibility of disease when the flow is discharged.
- **Sludge Treatment**—stabilizes the solids removed from wastewater during treatment, inactivates pathogenic organisms, and/or reduces the volume of the sludge by removing water.

The various treatment processes described earlier will be discussed in detail later.

13.5.1 Wastewater Sources

The principal sources of domestic wastewater in a community are the residential areas and commercial districts. Other significant sources include institutional and recreational facilities, storm water (runoff), and groundwater (infiltration). Each source produces wastewater with specific characteristics. In this section, wastewater sources and the particular characteristics of wastewater are described.

Wastewater is generated by five primary sources: human and animal wastes, household wastes, industrial wastes, stormwater runoff, and groundwater infiltration.

1. *Human and animal wastes*—contain the solid and liquid discharges of humans and animals and are considered by many to be the most dangerous from a human health viewpoint. The primary health hazard is presented by

the millions of bacteria, viruses, and other microorganisms (some of which may be pathogenic) present in the waste stream.

2. *Household wastes*—are wastes, other than human and animal wastes, discharged from the home. Household wastes usually contain paper, household cleaners, detergents, trash, garbage, and other substances that homeowners discharge into the sewer system.
3. *Industrial wastes*—include industry-specific materials, which can be discharged from industrial processes into the collection system. Typically contain chemicals, dyes, acids, alkalis, grit, detergents, and highly toxic materials.
4. *Stormwater runoff*—many collection systems are designed to carry both the wastes of the community and stormwater runoff. In this type of system, when a storm event occurs, the waste stream can contain large amounts of sand, gravel, and other grit as well as excessive amounts of water.
5. *Groundwater infiltration*—groundwater will enter older, improperly sealed collection systems through cracks or unsealed pipe joints. Not only can this add large amounts of water to wastewater flows, but it can also add additional grit.

Note that wastewater can be classified according to the sources of flows: domestic, sanitary, industrial, combined, and stormwater.

1. *Domestic (sewage) Wastewater*—mainly contains human and animal wastes, household wastes, small amounts of groundwater infiltration, and small amounts of industrial wastes.
2. *Sanitary Wastewater*—consists of domestic wastes and significant amounts of industrial wastes. In many cases, the industrial wastes can be treated without special precautions. However, in some cases, the industrial wastes will require special precautions or a pretreatment program to ensure the wastes do not cause compliance problems for the wastewater treatment plant.
3. *Industrial wastewater*—consists of industrial wastes only. Often, industries will determine that it is safer and more economical to treat their waste independently of domestic waste.
4. *Combined wastewater* is the combination of sanitary wastewater and stormwater runoff. All the wastewater and stormwater from the community are transported through one system to the treatment plant.
5. *Storm water*—a separate collection system (no sanitary waste) that carries storm water runoff, including street debris, road salt, and grit.

13.5.2 Wastewater Characteristics

Wastewater contains many different substances that can be used to characterize it. The specific substances and their amounts or concentrations will vary, depending on the source. Therefore, it is difficult to "precisely" characterize wastewater. Instead, wastewater characterization is typically based on and applied to average domestic wastewater. Wastewater is characterized in terms of its physical, chemical, and biological characteristics.

Note: Keep in mind that other sources and types of wastewaters can dramatically change the characteristics.

13.5.2.1 Physical Characteristics

The *physical characteristics* of wastewater are based on color, odor, temperature, and flow.

- **Color**—fresh wastewater is usually a light brownish-gray color. However, typical wastewater is gray and has a cloudy appearance. The color of the wastewater will change significantly if allowed to go septic (if travel time in the collection system increases). Typical septic wastewater will have a black color.
- **Odor**—odors in domestic wastewater are usually caused by gases produced by the decomposition of organic matter or by other substances added to the wastewater. Fresh domestic wastewater has a musty odor. If wastewater becomes septic, its odor typically shifts to a smell similar to rotten eggs, which is caused by the production of hydrogen sulfide (H_2S).
- **Temperature**—the temperature of wastewater is commonly higher than that of the water supply because of the addition of warm water from households and industrial plants. However, significant amounts of infiltration or stormwater flow can cause major temperature fluctuations.
- **Flow**—the actual volume of wastewater is commonly used as a physical characterization of sewage and is normally expressed in terms of gallons per person per day. Most treatment plants are designed using an expected flow of 100 to 200 gallons per person per day. This figure may have to be revised to reflect the degree of infiltration or storm flow the plant receives. Flow rates will vary throughout the day. This variation, which can be as much as 50% to 200% of the average daily flow, is known as the *diurnal flow variation.*

Note: *Diurnal* means occurring in a day or each day or within a day.

13.5.2.2 Chemical Characteristics

When discussing the chemical characteristics of wastewater, the discussion generally includes topics such as organic matter, inorganic matter, and gases. For the sake of simplicity, this handbook specifically describes chemical characteristics in terms of alkalinity, biochemical oxygen demand (BOD), COD, dissolved gases, nitrogen compounds, pH, phosphorus, solids (organic, inorganic, suspended, and dissolved solids), and water.

- *Alkalinity*—a measure of the wastewater's capability to neutralize acids. It is measured in terms of bicarbonate, carbonate, and hydroxide alkalinity.

Alkalinity is essential to buffer (maintain the neutral pH) wastewater during the biological treatment processes.

- *Biochemical Oxygen Demand (BOD)*—a measure of the amount of biodegradable matter in wastewater. It is usually measured by a five-day test conducted at 20 °C. The BOD_5 of domestic waste is generally in the range of 100 to 300 mg/l.
- *COD*—a measure of the amount of oxidizable matter present in a sample. The COD is generally in the range of 200 to 500 mg/l. The presence of industrial wastes can increase this significantly.
- *Dissolved gases*—gases that are dissolved in wastewater. The specific gases and their normal concentrations are based on the composition of the wastewater. Typical domestic wastewater contains oxygen in relatively low concentrations, carbon dioxide, and hydrogen sulfide (if septic conditions exist).
- *Nitrogen compounds*—the type and amount of nitrogen present will vary from the raw wastewater to the treated effluent. Nitrogen follows a cycle of oxidation and reduction. Most of the nitrogen in untreated wastewater will be in the form of organic nitrogen and ammonia nitrogen. Laboratory tests exist for the determination of both forms. The sum of these two forms of nitrogen is also measured and is known as *Total Kjeldahl Nitrogen*. Wastewater normally contains between 20 and 85 mg/l of nitrogen. Organic nitrogen will normally be in the range of 8 to 35 mg/l, and ammonia nitrogen will be in the range of 12 to 50 mg/l.
- *pH*—a method of expressing the acidic condition of wastewater. It is measured on a scale of 1 to 14. For proper treatment, wastewater pH should normally be in the range of 6.5 to 9.0 (ideal—6.5 to 8.0).
- *Phosphorus*—essential for biological activity and must be present in at least minimum quantities for secondary treatment processes to perform. Excessive amounts of phosphorus can cause stream damage and excessive algal growth. Phosphorus will normally be in the range of 6 to 20 mg/l. The removal of phosphate compounds from detergents has had a significant impact on the amounts of phosphorus in wastewater.
- *Solids*—most pollutants found in wastewater can be classified as solids. Wastewater treatment is generally designed to remove solids or convert them to a form that is more stable and can be removed. Solids can be classified by their chemical composition (organic or inorganic) or by their physical characteristics (settleable, floatable, and colloidal). The concentration of total solids in wastewater is normally in the range of 350 to 1,200 mg/l.
  - *Organic solids*—consist of carbon, hydrogen, oxygen, and nitrogen and can be converted to carbon dioxide and water by ignition at 550 °C. They are also known as fixed solids or loss on ignition.
  - *Inorganic solids*—mineral solids that are unaffected by ignition. They are also known as fixed solids or ash.
  - *Suspended solids*—will not pass through a glass fiber filter pad. They can be further classified as Total Suspended Solids (TSS), Volatile Suspended Solids, and/or Fixed Suspended Solids. It can also be separated into three components based on settling characteristics: settleable

solids, floatable solids, and colloidal solids. TSS in wastewater are generally in the range of 100 to 350 mg/l.

- *Dissolved solids*—will pass through a glass fiber filter pad. It can also be classified as Total Dissolved Solids (TDS), volatile dissolved solids, and fixed dissolved solids. TDS are generally in the range of 250 to 850 mg/l.

- *Water*—always the major constituent of wastewater. In most cases, water makes up 99.5% to 99.9% of the wastewater. Even in the strongest wastewater, the total amount of contamination present is less than 0.5% of the total, and in average-strength waste, it is usually less than 0.1%.

13.5.2.3 Biological Characteristics and Processes

(Note: The biological characteristics of water were discussed in detail earlier in this text.) After undergoing physical aspects of treatment (i.e., screening, grit removal, and sedimentation) in preliminary and primary treatment, wastewater still contains some suspended solids and other solids that are dissolved in the water. In a natural stream, such substances are a source of food for protozoa, fungi, algae, and several varieties of bacteria. In secondary wastewater treatment, these same microscopic organisms (which are one of the main reasons for treating wastewater) are allowed to work as fast as they can to biologically convert the dissolved solids to suspended solids, which will physically settle out at the end of secondary treatment.

Raw wastewater influent typically contains millions of organisms. Most of these organisms are non-pathogenic; however, several pathogenic organisms may also be present (these may include the microorganisms responsible for diseases such as typhoid, tetanus, hepatitis, dysentery, and gastroenteritis). Many of the organisms found in wastewater are microscopic (microorganisms); they include algae, bacteria, protozoans (such as amoeba, flagellates, free-swimming ciliates, and stalked ciliates), rotifers, and viruses. Table 13.1 is a summary of typical domestic wastewater characteristics.

TABLE 13.1
Typical Domestic Wastewater Characteristics

Characteristic	Typical Characteristic
Color	Gray
Odor	Musty
Dissolved oxygen	>1.0 mg/l
pH	6.5–9.0
TSS	100–350 mg/l
BOD_5	100–300 mg/l
COD	200–500 mg/l
Flow	100–200 gallons/person/day
Total nitrogen	20–85 mg/l
Total phosphorus	6–20 mg/l
Fecal coliform	500,000–3,000,000 MPN/100 ml

SIDEBAR 13.1 A MONSTER IN THE WASTEWATER COLLECTION SYSTEM

On occasion, a growing monster, a hidden menace in a sewer or wastewater collection system, best described as a nasty, blocking, jamming, obstructing hunk of rancid, foul-smelling fat-oil-grease (aka FOG) attached to sanitary napkins, condoms, other sanitary products, wet wipes, and other nasty ingredients, finds residence in sewer systems. This odd aberration of filth is commonly called a ***fatberg***—the well-known and unwanted collection and interceptor pipe blocker. From experience, I have seen clusters of congealed masses, some as firm as cement, some as long as a torpedo, and requiring expensive special equipment to remove. Unmentionables, such as raw sewage attached to wet wipes and sanitary napkins, mainly form blobs, or fatbergs, in underground sewage conveyance systems. Vegetable oils and animal fats share common physicochemical properties with petroleum products and produce similar environmental effects. Historically, petroleum oils entered sewer systems, but now FOG (fats, oils, and grease) from restaurants and food processing plants is increasingly prevalent. This fat, oil, and grease mix, fats and oils of chicken, pork, palm olein, soybean, olive, and coconut, forms a cocktail, a mix, that gloms onto smooth interceptor lines and rough lines with high turbulence and waits for the arrival of those "supposedly" disposable wet hands and body wipes, sanitary napkins, and globs of toilet paper. The icing of this disgusting cake happens when the uninformed take the used grease from their deep-fried foods and pour it down the toilet—out of sight, out of mind. Well, sort of, because when fatbergs form in a wastewater collection system, the result usually amounts to a visible experience such as sewer overflow.

Okay, this is a book about the impact of AI in Water Management Systems. So, there is literally no way to monitor or control the disposal of FOG in kitchen sinks and toilets. However, there are methods of monitoring the formation of fatbergs in wastewater/sewage treatment, and advancements highlight the growing role of AI in managing and preventing fatbergs in wastewater systems. AI is being used in various ways to take on fatbergs. Here are some key applications where AI is used:

- Detection and prevention is where AI systems are being used to detect fatbergs in sewer systems, helping to prevent blockages, overflows, and flooding.
- AI is used in wastewater treatment systems developed to tackle fatbergs at their source, reducing the need for costly clean-out efforts.
- AI-robotics can autonomously inspect and potentially repair pipelines, breaking up fatbergs and preventing their formation.
- AI and image processing are being researched to make the cleanup process more efficient and to mitigate the environmental impact of fatbergs.

13.6 WASTEWATER COLLECTION SYSTEMS

Wastewater collection systems collect and convey wastewater to the treatment plant. The complexity of the system depends on the size of the community and the type of system selected. Methods of collection and conveyance of wastewater include gravity systems, force main systems, vacuum systems, and combinations of all three types of systems.

13.6.1 Gravity Collection System

In a *gravity collection system*, the collection lines are sloped to permit the flow to move through the system with minimal pumping. The slope of the lines must keep the wastewater moving at a velocity (speed) of 2–4 feet per second. Otherwise, at lower velocities, solids will settle out, causing clogged lines, overflows, and offensive odors. To keep collection system lines at a reasonable depth, wastewater must be lifted (pumped) periodically so that it can continue flowing downhill to the treatment plant. Pump stations are installed at selected points within the system for this purpose.

13.6.2 Force Main Collection System

In a typical *force main collection system*, wastewater is collected at central points and pumped under pressure to the treatment plant. The system is typically used for conveying wastewater long distances. The use of the force main system allows the wastewater to flow to the treatment plant at the desired velocity without using sloped lines. It should be noted that the pump station discharge lines in a gravity system are force mains since the content of the lines is under pressure.

Note: Extra care must be taken when performing maintenance on force central systems since the content of the collection system is under pressure.

13.6.3 Vacuum System

In a *vacuum collection system*, wastewaters are collected at central points and then drawn toward the treatment plant under vacuum. The system consists of a large amount of mechanical equipment and requires extensive maintenance to perform correctly. Generally, the vacuum-type collection systems are not economically feasible.

13.6.4 Pumping Stations

Pumping stations provide the motive force (energy) to keep the wastewater moving at the desired velocity. They are used in both force main and gravity systems. They are designed in several different configurations and may use different sources of energy to move the wastewater (i.e., pumps, air pressure, or vacuum). One of the more commonly used types of pumping station designs is the wet well/dry well design.

13.6.4.1 Wet-Well/Dry-Well Pumping Stations

The wet well/dry well pumping station consists of two separate spaces or sections separated by a common wall. Wastewater is collected in one section (wet well section), and the pumping equipment (and in many cases, the motors and controllers) are located in a second section known as the dry well. There are many different designs for this type of system, but in most cases, the pumps selected for this system are of a centrifugal design. There are a couple of primary considerations in selecting a centrifugal design: (1) It allows for the separation of mechanical equipment (pumps, motors, controllers, wiring, etc.) from the potentially corrosive atmosphere (sulfides) of the wastewater; and (2) this type of design is usually safer for workers because they can monitor, maintain, operate, and repair equipment without entering the pumping station wet well.

Note: Most pumping station wet wells are confined spaces. To ensure safe entry into such spaces, compliance with OSHA's 29 CFR 1910.146 (Confined Space Entry Standard) is required.

13.6.4.2 Wet-Well Pumping Stations

Another type of pumping station design is the *wet well* type. This type consists of a single compartment that collects the wastewater flow. The pump is submerged in wastewater, with motor controls either in the same space or housed above the wet well in a weatherproof enclosure. In this type of station, a submersible centrifugal pump is normally used.

13.6.4.3 Pneumatic Pumping Stations

The *pneumatic pumping station* consists of a well and a control system that which controls the inlet and outlet valve operations and provides pressurized air to force or "push" the wastewater through the system. The exact method of operation depends on the system design. When operating, wastewater in the wet well reaches a predetermined level and activates an automatic valve, which closes the influent line. The tank (wet well) is then pressurized to a predetermined level. When the pressure reaches the predetermined level, the effluent line valve is opened and the waste stream is pushed out through the discharge line.

13.7 PRELIMINARY TREATMENT

The initial stage in the wastewater treatment process (following collection and influent pumping) is *preliminary treatment.* Raw influent entering the treatment plant may contain a wide variety of materials (trash). The purpose of preliminary treatment is to protect plant equipment by removing these materials, which could cause clogs, jams, or excessive wear to plant machinery. In addition, the removal of various materials at the beginning of the treatment process saves valuable space within the treatment plant.

Preliminary treatment may include many different processes, each designed to remove a specific type of material that could potentially be a problem for the treatment process. Processes include wastewater collection, influent pumping, screening, shredding, grit removal, flow measurement, pre-aeration, chemical addition, and flow equalization. The primary processes are shown in Figure 13.1. In this section, we will describe and discuss each of these processes and their importance in the treatment process.

Note: As mentioned, not all treatment plants will include all of the processes shown in Figure 13.1. Specific processes have been included to facilitate discussion of major potential problems with each process and its operation. This information may be important to the wastewater operator.

13.7.1 Screening

The purpose of *screening* is to remove large solids such as rags, cans, rocks, branches, leaves, and roots from the flow before it moves on to downstream processes.

Note: Typically, a treatment plant will remove anywhere from 0.5 to 12 ft^3 of screenings for each million gallons of influent received.

A *bar screen* traps debris as wastewater influent passes through. Typically, a bar screen consists of a series of parallel, evenly spaced bars or a perforated screen placed in a channel (see Figure 13.2). The waste stream passes through the screen, and the large solids (*screenings*) are trapped on the bars for removal.

Note: The screenings must be removed frequently enough to prevent accumulation, which can block the screen and cause the water level in front of the screen to rise.

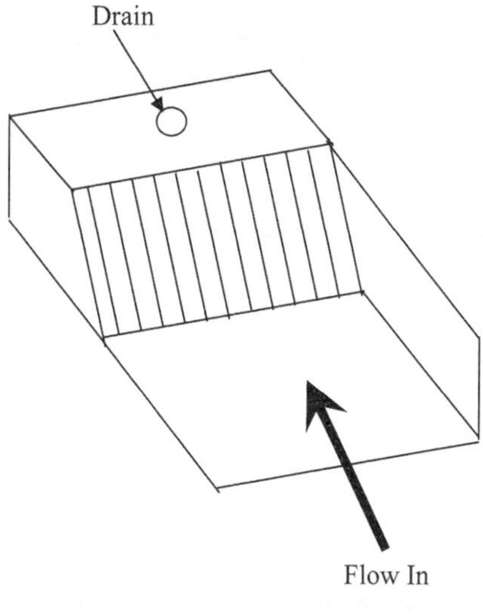

FIGURE 13.2 Basic bar screen.

The bar screen may have coarse (2–4-inch openings) or fine (0.75- to 2.0-inch openings). The bar screen can be cleaned manually, with bars or screens set at a 30-degree angle for easier solids removal (see Figure 13.2), or mechanically, with bars at a 45 to 60-degree angle for better mechanical cleaning.

The screening method employed depends on the design of the plant, the amount of solids expected, and whether the screen is for constant or emergency use only.

13.7.2 Shredding

As an alternative to screening, *shredding* can be used to reduce solids to a size that can enter the plant without causing mechanical problems or clogging. Shredding processes include comminution (comminute means to cut up) and barminution devices.

13.7.2.1 Comminution

The *comminutor* is the most common shredding device used in wastewater treatment. In this device, all the wastewater flows through the grinder assembly. The grinder consists of a screen or slotted basket, a rotating or oscillating cutter, and a stationary cutter. Solids pass through the screen and are chopped or shredded between the two cutters. The comminutor will not remove solids that are too large to fit through the slots, and it will not remove floating objects. These materials must be removed manually. Maintenance requirements for comminutors include aligning, sharpening, and replacing cutters, and corrective and preventive maintenance performed in accordance with the plant O&M Manual.

13.7.2.2 Barminution

In barminution, the *barminutor* uses a bar screen to collect solids, which are then shredded and passed through the bar screen for removal at a later process. In operation, each device's cutter alignment and sharpness are critical factors. Cutters must be sharpened or replaced, and alignment must be checked in accordance with the manufacturer's recommendations. Solids that are not shredded must be removed daily, stored in closed containers, and disposed of by burial or incineration. Barminutor operational problems are similar to those listed earlier for comminutors. Preventive and corrective maintenance, as well as lubrication, must be performed by qualified personnel and in accordance with the plant's O&M Manual. Due to higher maintenance requirements, the barminutor is less frequently used.

13.7.3 Grit Removal

The purpose of *grit removal* is to remove heavy inorganic solids that could cause excessive mechanical wear. Grit, which is heavier than organic solids, includes sand, gravel, clay, eggshells, coffee grounds, metal filings, seeds, and other similar materials. There are several processes or devices used for grit removal. All of the processes are based on the fact that grit is heavier than the organic solids, which should be kept in suspension for treatment in the following methods: Grit removal may be accomplished in grit chambers or by the centrifugal separation of sludge. Processes use gravity/velocity, aeration, or centrifugal force to separate the solids from the wastewater.

13.7.3.1 Gravity/Velocity-Controlled Grit Removal

Gravity/velocity-controlled grit removal is usually accomplished in a channel or tank where the velocity of the wastewater is controlled to about one foot per second (ideal) so that grit will settle while organic matter remains suspended. As long as the velocity is controlled in the range of 0.7 to 1.4 feet per second (fps), the grit removal will remain effective. Velocity is controlled by the volume of water flowing through the channel, the depth of the water in the channel, the width of the channel, or the cumulative width of channels in operation.

13.7.3.1.1 Aeration

Aerated grit removal systems use aeration to keep the lighter organic solids in suspension while allowing the heavier grit articles to settle out. Aerated grit removal may be manually or mechanically cleaned; however, most systems are mechanically cleaned. During regular operation, the rate of aeration controls actual grit removal. Adjusting the aeration rate produces the desired separation. This requires observation of mixing and aeration, and sampling of fixed suspended solids. The rate of aeration controls actual grit removal. If the rate is too high, all of the solids remain in suspension. If the rate is too low, both grit and organics will settle out. The operator monitors the same conditions as the gravity/velocity-controlled system and must also carefully check the air DS for correct function.

13.7.3.1.2 Centrifugal Force

The *cyclone degritter* uses a rapid spinning motion (centrifugal force) to separate heavy inorganic solids or grit from light organic solids. This unit process is typically used on primary sludge rather than the entire wastewater flow. The critical control factor for the process is the inlet pressure. If the pressure exceeds the recommendations of the manufacturer, the unit will flood, and grit will carry through with the flow. Grit is separated from flow, washed, and discharged directly to a storage container. Grit removal performance is determined by calculating the percent removal for inorganic (fixed) suspended solids. The operator observes the same kinds of conditions listed for the gravity/velocity-controlled and aerated grit removal systems, except for the air DS. Typical problems associated with grit removal include mechanical malfunctions and a rotten egg odor in the grit chamber (hydrogen sulfide formation), which can lead to metal and concrete corrosion problems. A low recovery rate of grit is another typical issue. Bottom scour, over-aerating, or not enough detention time can cause this. When these problems occur, the operator must make the required adjustments or repairs to correct the problem.

13.7.3.1.3 Preaeration

In the *preaeration process* (diffused or mechanical), we aerate wastewater to achieve and maintain an aerobic state (to freshen septic wastes), strip off hydrogen sulfide (to reduce odors and corrosion), agitate solids (to release trapped gases and improve solids separation and settling), and reduce BOD_5. All of this can be accomplished by aerating the wastewater for 10 to 30 minutes. To reduce BOD_5, preaeration must be conducted for 45 to 60 minutes.

13.7.4 Chemical Addition

Chemical addition to the waste stream is done (through either dry chemical metering or solution feed metering) to improve settling, reduce odors, neutralize acids or bases, reduce corrosion, reduce BOD_5, improve solids and grease removal, reduce loading on the plant, add or remove nutrients, add organisms, and/or aid subsequent downstream processes. The specific chemical and amount used depend on the desired result. Chemicals must be added at a point where sufficient mixing will occur to obtain maximum benefit. Chemicals typically used in wastewater treatment include chlorine, peroxide, acids and bases, mineral salts (ferric chloride, alum, etc.), bioadditives, and enzymes.

13.7.5 Equalization

The purpose of *flow equalization* (whether by surge, diurnal, or complete methods) is to reduce or remove the wide swings in flow rates normally associated with wastewater treatment plant loading. It minimizes the impact of storm flows. The process can be designed to prevent flows above the maximum plant design hydraulic capacity, reduce the magnitude of diurnal flow variations, and eliminate flow variations. Flow equalization is accomplished using mixing or aeration equipment, pumps, and flow measurement. Regular operation depends on the purpose and requirements of the flow equalization system. Equalized flows allow the plant to perform at optimum levels by providing stable hydraulic and organic loading. The downside to flow equalization is the additional costs associated with the construction and operation of the flow equalization facilities.

During normal operations, the operator must monitor all mechanical systems involved in flow equalization and watch for mechanical problems, taking appropriate corrective action. The operator also monitors dissolved oxygen levels, the impact of equalization on influent, and water levels in equalization basins, making necessary adjustments.

13.7.6 Aerated Systems

Aerated grit removal systems use aeration to keep the lighter organic solids in suspension while allowing the heavier grit particles to settle out. Aerated grit removal may be manually or mechanically cleaned; however, most systems are mechanically cleaned. In typical operation, the aeration rate is adjusted to produce the desired separation, which requires observation of mixing and aeration and sampling of fixed suspended solids. The rate of aeration controls actual grit removal. If the rate is too high, all of the solids remain in suspension. If the rate is too low, both the grit and the organics will settle out.

13.8 PRIMARY TREATMENT (SEDIMENTATION)

The purpose of primary treatment (primary sedimentation or primary clarification) is to remove settleable organic and floatable solids. Typically, each primary clarification unit can be expected to remove 90%–95% of settleable solids, 40%–60% of TSS, and 25%–35% of BOD_5.

Note: Performance expectations for settling devices used in other areas of plant operation are generally expressed as overall unit performance rather than settling unit performance.

Sedimentation may be used throughout the plant to remove settleable and floatable solids. It is used in primary treatment, secondary treatment, and advanced wastewater treatment processes. In this section, we focus on primary treatment or primary clarification, which uses large basins in which primary settling is achieved under relatively quiescent conditions (see Figure 13.1). Within these basins, mechanical scrapers collect the primary settled solids into a hopper, from which they are pumped to a sludge-processing area. Oil, grease, and other floating materials (scum) are skimmed from the surface. The effluent is discharged over weirs into a collection trough.

In primary sedimentation, wastewater enters a settling tank or basin. Velocity is reduced to approximately 1 ft/min.

Note: Here, velocity is measured in minutes, not seconds, as with the grit channels. A grit channel velocity of 1 ft/sec would be 60 ft/min.

Solids, which are heavier than water, settle to the bottom, while solids that are lighter than water float to the top. Settled solids are removed as sludge, and floating solids are removed as scum. Wastewater leaves the sedimentation tank over an effluent weir and goes on to the next step in treatment. Detention time, temperature, tank design, and condition of the equipment control the efficiency of the process.

- Primary treatment reduces the organic loading on downstream treatment processes by removing a large amount of settleable, suspended, and floatable materials.
- Primary treatment reduces the velocity of the wastewater through a clarifier to approximately 1 to 2 ft/min so that settling and flotation can take place. Slowing the flow enhances the removal of suspended solids in wastewater.
- Primary settling tanks remove floated grease and scum, remove the settled sludge solids, and collect them for pumped transfer to disposal or further treatment.
- Clarifiers used may be rectangular or circular. In rectangular clarifiers, wastewater flows from one end to the other, and the settled sludge is moved to a hopper at one end, either by flights set on parallel chains or by a single bottom scraper set on a traveling bridge. A surface skimmer collects floating material, mostly grease and oil.
- Wastewater in circular tanks typically enters centrally and moves outward. Settled sludge is pushed to a hopper at the bottom of the tank, and a surface skimmer removes floating material.

- Factors affecting primary clarifier performance include:
  - Rate of flow through the clarifier
  - Wastewater characteristics (strength, temperature, amount, and type of industrial waste, and the density, size, and shapes of particles)
  - Performance of pretreatment processes
  - Nature and amount of any wastes recycled to the primary clarifier

13.8.1 Types of Sedimentation Tanks

Sedimentation equipment includes septic tanks, two-story tanks, and plain settling tanks or clarifiers. All three devices may be used for primary treatment, while plain settling tanks are commonly used for secondary or advanced wastewater treatment processes.

13.8.1.1 Septic Tanks

Septic tanks are prefabricated tanks that serve as a combined settling and skimming tank and as an unheated-unmixed anaerobic digester. Septic tanks provide long settling times (six to eight hours or more) but do not separate decomposing solids from the wastewater flow. When the tank becomes full, solids will be discharged with the flow. The process is suitable for small facilities (i.e., schools, motels, and homes), but due to the long detention times and lack of control, it is not suitable for larger applications.

13.8.1.2 Two-Story (Imhoff) Tank

The *two-story or Imhoff tank*, named for German engineer Karl Imhoff (1876–1965), is similar to a septic tank in the removal of settleable solids and the anaerobic digestion of solids. The difference is that the two-story tank consists of a settling compartment where sedimentation is accomplished, a lower compartment where settled solids and digestion take place, and gas vents. Solids removed from the wastewater by settling pass from the settling tank compartment into the digestion compartment through a slot in the bottom of the settling tank compartment. The design of the slot prevents solids from returning to the settling compartment. Solids decompose anaerobically in the digestion section. Gases produced as a result of the solids' decomposition are released through the gas vents running along each side of the settling compartment.

13.8.1.3 Plain Settling Tanks (Clarifiers)

The plain settling tank or clarifier optimizes the settling process. Sludge is removed from the tank for processing in other downstream treatment units. Flow enters the tank, is slowed and distributed evenly across the width and depth of the unit, passes through the unit, and leaves over the effluent weir. Detention time within the primary settling tank is from one to three hours (two-hour average). Sludge removal is accomplished frequently on either a continuous or an intermittent basis. Continuous removal requires additional sludge treatment processes to remove the excess water resulting from the removal of sludge, which contains less than 2%–3% solids. Intermittent sludge removal involves pumping sludge from the tank often enough to prevent solids from surfacing, but not so frequently that solids content drops below 4%–8%.

Scum must be removed from the surface of the settling tank frequently. This is usually a mechanical process but may require manual intervention. The system should be operated frequently enough to prevent excessive buildup and scum carryover but not so often as to cause hydraulic overloading of the scum removal system. Settling tanks require housekeeping and maintenance. Baffles (which prevent floatable solids and scum from leaving the tank), scum troughs, scum collectors, effluent troughs, and effluent weirs require frequent cleaning to avoid heavy biological growth and solids accumulation. Mechanical equipment must be lubricated and maintained as specified in the manufacturer's recommendations or by procedures listed in the plant O&M Manual.

Process control, sampling, and testing are used to evaluate the performance of the settling process. Settleable solids, dissolved oxygen, pH, temperature, TSS, and BOD_5, as well as sludge solids and volatile matter testing, are routinely carried out.

13.8.1.3.1 Effluent From Settling Tanks

Upon completion of screening, degritting, and settling in sedimentation basins, large debris, grit, and many settleable materials have been removed from the waste stream. What is left is referred to as *Primary Effluent.* Usually cloudy and frequently gray in color, primary effluent still contains large amounts of dissolved food and other chemicals (nutrients). These nutrients are treated in the next step in the treatment process (secondary treatment), which is discussed in the next section.

***Note*:** Two of the most essential nutrients left to remove are phosphorus and ammonia. While we want to remove these two nutrients from the waste stream, we do not want to remove too much. Carbonaceous microorganisms in secondary treatment (biological treatment) need both phosphorus and ammonia.

13.9 SECONDARY TREATMENT

The primary purpose of *secondary treatment* (sometimes referred to as biological treatment) is to provide BOD removal beyond what is achievable by primary treatment. There are three commonly used approaches, all of which take advantage of the ability of microorganisms to convert organic wastes (via biological treatment) into stabilized, low-energy compounds. Two of these approaches, the *trickling filter* [and/or its variation, the *rotating biological contactor (RBC)*] and the *activated sludge* process, follow regular primary treatment sequentially. The third approach, *ponds* (oxidation ponds or lagoons), however, can provide equivalent results without preliminary treatment. In this section, we provide a brief overview of the secondary treatment process, followed by a detailed discussion of wastewater treatment ponds (primarily used in smaller treatment plants), trickling filters, and RBCs. We then shift focus to the activated sludge process—the secondary treatment process primarily used in large installations, which is the main focus of the handbook.

Secondary treatment refers to treatment processes that use biological processes to convert dissolved, suspended, and colloidal organic wastes into more stable solids, which can be removed by settling or discharged into the environment without causing harm. What is secondary treatment? As defined by the Clean Water Act (CWA), secondary treatment produces an effluent with no more than 30 mg/l BOD_5 and 30 mg/l TSS.

Note: The CWA also states that ponds and trickling filters will be included in the definition of secondary treatment even if they do not meet the effluent quality requirements continuously.

Most secondary treatment processes decompose solids aerobically, producing carbon dioxide, stable solids, and more organisms. Since solids are created, all of the biological processes must include some form of solids removal (settling tank, filter, etc.). Secondary treatment processes can be separated into two large categories: fixed film systems and suspended growth systems.

Fixed film systems are processes that use biological growth (biomass or slime). When the wastewater and slime are in contact, the organisms remove and oxidize the organic solids. The media may be stone, redwood, synthetic materials, or any other substance that is durable (capable of withstanding weather conditions for many years), provides a large area for slime growth while providing open space for ventilation, and is not toxic to the organisms in the biomass. Fixed film devices include trickling filters and rotating biological contactors (RBCs). *Suspended growth systems* are processes that use biological growth, which is mixed with the wastewater. Typical suspended growth systems consist of various modifications of the activated sludge process.

13.9.1 Treatment Ponds

Wastewater treatment can be accomplished using *ponds* (also known as *lagoons*). Ponds are relatively easy to build and manage; they can accommodate large fluctuations in flow and provide treatment that approaches conventional systems (producing a highly purified effluent) at a much lower cost. It is the cost (the economics) that drives many managers to decide on the pond option. The actual degree of treatment provided depends on the type and the number of ponds used. Ponds can be used as the sole type of treatment, or they can be used in conjunction with other forms of wastewater treatment; that is, other treatment processes followed by a pond or a pond followed by other treatment processes.

Stabilization ponds (also known as treatment ponds) have been used for the treatment of wastewater for over 3,000 years. The first recorded construction of a pond system in the United States was at San Antonio, Texas, in 1901. Today, over 8,000 wastewater treatment ponds are in place, involving more than 50% of the wastewater treatment facilities in the United States (CWNS, 2000). Facultative ponds account for 62%, aerated ponds 25%, anaerobic ponds 0.04%, and total containment 12% of the pond treatment systems. They treat a variety of wastewaters from domestic wastewater to complex industrial wastes, and they function under a wide range of weather conditions, from tropical to arctic. Ponds can be used alone or in combination with other wastewater treatment processes. As our understanding of pond operating mechanisms has increased, different types of ponds have been developed for application to specific kinds of wastewaters under local environmental conditions. This handbook focuses on municipal wastewater treatment pond systems.

DID YOU KNOW?

A pond can be considered an "Attractive Nuisance." The term "Attractive Nuisance" is a legal expression that implies the pond could be attractive to potential users, such as duck hunters, fisherman, or playing children. Since ponds have fairly steep slopes, the potential for someone falling in and drowning is a significant legal problem that must be a concern. It is important that adequate fencing and signage be provided.

While the tendency in the United States has been for smaller communities to build ponds, in other parts of the world, including Australia, New Zealand, Mexico, Latin America, Asia, and Africa, treatment ponds have been built for large cities. As a result, our understanding of the biological, biochemical, physical, and climatic factors that interact to transform the organic compounds, nutrients, and pathogenic organisms found in sewage into less harmful chemicals and unviable microorganisms (i.e., dead or sterile) has grown since 1983. A wealth of experience has been built up by civil, sanitary, and environmental engineers, operators, public works managers, and public health and environmental agencies as they have gained more experience with these systems. While some of this information makes its way into technical journals and textbooks, there is a need for a less formal presentation of the subject for those working in the field every day (USEPA, 2011).

Ponds are designed to enhance the growth of natural ecosystems that are anaerobic (providing conditions for bacteria that grow in the absence of oxygen (O_2) environments), aerobic (promoting the growth of O_2-producing and/or requiring organisms, such as algae and bacteria), or facultative, which is a combination of the two. Ponds are managed to reduce concentrations of BOD, TSS, and coliform numbers (fecal or total) to meet water quality requirements.

Ponds can be classified based on their location in the system, by the type of waste they receive, and by the primary biological process occurring in the pond. First, we look at the types of ponds according to their location and the type of waste they receive: Raw Sewage Stabilization ponds (see Figure 13.3), oxidation ponds, and polishing ponds. Then, in the following section, we look at ponds classified by the type of processes occurring within the pond: *Aerobic Ponds*, *Anaerobic Ponds*, *Facultative Ponds*, and *Aerated Ponds*.

13.9.2 Trickling Filters

Trickling filters have been used to treat wastewater since the 1890s. It was found that if settled wastewater was passed over rock surfaces, slime grew on the rocks and the

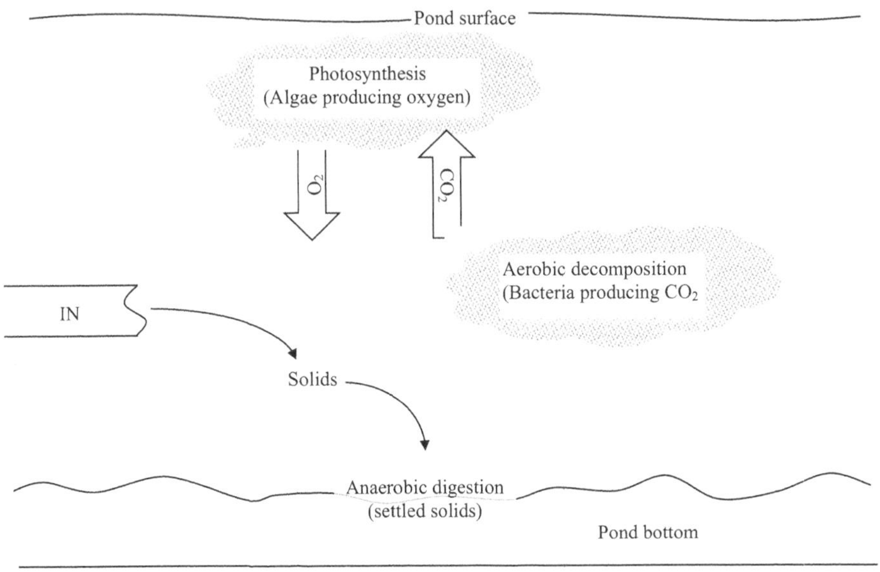

FIGURE 13.3 Stabilization pond processes.

water became cleaner. Today, we still use this principle, but in many installations, instead of stones, we use plastic media. In most wastewater treatment systems, the *trickling filter* follows primary treatment and includes a secondary settling tank or clarifier, as shown in Figure 13.4. Trickling filters are widely used for the treatment of domestic and industrial wastes. The process is a fixed-film biological treatment method designed to remove BOD_5 and suspended solids.

A trickling filter consists of a rotating distribution arm that sprays and evenly distributes liquid wastewater over a circular bed of fist-sized rocks, other coarse materials, or synthetic media (see Figure 13.5). The spaces between the media allow air to circulate easily so that aerobic conditions can be maintained. The spaces also allow wastewater to trickle down through, around, and over the media. A layer of biological slime that absorbs and consumes the wastes trickling through the bed covers the media material. The organisms aerobically decompose the solids, producing more organisms and stable wastes, which either become part of the slime or are discharged back into the wastewater flowing over the media. This slime consists mainly of bacteria, but it may also include algae, protozoa, worms, snails, fungi, and insect larvae. The accumulating slime occasionally sloughs off individual media materials (see Figure 13.6) and is collected at the bottom of the filter, along with the treated wastewater, and passed on to the secondary settling tank, where it is removed. The overall performance of the trickling filter is dependent on hydraulic and organic loading, temperature, and recirculation.

Trickling filters are classed as shown in Table 13.2.

The trickling filter DS is designed to spread wastewater evenly over the surface of the entire media. The most common system is the rotary distributor, which moves

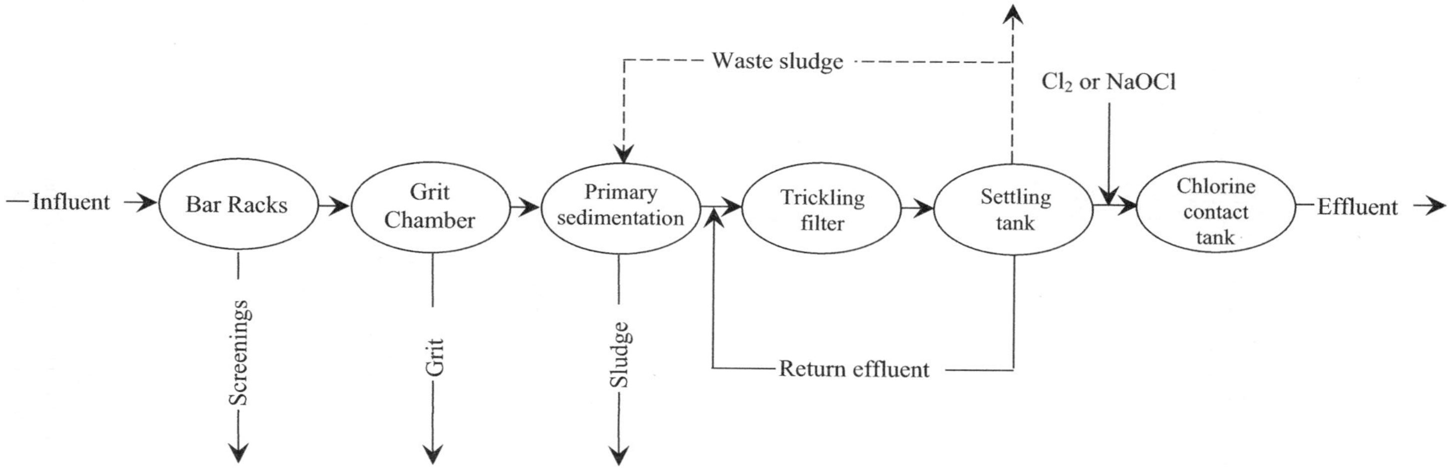

FIGURE 13.4 Simplified flow diagram of a trickling filter used for wastewater treatment.

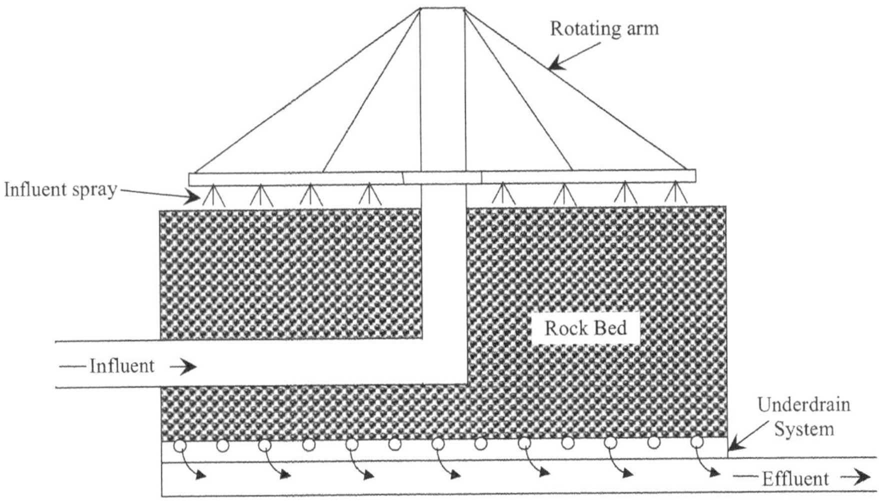

FIGURE 13.5 Schematic of the cross section of a trickling filter.

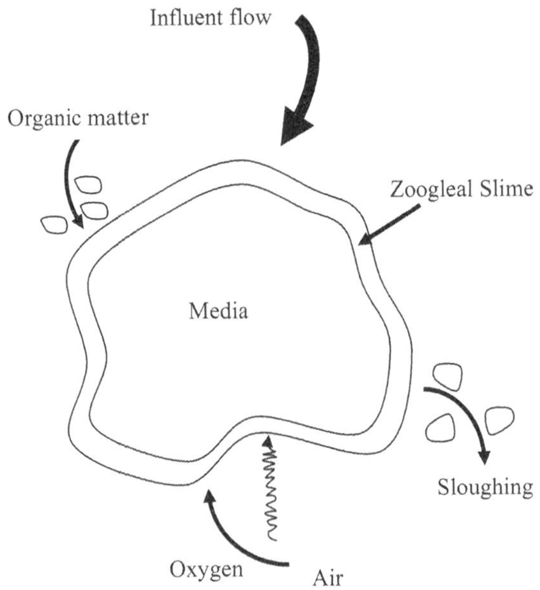

FIGURE 13.6 Filter media showing biological activities that take place on the surface area.

above the surface of the press and sprays the wastewater on the surface. The force of the water leaving the orifices drives the rotary system. The distributor arms usually have small plates below each orifice to spread the wastewater into a fan-shaped DS. The second type of distributor is the fixed nozzle system. In this system, the nozzles

TABLE 13.2
Trickling Filter Classification

Filter Class	Standard	Intermediate	High-Rate	Super High-Rate	Roughing
Hydraulic loading gpd/ft^2	25–90	90–230	230–900	350–2,100	>900
Organic loading BOD per 1,000 ft^3	5–25	15–30	25–300	Up to 300	>300
Sloughing frequency	Seasonal	Varies	Continuous	Continuous	Continuous
Distribution	Rotary	Rotary fixed	Rotary fixed	Rotary	Rotary fixed
Recirculation	No	Usually	Always	Usually	Not usually
Media depth, ft	6–8	6–8	3–8	Up to 40	3–20
Media type	Rock Plastic Wood	Rock Plastic Wood	Rock Plastic Wood	Plastic Plastic Wood	Rock
Nitrification	Yes	Some	Some	Limited	None
Filter flies	Yes	Variable	Variable	Very few	Not usually
BOD removal	80%–85%	50%–70%	65%–80%	65%–85%	40%–65%
TSS removal	80%–85%	50%–70%	65%–80%	65%–85%	40%–65%

are fixed in place above the media and are designed to spray the wastewater over a fixed portion of the media. This system is used frequently with deep-bed synthetic media filters.

Note: Trickling filters that use ordinary rock are usually only about three meters in depth because of structural problems caused by the weight of stones, which also requires the construction of beds that are quite wide. In many applications, beds can be up to 60 feet in diameter. When synthetic media is used, the bed can be much deeper.

No matter which type of media is selected, the primary consideration is that it must be capable of providing the desired film location for the development of the biomass. Depending on the type of media used and the filter classification, the press may be 3 to 20 feet or more in depth.

The underdrains are designed to support the media, collect the wastewater and sloughing, and carry them out of the filter, while also providing ventilation to the filter.

Note: To ensure sufficient airflow to the filter, the underdrains should never be allowed to be more than 50% full of wastewater.

The effluent channel is designed to carry the flow from the trickling filter to the secondary settling tank. The secondary settling tank provides two to four hours of detention time to separate the sloughing materials from the treated wastewater. Design, construction, and operation are similar to those of the primary settling tank. Longer detention times are provided because the sloughing materials are lighter and settle more slowly.

Recirculation pumps and piping are designed to recirculate (and thus improve the performance of the trickling filter or settling tank) a portion of the effluent back to be mixed with the filter influent. When recirculation is used, pumps and metering devices must be provided.

Trickling filters are classified based on hydraulic and organic loading. Moreover, the expected performance and the construction of the trickling filter are determined by the filter classification. Filter classifications include standard rate, intermediate rate, high rate, super high rate (plastic media), and roughing rate types. Standard rate, high rate, and roughing rate are the filter types most commonly used. The *standard rate filter* has a hydraulic loading (gpd/ft^3) of from 25 to 90; a seasonal sloughing frequency; does not employ recirculation; and typically has an 80%–85% BOD_5 removal rate and 80% to 85% TSS removal rate. The *high-rate filter* has a hydraulic loading (gpd/ft^3) of 230 to 900, a continuous sloughing frequency, always employs recirculation, and typically has a 65% to 80% BOD_5 removal rate and a 65% to 80% TSS removal rate. The *roughing filter* operates at a hydraulic loading above 900 gpd/ft^3, sloughs continuously, rarely uses recirculation, and achieves 40%–65% removal for both general contaminants and TSS.

13.9.3 Rotating Biological Contractors (RBCs)

The *RBC* is a biological treatment system (see Figure 13.7) and is a variation of the attached growth idea provided by the trickling filter. Still relying on microorganisms

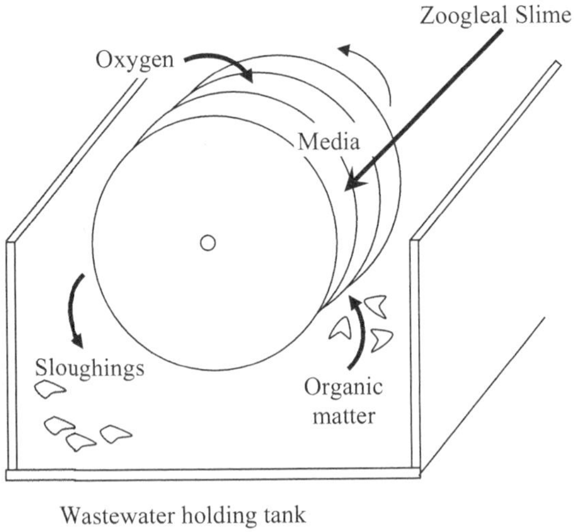

FIGURE 13.7 Rotating biological contactor (RBC) cross section and treatment system.

that grow on the surface of a medium, the RBC is instead a **fixed film** biological treatment device—the basic biological process, however, is similar to that occurring in the trickling filter. An RBC consists of a series of closely spaced (mounted side by side), circular, plastic (synthetic) disks that are typically about 3.5 m in diameter and attached to a rotating horizontal shaft (see Figure 13.8). Approximately 40% of each disk is submerged in a tank containing the wastewater to be treated. As the RBC rotates, the attached biomass film (zoogleal slime) that grows on the surface of the disk moves into and out of the wastewater. While submerged in the wastewater, the microorganisms absorb organics; while they are rotated out of the wastewater, they are supplied with needed oxygen for aerobic decomposition. As the zoogleal slime reenters the wastewater, excess solids and waste products are stripped off the media as sloughings. These sloughings are transported with the wastewater flow to a settling tank for removal.

Modular RBC units are placed in series (see Figure 13.8). Simply because a single contactor is not sufficient to achieve the desired level of treatment, the resulting treatment achieved exceeds conventional secondary treatment. Each contactor is called a stage, and the group is known as a train. Most RBC systems consist of two or more trains with three or more stages in each. The key advantage in using RBCs instead of trickling filters is that RBCs are easier to operate under varying load conditions, since it is easier to keep the solid medium wet at all times. Moreover, the level of nitrification, which an RBC system can achieve, is significant—especially when multiple stages are employed.

The equipment that makes up an RBC includes the rotating biological contactor (the media: either standard or high density), a center shaft, a drive system, a tank, baffles, a housing or cover, and a settling tank. The RBC consists of circular sheets of synthetic material (usually plastic) that are mounted side by side on a shaft. The sheets (media) contain large amounts of surface area for the growth of the biomass. The *center shaft* provides support for the disks of media and must be strong enough to support the weight of the media and the biomass. Experience has indicated that a significant problem has been the collapse of the support shaft. The *drive system* provides the motive force to rotate the disks and shaft. The drive system may be mechanical, air-driven, or a combination of each. When the drive system does not provide uniform movement of the RBC, major operational problems can arise.

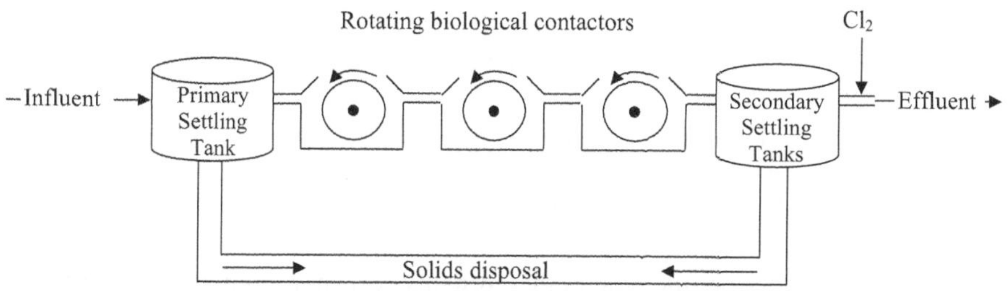

FIGURE 13.8 Rotating biological contactor (RBC) treatment system.

The *tank* holds the wastewater that the RBC rotates in. It should be large enough to permit variation of the liquid depth and detention time. *Baffles* are required to allow proper adjustment of the loading applied to each stage of the RBC process. Adjustments can be made to increase or decrease the submergence of the RBC. RBC stages are usually enclosed in some type of protective structure (*cover*) to prevent loss of biomass due to severe weather changes (snow, rain, temperature, wind, sunlight, etc.). In many instances, this housing greatly restricts access to the RBC. The *settling tank* is provided to remove the sloughing material created by the biological activity and is similar in design to the primary settling tank. The settling tank provides two- to four-hour detention times to allow for the settling of lighter biological solids.

During normal operation, operator vigilance is required to observe the RBC movement, slime color, and appearance. However, if the unit is covered, observations may be limited to the portion of the media that can be viewed through the access door. Slime color and appearance can indicate process condition; for example:

- Gray, shaggy slime growth—indicates normal operation
- Reddish-brown, golden shaggy growth—indicates nitrification
- White chalky appearance—indicates high sulfur concentrations
- No slime—indicates severe temperature or pH changes

Sampling and testing should be conducted daily for dissolved oxygen content and pH. BOD_5 and suspended solids testing should also be accomplished to assess performance.

The RBC typically produces a high-quality effluent with BOD_5 at 85%–95% and suspended solids removal at 85%–95%. The RBC treatment process may also significantly reduce (if designed for this purpose) the levels of organic nitrogen and ammonia nitrogen.

Rotating biological filter operation requires routine observation, process control, sampling and testing, troubleshooting, and process control calculations. Comparison of daily results with expected "normal" ranges is the key to identifying problems and appropriate corrective actions.

Note: If the RBC is covered, observations may be limited to the portion of the media that can be viewed through the access door.

13.10 ACTIVATED SLUDGE

The biological treatment systems discussed to this point [ponds, trickling filters, and rotating biological contactors (RBCs)] have been around for years. The trickling filter, for example, has been around and successfully used since the late 1800s. The problem with ponds, trickling filters, and RBCs is that they are temperature-sensitive, remove less BOD, and trickling filters, for example, cost more to build than the activated sludge systems that were later developed.

Note: Although trickling filters and other systems cost more to build than activated sludge systems, it is important to point out that activated sludge systems cost more to operate due to the energy required for pump and blower operation.

As shown in Figure 13.1, the activated sludge process follows primary settling. The basic components of an activated sludge sewage treatment system include an aeration tank and a secondary basin, settling basin, or clarifier (see Figure 13.9). Primary effluent is mixed with settled solids recycled from the secondary clarifier and is then introduced into the aeration tank. Compressed air is injected continuously into the mixture through porous diffusers located at the bottom of the tank, usually along one side.

Wastewater is fed continuously into an aerated tank, where the microorganisms metabolize and biologically flocculate the organics. Microorganisms (activated sludge) are settled from the aerated mixed liquor under quiescent conditions in the final clarifier and then returned to the aeration tank. Left uncontrolled, the number of organisms would eventually become too great; therefore, some must periodically be removed (wasted). A portion of the concentrated solids from the bottom of the settling tank must be removed from the process (waste activated sludge, or WAS). Clear supernatant from the final settling tank is the plant effluent.

To better understand the discussion of the activated sludge process presented in the following sections, you must understand the terms associated with the process. Some of these terms have been used and defined earlier in the text, but we list them here again to refresh your memory. Please review these terms and remember them. They are used throughout the discussion.

- *Adsorption*—taking in or reception of one substance into the body of another by molecular or chemical actions and distribution throughout the absorber.
- *Activated*—to speed up the reaction. When applied to sludge, it means that many aerobic bacteria and other microorganisms are in the sludge particles.
- *Activated sludge*—a floc or solid formed by the microorganisms. It includes organisms, accumulated food materials, and waste products from the aerobic decomposition process.

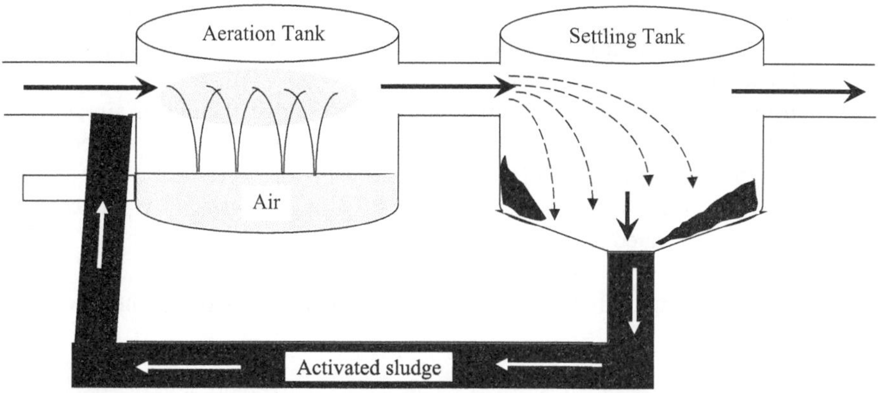

FIGURE 13.9 The activated sludge process.

- *Activated sludge process*—a biological wastewater treatment process in which a mixture of influent and activated sludge is agitated and aerated. The activated sludge is subsequently separated from the treated mixed liquor by sedimentation and is returned to the process as needed. The treated wastewater overflows the weir of the settling tank, where separation from the sludge takes place.
- *Adsorption*—the adherence of dissolved, colloidal, or finely divided solids to the surface of solid bodies when they are brought into contact.
- *Aeration*—mixing air and a liquid by one of the following methods: spraying the liquid in the air, diffusing air into the liquid, or agitating the liquid to promote surface adsorption of air.
- *Aerobic*—a condition in which "free" or dissolved oxygen is present in the aquatic environment. Aerobic organisms must be in the presence of dissolved oxygen to be active.
- *Bacteria*—single-cell plants that play a vital role in the stabilization of organic waste.
- *BOD*—a measure of the amount of food available to the microorganisms in a particular waste. It is measured by the amount of dissolved oxygen used up during a specific period (usually five days, expressed as BOD_5).
- *Biodegradable*—from "degrade" (to wear away or break down chemically) and "bio" (by living organisms). Put it all together, and you have a "substance, usually organic, which can be decomposed by biological action."
- *Bulking*—a problem in activated sludge plants that results in poor settleability of sludge particles.
- *Coning*—a condition that may be established in a sludge hopper during sludge withdrawal, when part of the sludge moves toward the outlet while the remainder tends to stay in place. Development of a cone or channel of moving liquids surrounded by relatively stationary sludge.
- *Decomposition*—generally, in waste treatment, decomposition refers to the changing of waste matter into simpler, more stable forms that will not harm the receiving stream.
- *Diffuser*—a porous plate or tube through which air is forced and divided into tiny bubbles for distribution in liquids. Commonly made of carborundum, aluminum, or silica sand.
- *Diffused air aeration*—a diffused air activated sludge plant takes air, compresses it, then discharges the air below the water surface to the aerator through some type of air diffusion device.
- *Dissolved oxygen*—atmospheric oxygen dissolved in water or wastewater, usually abbreviated as DO.

Note: The typical required DO for a well-operated activated sludge plant is between 2.0 and 2.5 mg/l.

- *Facultative*—facultative bacteria can use either molecular (dissolved) oxygen or oxygen obtained from food materials. In other words, facultative bacteria can live under aerobic or anaerobic conditions.
- *Filamentous bacteria*—organisms that grow in a thread or filamentous form.
- *Food-to-microorganisms ratio*—a process control calculation used to evaluate the amount of food (BOD or COD) available per pound of mixed liquor volatile suspended solids.
- *Fungi*—multicellular aerobic organisms.
- *Gould sludge age*—a process control calculation used to evaluate the amount of influent suspended solids available per pound of mixed liquor suspended solids.
- *Mean cell residence time* (MCRT)—the average length of time a mixed liquor suspended solids particle remains in the activated sludge process. This is usually written as MCRT and may also be referred to as *sludge retention rate* (STR).
- *Mixed liquor*—the combination of return activated sludge and wastewater (either influent or primary effluent) that flows into the aeration tank.
- *Mixed liquor suspended solids (MLSS)*—the suspended solids concentration of the mixed liquor. Many references use this concentration to represent the number of organisms in the activated sludge process. This is usually written as MLSS.
- *Mixed liquor volatile suspended solids* (MLVSS)—the organic matter in the mixed liquor suspended solids. This can also be used to represent the number of organisms in the process. This is usually written as MLVSS.
- *Nematodes*—microscopic worms that may appear in biological waste treatment systems.
- *Nutrients*—substances required to support plant organisms. Primary nutrients are carbon, hydrogen, oxygen, sulfur, nitrogen, and phosphorus.
- *Protozoa*—single-cell animals that can be easily observed under the microscope at a magnification of 100×. Bacteria and algae are prime sources of food for advanced forms of protozoa.
- *Return activated sludge*—the solids returned from the settling tank to the head of the aeration tank. This is typically written as RAS.
- *Rising sludge*—rising sludge occurs in the secondary clarifiers or activated sludge plant when the sludge settles to the bottom of the clarifier, is compacted, and then rises to the surface in a relatively short time.
- *Rotifers*—multicellular animals with flexible bodies and cilia near their mouths used to attract food. Bacteria and algae are their primary source of food.
- *Secondary treatment*—a wastewater treatment process used to convert dissolved or suspended materials into a form that can be removed.
- *Settleability*—a process control test used to evaluate the settling characteristics of the activated sludge. Readings taken at 30 to 60 minutes are used to calculate the settled sludge volume (SSV) and the sludge volume index (SVI).

- *Settled sludge volume*—the volume of ml/l (or percent) occupied by an activated sludge sample after 30 or 60 minutes of settling. It is usually written as SSV with a subscript to indicate the time of the reading used for calculation (SSV_{30} or SSV_{60}).
- *Shock load*—the arrival at a plant of a waste toxic to organisms, in sufficient quantity or strength to cause operating problems, such as odor or sloughing off of the growth of slime on the trickling filter media. Organic overloads can also cause a shock load.
- *Sludge volume index*—a process control calculation used to evaluate the settling quality of the activated sludge. It requires the results of the SSV_{30} and mixed liquor suspended solids tests to calculate.
- *Solids*—material in the solid state

  Dissolved—solids present in solution; solids that will pass through a glass fiber filter.

  Fixed—solids, also known as the inorganic solids, are the solids that are left after a sample is ignited at 550° C for 15 minutes.

  Floatable solids—solids that will float to the surface of still water, sewage, or other liquids. They are usually composed of greased particles, oils, light plastic material, etc. and also called *scum*.

  Non-settleable—finely divided suspended solids that will not sink to the bottom in still water, sewage, or other liquids within a reasonable period, usually two hours. Non-settleable solids are also known as colloidal solids.

  Suspended—solids that will not pass through a glass fiber filter.

  Total—solids in water, sewage, or other liquids; it includes suspended solids and dissolved solids.

  Volatile—organic solids. Measured as the solids lost on ignition of the dry solids at 550°C.

  WAS—solids being removed from the activated sludge process. This is typically written as WAS.

The equipment requirements for the activated sludge process are more complex than those of other processes discussed. Equipment includes an *aeration tank, an aeration system, a settling tank, a return sludge system, and a waste sludge system.* These are discussed as follows:

13.10.1 Aeration Tank

The *aeration tank* is designed to provide the required detention time (depending on the specific modification) and ensure that the activated sludge and the influent wastewater are thoroughly mixed. Tank design typically attempts to ensure that no dead spots are created.

13.10.2 Aeration

Aeration can be mechanical or diffused. Mechanical aeration systems use agitators or mixers to mix air and mixed liquor. Some systems use a sparge ring to release air directly into the mixer. Diffused aeration systems use pressurized air released through diffusers near the bottom of the tank. Efficiency is directly related to the size of the air bubbles produced. Fine bubble systems have higher efficiency. The diffused air system has a blower to produce large volumes of low-pressure air (5 to 10 psi), air lines to carry the air to the aeration tank, and headers to distribute the air to the diffusers, which release the air into the wastewater.

13.10.3 Settling Tank

Activated sludge systems are equipped with plain *settling tanks* designed to provide two to four hours of hydraulic detention time.

13.10.4 Return Sludge

The return sludge system includes pumps, a timer or variable speed drive to regulate pump delivery, and a flow measurement device to determine actual flow rates.

13.10.5 Waste Sludge

In some cases, the waste-activated *sludge* withdrawal is accomplished by adjusting valves on the return system. When a separate system is used, it includes pump(s), a timer or variable speed drive, and a flow measurement device.

The activated sludge process is a treatment technique in which wastewater and reused biological sludge, full of living microorganisms, are mixed and aerated. The biological solids are then separated from the treated wastewater in a clarifier and are returned to the aeration process or wasted. The microorganisms are mixed thoroughly with the incoming organic material, and they grow and reproduce by using the organic material as food. As they grow and are combined with air, the individual organisms cling together (flocculate). Once flocculated, they more readily settle in the secondary clarifiers.

The wastewater being treated flows continuously into an aeration tank where air is injected to mix the wastewater with the returned activated sludge and to supply the oxygen needed by the microbes to live and feed on the organics. Aeration can be provided by injection through air diffusers in the bottom of the tank or by mechanical aerators located at the surface. The mixture of activated sludge and wastewater in the aeration tank is called the "mixed liquor." The mixed liquor flows to a secondary clarifier where the activated sludge is allowed to settle.

The activated sludge is constantly growing, and more is produced than can be returned for use in the aeration basin. Some of this sludge must, therefore, be wasted to a sludge handling system for treatment and disposal. The volume of sludge returned to the aeration basins usually is 40% to 60% of the wastewater flow. The rest is wasted.

One of the most common types of modified activated sludge processes, which provides biological treatment for the removal of biodegradable organic waste under aerobic conditions, is the extended aeration process called the package plant. Package plants are pre-manufactured treatment facilities used to treat wastewater in small communities or on individual properties. According to manufacturers, package plants can be designed to treat flows as low as 0.002 MGD or as high as 0.5 MGD, although they more commonly treat flows between 0.01 and 0.25 MGD (Metcalf & Eddy, 2003).

In operation, air may be supplied to the extended aeration package plant by mechanical or diffused aeration to provide the oxygen required to sustain the aerobic biological process. Mixing must be supplied by aeration or mechanical means to maintain the microbial organisms in contact with the dissolved organics. In addition, the pH must be controlled to optimize the biological process, and essential nutrients must be present to facilitate biological growth and the continuation of biological degradation. Wastewater enters the treatment system and is typically screened immediately to remove large suspended, settleable, or floating solids that could interfere with or damage equipment downstream in the process. Wastewater may then pass through a grinder to reduce large particles that are not captured in the screening process. If the plant requires the flow to be regulated, the effluent will then flow into equalization basins, which regulate peak wastewater flow rates. Wastewater then enters the aeration chamber, where it is mixed and oxygen is provided to the microorganisms. The mixed liquor then flows to a clarifier or settling chamber, where most microorganisms settle to the bottom of the clarifier, and a portion is pumped back to the incoming wastewater at the beginning of the plant. This returned material is the retune activated sludge (RAS). The material that is not returned, the WAS, is removed for treatment and disposal. The clarified wastewater then flows over a weir and into a collection channel before being diverted to the disinfection system (USEPA, 2000)

Typically located in small municipalities, suburban subdivisions, apartment complexes, highway rest areas, trailer parks, small institutions, and other sites where flow rates are below 0.1 MGD, extended aeration package plants consist of a steel tank that is compartmentalized into flow equalization, aeration, clarification, disinfection, and aerated sludge holding/digestion segments. Extended aeration systems are typically manufactured to treat wastewater flow rates between 0.002 and 0.1 MGD. Using concrete tanks may be preferable for large sizes (Sloan Equipment, 1999).

Extended aeration plants are usually started up using “seed sludge” from another sewage plant. It may take as many as two to four weeks from the time it is seeded for the plant to stabilize (Sloan Equipment, 1999). These systems are also suitable for areas requiring nitrification.

Key internal components of extended aeration treatment package plants consist of the following: transfer pumps to move wastewater between the equalization and aeration zones; a bar screen and/or grinder to decrease the size of large solids; an aeration system consisting of blowers and diffusers for the equalization, aeration, and sludge holding zones; transfer pumps to move wastewater between the equalization and aeration zones; an airlift pump for returning sludge; a skimmer and effluent weir for the clarifier; and UV, liquid hypochlorite, or tablet modules used in the disinfection zone. Blowers and the control panel containing switches, lights, and motor starters are typically attached to either the top or one side of the package plant (Sloan Equipment, 1999).

13.11 DISINFECTION OF WASTEWATER

Like drinking water, liquid wastewater effluent is disinfected. Unlike drinking water, wastewater effluent is not directly disinfected (direct end-of-pipe connection) to protect a drinking water supply, but instead is treated to protect public health in general. This is particularly important when the secondary effluent is discharged into a body of water used for swimming or as a water supply for downstream water supplies. In the treatment of water for human consumption, treated water is typically chlorinated (although ozonation is also currently being applied in many cases). Chlorination is the preferred method of disinfection in potable water supplies because of chlorine's unique ability to provide a residual. This chlorine residual is essential because when treated water leaves the waterworks facility and enters the DS, the possibility of contamination increases. The residual works to disinfect water right up to the consumer's tap continuously.

In this section, we will discuss basic chlorination and dichlorination. Additionally, we describe ultraviolet (UV) irradiation, ozonation, bromine, chlorine, and no disinfection. Keep in mind that much of the chlorination material presented in the following is similar to the chlorination information presented earlier.

13.11.1 Chlorine Disinfection

Chlorination for disinfection, as shown in Figure 13.1, follows all other steps in conventional wastewater treatment. The purpose of chlorination is to reduce the population of organisms in the wastewater to levels low enough to ensure that pathogenic microorganisms will not be present in sufficient quantities to cause disease when discharged.

Note: Chlorine gas is heavier than air (vapor density of 2.5). Therefore, exhaust from a chlorinator room should be taken from floor level.

Note: The safest action to take in the event of a major chlorine container leak is to call the fire department.

Note: The use of chlorination in critical waters, such as natural trout streams, is not a standard practice. This practice is strictly prohibited because chlorine and its by-products (i.e., chloramines) are highly toxic to aquatic organisms.

Remember that there are several terms used in the discussion of disinfection by chlorination. Because it is essential for the operator to be familiar with these terms, we repeat key terms.

- *Chlorine*—a strong oxidizing agent that has strong disinfecting capabilities. It is a yellow-green gas that is highly corrosive and toxic to humans in extremely low concentrations in the air.
- *Contact Time*—the length of time the disinfecting agent and the wastewater remain in contact.

- *Demand*—the chemical reactions that must be satisfied before a residual or excess chemical will appear.
- *Disinfection*—refers to the selective destruction of disease-causing organisms. Not all organisms are destroyed during the process. This differentiates disinfection from sterilization, which is the destruction of all microorganisms.
- *Dose*—the amount of chemical being added in milligrams/liter.
- *Feed Rate*—the amount of chemical being added in pounds per day.
- *Residual*—the amount of disinfecting chemical remaining after the demand has been satisfied.
- *Sterilization*—the removal of all living organisms.

13.11.1.1 Chlorine Facts

- Elemental chlorine (Cl_2—gaseous) is a yellow-green gas, 2.5 times heavier than air.
- The most common use of chlorine in wastewater treatment is for disinfection. Other uses include odor control and activated sludge bulking control. Chlorination takes place before the discharge of the final effluent to the receiving waters (see Figure 13.1).
- Chlorine may also be used for nitrogen removal through a process called *breakpoint chlorination*. For nitrogen removal, enough chlorine is added to the wastewater to convert all the ammonium nitrogen gas. To do this, approximately 10 mg/l of chlorine must be added for every 1 mg/l of ammonium nitrogen in the wastewater.
- For disinfection, chlorine is fed manually or automatically into a chlorine contact tank or basin, where it comes into contact with flowing wastewater for at least 30 minutes to destroy disease-causing microorganisms (pathogens) found in treated wastewater.
- Chlorine may be applied as a gas, a solid, or in liquid hypochlorite form.
- Chlorine is a very reactive substance. It has the potential to react with many different chemicals (including ammonia), as well as with organic matter. When chlorine is added to wastewater, several reactions occur:
  (1) Chlorine will react with any reducing agent (i.e., sulfide, nitrite, iron, and thiosulfate) present in wastewater. These reactions are known as the *chlorine demand*. The chlorine used for these reactions is not available for disinfection.
  (2) Chlorine also reacts with organic compounds and ammonia compounds to form chlor-organics and chloramines. Chloramines are part of the group of chlorine compounds that have disinfecting properties and are detected in the chlorine residual test.
  (3) After all of the chlorine demands are met, the addition of more chlorine will produce free residual chlorine. Producing free residual chlorine in wastewater requires substantial additions of chlorine.

13.11.1.2 Hypochlorite Facts

Hypochlorite, though there are some minor hazards associated with its use (skin irritation, nose irritation, and burning eyes), is relatively safe to work with. It is usually available in dry form as a white powder, pellet, or tablet, or in liquid form. It can be added directly using a dry chemical feeder or dissolved and fed as a solution.

Note: In most wastewater treatment systems, disinfection is accomplished using a combined residual.

Chlorine is a very reactive substance. It is added to wastewater to meet all chemical demands, reacting with certain chemicals (such as sulfide, sulfite, and ferrous iron). Once these initial chemical demands are met, chlorine will react with substances like ammonia to form chloramines and other substances, which, although not as effective as chlorine, have disinfecting capability. This creates a combined residual, measurable using residual chlorine test methods. If additional chlorine is added, free residual chlorine can be produced. Due to the chemicals typically found in wastewater, chlorine residuals are usually combined rather than free residuals. Control of the disinfection process is usually based on maintaining a total residual chlorine of at least 1.0 mg/l for a contact time of at least 30 minutes at design flow.

Note: Residual level, contact time, and effluent quality affect disinfection. Failure to maintain the desired residual levels for the required contact time will result in lower efficiency and an increased probability that disease organisms will be discharged.

On the basis of water quality standards, the **total residual limitations** on chlorine are:

- Fresh Water—Less than 11 ppb total residual chlorine.
- Estuaries—Less than 7.5 ppb for halogen-produced oxidants.
- Endangered Species—Use of chlorine is prohibited.

Depending on the form of hypochlorite selected for use, special equipment, which will control the addition of hypochlorite to the wastewater, is required. Liquid forms require the use of metering pumps, which can deliver varying flows of hypochlorite solution. Dry chemicals require the use of a feed system designed to provide variable doses of the form used. The tablet form of hypochlorite involves the use of a tablet chlorinator, explicitly intended to give the desired dose of chlorine. The hypochlorite solution or dry feed systems dispense the hypochlorite, which is then mixed with the flow. The treated wastewater then enters the contact tank to provide the required contact time.

Because of the potential hazards associated with the use of chlorine, the equipment requirements are significantly greater than those associated with hypochlorite use. The system most widely used is a *solution feed system*. In this system, chlorine is removed from the container at a flow rate controlled by a variable orifice. Water moving through the chlorine injector creates a vacuum, which draws the chlorine gas to the injector and mixes it with the water. The chlorine gas reacts with the water to form hypochlorous and hydrochloric acid. The solution is then piped to the chlorine contact tank and dispersed into the wastewater through a diffuser. Larger facilities may withdraw the liquid form of chlorine and use evaporators (heaters) to convert it to the gas form. Small facilities will usually draw the gas form of chlorine from the cylinder. As gas is withdrawn, liquid will be converted to gas. This requires heat energy and may result in the chlorine line freezing if the withdrawal rate exceeds the available energy levels.

In either type of system, regular operation requires adjusting feed rates to ensure the necessary residual levels are maintained. This usually involves chlorine residual testing and adjustment based on the test results. Other activities include removal of accumulated solids from the contact tank, collection of bacteriological samples to evaluate process performance, and maintenance of safety equipment (respirator-air pack, safety lines, etc.). Hypochlorite operation may also include make-up solution (solution feed systems), adding powder or pellets to the dry chemical feeder, or tablets to the tablet chlorinator.

Chlorine operations include adjustment of chlorinator feed rates, inspection of mechanical equipment, testing for leaks using an ammonia swab (white smoke means leaks), changing containers (requires more than one person for safety), and adjusting the injector water feed rate when needed. Chlorination requires routine testing of plant effluent for total residual chlorine and may also require collection and analysis of samples to determine the fecal coliform concentration in the effluent.

13.11.2 Ultraviolet Irradiation

Although ultraviolet (UV) disinfection was recognized as a method for achieving disinfection in the late nineteenth century, its application virtually disappeared with the evolution of chlorination technologies. However, in recent years, there has been a resurgence in its use in the wastewater field, mainly as a consequence of concern for the discharge of toxic chlorine residual. Even more recently, UV has gained more attention because of the strict new regulations on chlorine use imposed by both OSHA and USEPA. Due to this relatively recent increased regulatory pressure, many facilities are actively engaged in substituting chlorine with other disinfection alternatives. Moreover, UV technology itself has made many improvements, which now make UV attractive as a disinfection alternative. Ultraviolet light has excellent germicidal properties and is very effective in destroying microorganisms. It is used in hospitals, biological testing facilities, and many other similar locations. In wastewater treatment, the plant effluent is exposed to ultraviolet light of a specified wavelength and intensity for a specified contact period. The effectiveness of the process depends on the following factors:

- UV light intensity
- Contact time
- Wastewater quality (turbidity)
- For any treatment plant, disinfection success is directly related to the concentration of colloidal and particulate constituents in the wastewater.

The Achilles' heel of UV for disinfecting wastewater is turbidity. If the wastewater quality is poor, the ultraviolet light will be unable to penetrate the solids, and the effectiveness of the process decreases dramatically. For this reason, many states limit the use of UV disinfection to facilities that can reasonably be expected to produce an effluent containing ≤30 mg/l of BOD_5 and TSS.

The main components of a UV disinfection system are mercury arc lamps, a reactor, and ballasts. The source of UV radiation is either the low-pressure or medium-pressure mercury arc lamp with low or high intensities. Note that in the operation of UV systems, UV lamps must be readily available when replacements are required. The ideal lamps have a lifespan of at least 7,500 hours and emit minimal ozone or hydrogen peroxide. The lamps must also meet technical specifications for intensity, output, and arc length. If the UV light tubes are submerged in the waste stream, they must be protected inside quartz tubes, which not only protect the lights but also make cleaning and replacement easier.

Contact tanks must be used with UV disinfection. The UV lights must be designed with banks in a horizontal position, either parallel or perpendicular to the flow, or with banks of lights placed in a vertical position perpendicular to the flow.

Note: The contact tank must provide, at a minimum, 10-second exposure time.

It was stated earlier that turbidity problems have been a problem with using UV in wastewater treatment, and this is the case. However, if turbidity is its Achilles' heel, then the need for increased maintenance (compared to other disinfection alternatives) is the toe of the same foot. UV maintenance requires that the tubes be cleaned regularly or as needed. In addition, periodic acid washing is also necessary to remove chemical buildup.

Routine monitoring is required. Monitoring is necessary to check for bulb burnout, buildup of solids on quartz tubes, and UV light intensity.

Note: UV light is highly hazardous to the eyes. Never enter an area where UV lights are in operation without proper eye protection. Never look directly into the ultraviolet light.

This eliminates the need to generate, handle, transport, or store toxic/hazardous or corrosive chemicals.

13.11.3 Ozonation

Ozone is a potent oxidizing gas that reacts with most organic and many inorganic molecules. It is produced when oxygen molecules separate, collide with other oxygen atoms, and form a molecule consisting of three oxygen atoms. For high-quality effluents, ozone is a very effective disinfectant. Current regulations for domestic treatment systems limit the use of ozonation to filtered effluents unless the system's effectiveness can be demonstrated before installation.

Note: Effluent quality is the key performance factor for ozonation.

For ozonation of wastewater, the facility must have the capability to generate pure oxygen along with an ozone generator. A contact tank with a ≥10-minute contact time at the design average daily flow is required. Off-gas monitoring for process control is also needed. In addition, safety equipment capable of monitoring ozone layers in the atmosphere and a ventilation system to prevent ozone levels from exceeding 0.1 ppm are required.

The actual operation of the ozonation process consists of monitoring and adjusting the ozone generator, as well as monitoring the control system to maintain the required ozone concentration in the off-gas. The process must also be evaluated periodically using biological testing to assess its effectiveness.

Note: Ozone is an extremely toxic substance. Concentrations in air should not exceed 0.1 ppm. It also has the potential to create an explosive atmosphere. Sufficient ventilation and purging capabilities should be provided.

Note: Ozone has certain advantages over chlorine for disinfection of wastewater: (1) Ozone increases DO in the effluent; (2) ozone has a briefer contact time; (3) ozone has no undesirable effects on marine organisms; and (4) ozone decreases turbidity and odor.

13.11.4 Bromine Chloride

Bromine chloride is a mixture of bromine and chlorine. It forms hydrocarbons and hydrochloric acid when mixed with water. Bromine chloride is an excellent disinfectant that reacts quickly and usually does not produce any long-term residuals.

Note: Bromine chloride is a highly corrosive compound in the presence of low concentrations of moisture.

The reactions occurring when bromine chloride is added to the wastewater are similar to those occurring when chlorine is added. The significant difference is the production of bromamine compounds rather than chloramines. The bromamine compounds are excellent disinfectants but are less stable and dissipate quickly. In most cases, the bromamines decay into other, less toxic compounds rapidly and are undetectable in the plant effluent. The factors that affect performance are similar to those affecting the performance of the chlorine disinfection process. Effluent quality, contact time, etc. have a direct impact on the performance of the process.

13.11.5 No Disinfection

In a minimal number of cases, treated wastewater discharges without disinfection are permitted. These are approved on a case-by-case basis. Each request must be evaluated based on the point of discharge, the quality of the discharge, the potential for human contact, and many other factors.

13.12 ADVANCED WASTEWATER TREATMENT

Advanced wastewater treatment is defined as the method(s) and/or process(es) that remove more contaminants (suspended and dissolved substances) from wastewater than are taken out by conventional biological treatment. Put another way, advanced wastewater treatment is the application of a process or system that follows secondary treatment or that includes phosphorus removal or nitrification in conventional secondary treatment.

Advanced wastewater treatment is used to augment conventional secondary treatment because secondary treatment typically removes only between 85% and 95% of the BOD and TSS in raw sanitary sewage. Generally, this leaves 30 mg/l or less of BOD and TSS in the secondary effluent. This level of BOD and TSS in secondary effluent may still cause water-quality violations, preventing the plant from meeting permit requirements. Thus, advanced wastewater treatment is often used to remove additional pollutants from treated wastewater.

In addition to meeting or exceeding the requirements of water-quality standards, treatment facilities use advanced wastewater treatment for other reasons as well. For example, sometimes, conventional secondary wastewater treatment is not sufficient to protect the aquatic environment. In a stream, for example, when periodic flow events occur, the stream may not provide the necessary level of effluent dilution to maintain the required dissolved oxygen (DO) levels for aquatic organism survival.

Secondary treatment has other limitations. It does not significantly reduce the effluent concentration of nitrogen and phosphorus (essential plant nutrients) in sewage. An overabundance of these nutrients can overstimulate plant and algae growth, leading to water quality problems. For example, if discharged into lakes, these nutrients contribute to algal blooms and accelerated eutrophication (lake aging). Also, the nitrogen in the sewage effluent may be present mostly in the form of ammonia compounds. If present in high enough concentrations, ammonia compounds can be toxic to aquatic organisms. Yet another problem with these compounds is that they exert a *nitrogenous* oxygen demand in the receiving water, as they convert to nitrates. This process is called nitrification.

Note: The term *tertiary treatment* is commonly used as a synonym for advanced wastewater treatment. However, these two terms do not have precisely the same meaning. Tertiary suggests a third step that is applied after primary and secondary treatment.

Advanced wastewater treatment can remove more than 99% of the pollutants from raw sewage and can produce an effluent of almost potable (drinking) water quality. However, obviously, advanced treatment is not cost-free. The cost of advanced therapy, for operation and maintenance as well as for retrofit of present conventional processes, is very high (sometimes doubling the cost of secondary treatment). Therefore, a plan to install advanced treatment technology calls for careful study as the benefit-to-cost ratio is not always big enough to justify the additional expense.

Even considering the expense, application of some form of advanced treatment is not uncommon. These treatment processes can be physical, chemical, or biological.

The specific method used is based on the purpose of the treatment and the quality of the effluent desired.

13.12.1 Chemical Treatment

The purpose of chemical treatment is to remove the following:

- BOD
- TSS
- Phosphorus
- Heavy metals
- Other substances that can be chemically converted to a settleable solid

Chemical treatment is often accomplished as an "add-on" to existing treatment systems or using separate facilities specifically designed for chemical addition. In each case, the basic process necessary to achieve the desired results remains the same:

- Chemicals are thoroughly mixed with the wastewater.
- The chemical reactions occur from solids (coagulation).
- The solids are mixed to increase particle size (flocculation).
- After settling and/or filtration (separation), the solids are removed.

The specific chemical used depends on the pollutant to be removed and the characteristics of the wastewater. Chemicals may include the following:

- Lime
- Alum (aluminum sulfate)
- Aluminum salts
- Ferric or ferrous salts
- Polymers
- Bioadditives

13.12.2 Microscreening

Microscreening (also called *microstraining*) is an advanced treatment process used to reduce suspended solids. The microscreens are composed of specially woven steel wire fabric mounted around the perimeter of a large revolving drum. The steel wire cloth acts as a fine screen, with openings as small as 20 μm (or millionths of a meter)—small enough to remove microscopic organisms and debris. The rotating drum is partially submerged in the secondary effluent, which must flow into the drum and then outward through the microscreen. As the drum rotates, captured solids are carried to the top, where a high-velocity water spray flushes them into a hopper or backwash tray mounted on the hollow axle of the drum. Backwash solids are recycled to the plant influent for treatment. These units have found the most excellent application in the treatment of industrial waters and final polishing filtration of wastewater effluents. Expected performance for suspended solids removal is 95%–99%, but the typical suspended-solids removal achieved with these units is about 55%. The normal range is from 10% to 80%.

According to Metcalf & Eddy (2003), the functional design of the microscreen unit involves the following considerations: (1) the characterization of the suspended solids with respect to the concentration and degree of flocculation; (2) the selection of unit design parameter values that will not only ensure capacity to meet maximum hydraulic loadings with critical solids characteristics but also provide desired design performance over the expected range of hydraulic and solids loadings; (3) the provision of backwash and cleaning facilities to maintain the capacity of the screen.

13.12.3 Filtration

The purpose of *filtration* processes used in advanced treatment is to remove suspended solids. The specific operations associated with a filtration system are dependent on the equipment used. Wastewater flows to a filter (gravity or pressurized). The filter can be single, dual, or multimedia. Wastewater flows through the media, which removes solids. The solids remain in the filter. Backwashing the filter as needed removes trapped solids. Backwash solids are returned to the plant for treatment. Processes typically remove 95%–99% of the suspended matter.

13.12.4 Membrane Bioreactors (MBRs)

The most commonly used technologies for performing secondary treatment of municipal wastewater rely on microorganisms suspended in the wastewater to treat it. These are good technologies that work well in many situations. Still, they have several drawbacks, including the difficulty of growing the right types of microorganisms and the physical requirement of a large plant site. In the past 20 years, the use of microfiltration membrane bioreactors (MBRs) has overcome many of the limitations of conventional systems. The advantage of MBRs is the combination of suspended growth biological treatment with solids removal via filtration. The membranes can be designed for and operated in small spaces and with high removal efficiency of contaminants such as nitrogen, phosphorus, bacteria, biological oxygen demand, and TSS. In effect, the membranes can be designed to replace secondary clarifiers and sand filters in a typical activated sludge treatment system.

The bottom line: Membrane filtration allows a higher biomass concentration to be maintained, thereby allowing smaller bioreactors to be used (USEPA, 2007a).

13.12.5 Biological Nitrification

Biological nitrification is a microbial process in which reduced nitrogen compounds, primarily ammonia, are sequentially oxidized to nitrite and nitrate. Nitrification occurs naturally in drinking water or during secondary disinfection when ammonia is added to form chloramines.

13.12.6 Biological Denitrification

Biological denitrification removes nitrogen from the wastewater. When bacteria come in contact with a nitrified element in the absence of oxygen, they reduce the nitrates to nitrogen gas, which escapes the wastewater. The denitrification process can be done in either an anoxic-activated sludge system (suspended growth) or a column system (fixed growth). The denitrification process can remove up to 85% or more of nitrogen. After effective biological treatment, a small amount of oxygen-demanding material is left in the wastewater when it reaches the denitrification process. The denitrification reaction will only occur if an oxygen demand source exists when no dissolved oxygen is present in the wastewater. An oxygen demand source is usually added to reduce the nitrates quickly. The most common demand source added is soluble BOD or methanol. Approximately 3 mg/l of methanol is added for every 1 mg/l of nitrate-nitrogen. Suspended growth denitrification reactors are mixed mechanically, but only enough to keep the biomass from settling without adding unwanted oxygen. Submerged filters of different types of media may also be used to provide denitrification. A fine media downflow filter is sometimes used to give both denitrification and effluent filtration. A fluidized sand bed where wastewater flows upward through a medium of sand or activated carbon at a rate that fluidizes the bed may also be used. Denitrification bacteria grow in the media.

13.12.7 Carbon Adsorption

The primary purpose of *carbon adsorption* used in advanced treatment processes is the removal of refractory organic compounds (non-BOD_5) and soluble organic material that are difficult to eliminate by biological or physical/chemical treatment. In the carbon adsorption process, wastewater passes through a container filled with either carbon powder or carbon slurry. Organics adsorb onto the carbon (i.e., organic molecules are attracted to the activated carbon surface and are held there) with sufficient contact time. A carbon system usually has several columns or basins used as contactors. Most contact chambers are either open concrete gravity-type systems or steel pressure containers applicable to either upflow or downflow operation. With use, carbon loses its adsorptive capacity. The carbon must then be regenerated or replaced with fresh carbon. As head loss develops in carbon contactors, they are backwashed with clean effluent in much the same way the effluent filters are backwashed. Carbon used for adsorption may be in a granular or powdered form.

Note: Powdered carbon is too fine for use in columns and is usually added to the wastewater and then later removed by coagulation and settling.

13.12.8 Land Application

The application of secondary effluent onto a land surface can provide an effective alternative to the expensive and complicated advanced treatment methods discussed previously and the biological nutrient removal (BNR) system discussed later. A high-quality polished effluent (i.e., effluent with high levels of TSS, BOD, phosphorus, and nitrogen compounds, as well as refractory organics) can be obtained by the natural processes that occur as the effluent flows over the vegetated ground surface and percolates through the soil. Limitations are involved with the land application of wastewater effluent. For example, the process needs large land areas. Soil type and climate are also critical factors in controlling the design and feasibility of a land treatment process.

13.12.9 Biological Nutrient Removal (BNR)

Nitrogen and phosphorus are the primary causes of cultural eutrophication (i.e., nutrient enrichment due to human activities) in surface waters. The most recognizable manifestations of this eutrophication are algal blooms that occur during the summer. Chronic symptoms of over-enrichment include low dissolved oxygen, fish kills, murky water, and depletion of desirable flora and fauna. In addition, the increase in algae and turbidity increases the need to chlorinate drinking water, which, in turn, leads to higher levels of disinfection by-products that have been shown to increase the risk of cancer (USEPA, 2007c). Excessive amounts of nutrients can also stimulate the activity of microbes, such as *Pfiesteria*, which may be harmful to human health (USEPA, 2007d).

Approximately 25% of all water body impairments are due to nutrient-related causes (e.g., nutrients, oxygen depletion, algal growth, ammonia, harmful algal blooms, biological integrity, and turbidity) (USEPA, 2007d). In efforts to reduce the number of nutrient impairments, many point source discharges have received more stringent effluent limits for nitrogen and phosphorus. To achieve these new, lower effluent limits, facilities have begun to look beyond traditional treatment technologies.

Recent experience has reinforced the concept that biological nutrient removal (BNR) systems are reliable and effective in removing nitrogen and phosphorus. The process is based on the principle that, under specific conditions, microorganisms will remove more phosphorus and nitrogen than is required for biological activity; thus, treatment can be accomplished without the use of chemicals. Not having to use and therefore having to purchase chemicals to remove nitrogen and phosphorus potentially has numerous cost-benefit implications. In addition, because chemicals are not required to be used, chemical waste products are not produced, reducing the need to handle and dispose of waste. Several patented processes are available for this purpose. Performance depends on the biological activity and the process employed.

13.12.10 Enhanced Biological Nutrient Removal (EBNR)

Removing phosphorus from wastewater in secondary treatment processes has evolved into innovative *enhanced biological nutrient removal* (EBNR) technologies. An ENBR treatment process promotes the production of phosphorus-accumulating organisms, which utilize more phosphorus in their metabolic processes than a conventional secondary biological treatment process (USEPA, 2007b). The average total phosphorus concentration in raw domestic wastewater is usually between 6 and 8 mg/l, and the total phosphorus concentration in municipal wastewater after conventional secondary treatment is routinely reduced to 3 or 4 mg/l. Incorporating

EBR into the secondary treatment system can often reduce total phosphorus concentrations to 0.3 mg/l or less. Facilities using EBNR have significantly reduced the amount of phosphorus that needs to be removed through subsequent chemical addition and tertiary filtration processes. This has improved the efficiency of the tertiary process and significantly reduced the costs of chemicals used to remove phosphorus. Facilities using EBNR reported that their chemical dosing was cut in half after EBNR was installed to remove phosphorus (USEPA, 2007b).

Treatment provided by these WWTPs also removes other pollutants that commonly affect water quality to very low levels (USEPA, 2007b). BOD and TSS are routinely less than 2 mg/l, and fecal coliform bacteria are less than 10 fcu/100 ml. Turbidity of the final effluent is very low, which allows for effective disinfection using ultraviolet light, rather than chlorination. Recent studies report that wastewater treatment plants using EBNR also significantly reduce the amount of pharmaceuticals and personal healthcare products from municipal wastewater, compared to the removal accomplished by conventional secondary treatment.

13.13 SOLIDS (SLUDGE/BIOSOLIDS) HANDLING

The wastewater treatment unit processes described up to this point remove solids and BOD from the waste stream before the liquid effluent is discharged to its receiving waters. What remains to be disposed of is a mixture of solids and wastes, called *process residuals*, which are more commonly referred to as *sludge* or *biosolids*.

Note: *Sludge* is the commonly accepted name for wastewater solids. However, if wastewater sludge is used for beneficial reuse (e.g., as a soil amendment or fertilizer), it is commonly called *biosolids*.

The most costly and complex aspect of wastewater treatment can be the collection, processing, and disposal of sludge. This is the case because the quantity of sludge produced may be as high as 2% of the original volume of wastewater, depending somewhat on the treatment process being used.

Because sludge can contain as much as 97% water content, and because the cost of disposal is related to the volume of sludge being processed, one of the primary purposes or goals (along with stabilizing it so it is no longer objectionable or environmentally damaging) of sludge treatment is to separate as much water from the solids as possible. Sludge treatment methods may be designed to accomplish both of these purposes.

Note: Sludge treatment methods are generally divided into three major categories: thickening, stabilization, and dewatering. Many of these processes include complex sludge treatment methods (i.e., heat treatment, vacuum filtration, and incineration).

When we speak of *sludge* or *biosolids*, we are speaking of the same substance or material; each is defined as the suspended solids removed from wastewater during sedimentation and then concentrated for further treatment, disposal, or reuse. The difference between the terms *sludge* and *biosolids* is determined by the way they are managed.

Note: The task of disposing of, treating, or reusing wastewater solids is called *sludge* or *biosolids management.*

Sludge is typically seen as wastewater solids that are disposed of. Biosolids are the same substance managed for reuse—commonly called beneficial reuse (e.g., for land application as a soil amendment, such as biosolids compost). Note that even as wastewater treatment standards have become more stringent because of increasing environmental regulations, the volume of wastewater sludge has increased. Also, note that before sludge can be disposed of or reused, it requires some form of treatment to reduce its volume, to stabilize it, and to inactivate pathogenic organisms.

Sludge forms initially as a 3%–7% suspension of solids, and with each person typically generating about four gallons of sludge per week, the total quantity generated each day, week, month, and year is significant. Due to the volume and nature of the material, sludge management is a significant factor in the design and operation of all water pollution control plants.

Note: Wastewater solids treatment, handling, and disposal account for more than half of the total costs in a typical secondary treatment plant.

13.13.1 Sources of Sludge

Wastewater sludge is generated in primary, secondary, and chemical treatment processes. In primary treatment, the solids that float or settle are removed. The floatable material makes up a portion of the solid waste known as scum. Scum is not normally considered sludge; however, it should be disposed of in an environmentally sound way. The settleable material that collects on the bottom of the clarifier is known as *primary sludge.* Primary sludge can also be referred to as raw sludge because it has not undergone decomposition. Raw primary sludge from a typical domestic facility is quite objectionable and has a high percentage of water, two characteristics that make handling difficult.

Solids not removed in the primary clarifier are carried out of the primary unit. These solids are known as *colloidal suspended solids.* The secondary treatment system (i.e., trickling filter, activated sludge) is designed to change those colloidal solids into settleable solids that can be removed. Once in a settleable form, these solids are removed in the secondary clarifier. The sludge at the bottom of the secondary clarifier is called *secondary sludge.* Secondary sludges are light and fluffy and more difficult to process than primary sludges. In short, secondary sludges do not de-water well.

The addition of chemicals and various organic and inorganic substances before sedimentation and clarification may increase the solids capture and reduce the amount of solids lost in the effluent. This *chemical addition* results in the formation of heavier solids, which trap the colloidal solids or convert dissolved solids to settleable solids. The resultant solids are known as *chemical sludges.* As chemical usage increases, so does the quantity of sludge that must be handled and disposed of. Chemical sludges can be very difficult to process; they do not de-water well and contain lower percentages of solids.

The composition and characteristics of sewage sludge vary widely and can change considerably with time. Notwithstanding these facts, the basic components of wastewater sludge remain the same. The only variations occur in the quantity of the various components as the type of sludge and the process from which it originated change. The main element of all sludges is *water*. Before treatment, most sludges contain 95%–99+% water (see Table 13.3). This high-water content makes sludge handling and processing extremely costly in terms of both money and time. Sludge handling may represent up to 40% of the capital cost and 50% of the operation cost of a treatment plant. Therefore, the importance of optimal design for handling and disposal of sludge cannot be overstated. The water content of the sludge is present in various forms. Some forms can be removed by different sludge treatment processes, providing flexibility in selecting the optimum sludge treatment and disposal method.

The various forms of water and their approximate percentages for a typical activated sludge are shown in Table 13.4. The forms of water associated with sludges include the following:

- *Free water*—water that is not attached to sludge solids in any way. This can be removed by simple gravitational settling.
- *Floc water*—water that is trapped within the floc and travels with it. Its removal is possible by mechanical dewatering.
- *Capillary water*—water that adheres to the individual particles and can be squeezed out of shape and compacted.
- *Particle water*—water that is chemically bound to individual particles and can't be removed without inclination.

Table 13.3 lists typical water content in sludge.

From a public health point of view, the second and probably more critical component of sludge is the *solid matter*. Representing from 1% to 8% of the total mixture, these solids are volatile. Wastewater solids can be classified into two categories based on their origin: organic and inorganic. *Organic solids* in wastewater, simply put, are

TABLE 13.3
Typical Water Content of Sludges

Water Treatment Process	% Moisture of Sludge	Generated lb Water/lb Sludge Solids
Primary sedimentation	95	19
Trickling filter		
Humus—low rate	93	13.3
Humus—high rate	97	32.3
Activated sludge	99	99

Source: USEPA's *Operational Manual: Sludge Handling and Conditioning*. EPA-430/9-78-002, 1978. Washington, DC

TABLE 13.4
Distribution of Water in an Activated Sludge

Water Type	% Volume
Free water	75
Floc water	20
Capillary water	2
Particle water	2.5
Solids	0.5
Total	**100**

Source: USEPA's *Operations Manual, Sludge Handling and Conditioning. EPA-430/9-78-002, 1978.* Washington, DC.

materials that were at one time alive and that will burn or volatilize at 550°C after 15 minutes in a muffle furnace. The percentage of organic material within the sludge will determine its stability.

The inorganic material within sludge will determine its stability. The *inorganic solids* are those solids that were never alive and will not burn or volatilize at 550°C after 15 minutes in a muffle furnace. Inorganic solids are generally not subject to breakdown by biological action and are considered stable. Certain inorganic solids, however, can create problems when related to the environment, for example, heavy metals such as copper, lead, zinc, mercury, and others. These can be extremely harmful if discharged.

Organic solids may be subject to biological decomposition in either an aerobic or anaerobic environment. Decomposition of organic matter (with its production of objectionable by-products) and the possibility of toxic organic solids within the sludge compound the problems of sludge disposal.

The pathogens in domestic sewage are primarily associated with insoluble solids. Primary wastewater treatment processes concentrate these solids into sewage sludge, so untreated or raw primary sewage sludges have higher quantities of pathogens than the incoming wastewater. Biological wastewater treatment processes such as lagoons, trickling filters, and activated sludge treatment may substantially reduce the number of pathogens in sewage (USEPA, 1989). These processes may also reduce the number of pathogens in sewage sludge by creating adverse conditions for pathogen survival.

Nevertheless, the resulting biological sewage sludges may still contain sufficient levels of pathogens to pose a public health and environmental concern. Moreover, insects, birds, rodents, and domestic animals may transport sewage sludge and pathogens from sewage sludge to humans and animals. Vectors are attracted to sewage sludge as a food source, and reducing the attraction of vectors to sewage sludge to prevent the spread of pathogens is a focus of current regulations. Sludge-borne pathogens and vector attraction are discussed in the following section.

A pathogen is an organism capable of causing disease. Pathogens infect humans through several different pathways, including ingestion, inhalation, and dermal contact. The infective dose, or the number of pathogenic organisms to which a human must be exposed to become infected, varies depending on the organism and the health status of the exposed individual.

Pathogens that propagate in the enteric or urinary system in humans and are discharged in feces or urine pose the most significant risk to public health regarding the use and disposal of sewage sludge. Pathogens are also found in the urinary and enteric systems of other animals and may propagate in non-enteric settings.

As mentioned earlier, the four major types of human pathogenic (disease-causing) organisms (bacteria, viruses, protozoa, and helminths) all may be present in domestic sewage. The actual species and quantity of pathogens present in the domestic wastewater from a particular municipality (and the sewage sludge produced when treating the domestic wastewater) depend on the health status of the local community and may vary substantially at different times. The level of pathogens present in treated sewage sludge (biosolids) also depends on the reductions achieved by the wastewater and sewage sludge treatment processes.

If improperly treated sewage sludge were illegally applied to land or placed on a surface disposal site, humans and animals could be exposed to pathogens directly by coming into contact with sewage sludge or indirectly by consuming drinking water or food contaminated by sewage sludge pathogens. Insects, birds, rodents, and even farm workers could contribute to these exposure routes by transporting sewage sludge and sewage sludge pathogens away from the site.

13.13.2 Sludge Thickening

The solids content of primary, activated, trickling-filter, or even mixed sludge (i.e., primary plus activated sludge) varies considerably, depending on the characteristics of the sludge. Note that the sludge removal and pumping facilities, as well as the method of operation, also affect the solids content. *Sludge thickening* (or *concentration*) is a unit process used to increase the solids content of the sludge by removing a portion of the liquid fraction. By increasing the solids content, a more economical treatment of the sludge can be achieved. Sludge thickening processes include the following:

- Gravity thickeners
- Flotation thickeners
- Solids concentrators

13.13.2.1 Gravity Thickening

Gravity thickening is the most effective part of the primary sludge. In operation, solids are withdrawn from primary treatment (and sometimes secondary treatment) and pumped to the thickener. The solids build up in the thickener and form a solid blanket on the bottom. The weight of the blanket compresses the solids on the bottom and "squeezes" the water out. By adjusting the blanket thickness, the percent solids in the underflow (solids withdrawn from the bottom of the thickener) can be increased or decreased. The supernatant (clear water), which rises to the surface, is returned to the wastewater flow for treatment. Daily operations of the thickening process include pumping, observation, sampling and testing, process control calculations, maintenance, and housekeeping.

Note: The equipment employed for thickening depends on the specific thickening processes used.

Equipment used for gravity thickening consists of a thickening tank, which is similar in design to the settling tank used in primary treatment. Generally, the tank is circular and provides equipment for continuous solids collection. The collector mechanism uses heavier construction than that in a settling tank because the solids being moved are more concentrated. The gravity thickener pumping facilities (i.e., pump and flow measurement) are used for the withdrawal of thickened solids.

Solids concentrations achieved by gravity thickening are typically 8%–10% solids from primary underflow, 2%–4% solids from WAS, 7%–9% solids from trickling filter residuals, and 4%–9% from combined primary and secondary residuals. The performance of gravity thickening processes depends on various factors, which include the following:

- Type of sludge
- Condition of influent sludge
- Temperature
- Blanket depth
- Solids loading
- Hydraulic loading
- Solids retention time
- Hydraulic detention time

13.13.2.2 Flotation Thickening

Flotation thickening is most efficiently used for waste sludges from suspended-growth biological treatment processes, such as the activated sludge process. In operation, recycled water from the flotation thickener is aerated under pressure. During this process, the water absorbs more air than it would under normal pressure. The recycled flow, along with any chemical additives (if used), is mixed with the flow. When the mixture enters the flotation thickener, the excess air is released in the form of fine bubbles. These bubbles become attached to the solids and lift them toward the surface. The accumulation of solids on the surface is called the **float cake**. As more solids are added to the bottom of the float cake, it becomes thicker, and water drains from the upper levels of the cake. The solids are then moved up an inclined plane by a scraper and discharged. The supernatant leaves the tank below the surface of the float solids and is recycled or returned to the waste stream for treatment. Typically, flotation thickener performance is 3%–5% solids for WAS with polymer addition and 2%–4% solids without polymer addition.

The flotation thickening process requires pressurized air, a vessel for mixing the air with all or part of the process residual flow, a tank for the flotation process to occur, solids collector mechanisms to remove the float cake (solids) from the top of the tank, and accumulated heavy solids from the bottom of the tank. Since the process typically requires chemicals to be added to improve separation, chemical mixing equipment, storage tanks, and metering equipment to dispense the chemicals at the desired dose are needed. The performance of the dissolved air-thickening process depends on various factors:

- Bubble size
- Solids loading
- Sludge characteristics
- Chemical selection
- Chemical dose

13.13.2.3 Solids Concentrators

Solids concentrators (belt thickeners) usually consist of a mixing tank, chemical storage, metering equipment, and a moving porous belt. In operation, the residual flow process is chemically treated and then spread evenly over the surface of the moving porous belt. As the flow is carried down the belt (similar to a conveyor belt), the solids are mechanically turned or agitated, and water drains through the belt. This process is primarily used in facilities where space is limited.

13.13.3 Sludge Stabilization

The purpose of sludge stabilization is to reduce volume, stabilize the organic matter, and eliminate pathogenic organisms to permit reuse or disposal. The equipment required for stabilization depends on the specific process used. Sludge stabilization processes include the following:

- Aerobic digestion
- Anaerobic digestion
- Composting
- Lime stabilization
- Wet air oxidation (heat treatment)
- Chemical oxidation (chlorine oxidation)
- Incineration

13.13.3.1 Aerobic Digestion

Equipment used for *aerobic digestion* consists of an aeration tank (digester), which is similar in design to the aeration tank used for the activated sludge process. Either diffused or mechanical aeration equipment is necessary to maintain the aerobic conditions in the tank. Solids and supernatant removal equipment are also required. In operation, process residuals (sludge) are added to the digester and aerated to maintain a dissolved oxygen (D.O.) concentration of 1.0 mg/l. Aeration also ensures that the tank contents are well mixed. Generally, aeration continues for approximately 20 days of retention time. Periodically, aeration is stopped, and the solids are allowed to settle. Sludge and the clear liquid supernatant are withdrawn as needed to provide more room in the digester. When no additional volume is available, mixing is stopped for 12–24 hours before solids

TABLE 13.5
Aerobic Digester Normal Operating Levels

Parameter	Normal Levels
Detention time, days	10–20
Volatile solids loading, lbs/ft^3/day	0.1–0.3
D.O., mg/l	1.0
pH	5.9–7.7
Volatile solids reduction	40%–50%

are withdrawn for disposal. Process control testing should include alkalinity, pH, % solids, % volatile solids for influent sludge, supernatant, digested sludge, and digester contents. Normal operating levels for an aerobic digester are listed in Table 13.5.

A typical operational problem associated with an aerobic digester is pH control. When the pH drops, for example, this may indicate irregular biological activity or low influent alkalinity. This problem is corrected by adding alkalinity (lime, bicarbonate, etc.).

13.13.3.2 Anaerobic Digestion

Anaerobic digestion is the traditional method of sludge stabilization. This process uses bacteria that grow without oxygen, works more slowly than aerobic digestion, but converts only a small fraction of waste into new bacterial cells. Instead, most of the organics are converted into carbon dioxide and methane gas.

Note: In an anaerobic digester, the entrance of air should be prevented because of the potential for air to mix with the gas produced in the digester, creating an explosive mixture.

Equipment used in anaerobic digestion includes a sealed digestion tank with either a fixed or a floating cover, heating and mixing equipment, gas storage tanks, solids and supernatant withdrawal equipment, and safety equipment (e.g., vacuum relief, pressure relief, flame traps, explosion-proof electrical equipment).

In operation, process residual (thickened or unthickened sludge) is pumped into the sealed digester. The organic matter digests anaerobically through a two-stage process. Sugars, starches, and carbohydrates are converted to volatile acids, carbon dioxide, and hydrogen sulfide. The volatile acids are then converted to methane gas. This operation can occur in a single tank (single stage) or two tanks (two stages). In a single-stage system, supernatant and/or digested solids must be removed whenever flow is added. In a two-stage operation, solids and liquids from the first stage flow into the second stage each time fresh solids are added. Supernatant is withdrawn from the second stage to provide additional treatment space. Periodically, solids are withdrawn for dewatering or disposal. The methane gas produced in the process may be used for many plant activities.

TABLE 13.6
Anaerobic Digester—Sludge Parameters

Raw Sludge Solids	Impact
<4% solids	Loss of alkalinity
	Decreased sludge retention time
	Increased heating requirements
	Decreased volatile acid: alk ratio
4%–8% solids	Normal operation
>8% solids	Poor mixing
	Organic overloading
	Decreased volatile acid: alk ratio

Note: The primary purpose of a secondary digester is to allow for solid separation.

Multiple operational factors influence the functioning of the anaerobic digester. For example, % volatile matter in raw sludge, digester temperature, mixing, volatile acids/alkalinity ratio, feed rate, % solids in raw sludge, and pH are all essential operational parameters that the operator must monitor.

Along with being able to recognize normal and abnormal anaerobic digester performance parameters, wastewater operators must also know and understand normal operating procedures. Standard operating procedures include sludge additions, supernatant withdrawal, sludge withdrawal, pH control, temperature control, mixing, and safety requirements. Essential performance parameters are listed in Table 13.6.

13.13.4 Other Sludge Stabilization Processes

In addition to aerobic and anaerobic digestion, other sludge stabilization processes include composting, lime stabilization, wet air oxidation, and chemical (chlorine) oxidation. These other stabilization processes are briefly described in this section.

13.13.4.1 Composting

The purpose of composting sludge is to stabilize the organic matter, reduce volume, and eliminate pathogenic organisms. In a *composting operation*, dewatered solids are usually mixed with a bulking agent (i.e., hardwood chips) and stored until biological stabilization occurs. The composting mixture is ventilated during storage to provide sufficient oxygen for oxidation and to prevent odors. After the solids are stabilized, they are separated from the bulking agent. The composted solids are then stored for curing and applied to farmlands or other beneficial uses. Expected performance of the composting operation for both percent volatile matter reduction and percent moisture reduction ranges from 40% to 60%.

13.13.4.1.1 Aerated Static Pile (ASP)

Three methods of composting wastewater biosolids are standard. Each technique involves mixing dewatered wastewater solids with a bulking agent to provide carbon and increase porosity. The resulting mixture is piled or placed in a vessel where microbial activity causes the temperatures of the mixture to rise during the "active composting" period. The specific temperatures that must be achieved and maintained for successful composting vary based on the method and use of the end product. After active composting, the material is cured and distributed. Again, there are three commonly employed composting methods, but we only described the aerated static pile (ASP) method because it is widely used. For an in-depth treatment of the other two methods, windrow and in-vessel, we refer you to F.R. Spellman's *Wastewater Biosolids to Compost* (1997). Boca Raton, FL: CRC Press.

DID YOU KNOW?

Aeration is a critical process control parameter in the aerated static pile composting system. Air is required to supply oxygen for the biological degradation of organic solids in the biosolids and wood chips. Aeration is also needed for the removal of heat generated by the biological activity in the compost pile and excess moisture from the compost mix. Fans are used to ensure that sufficient quantities of air are supplied to meet composting process requirements and to provide the process control flexibility necessary for optimizing operations.

The ASP Model Composting Facility uses a homogenized mixture of bulking agent (coarse hardwood wood chips) and dewatered biosolids. This mixture is piled by front-end loaders onto a large concrete composting pad where it is mechanically aerated using PVC plastic pipe embedded within the concrete slab. This ventilation procedure is part of the 26-day period of "active" composting when adequate air and oxygen are necessary to support aerobic biological activity in the compost mass and to reduce the heat and moisture content of the compost mixture. Keep in mind that a compost pile without a properly sized air DS can lead to the onset of anaerobic conditions and the appearance of putrefactive odors.

For illustration and discussion purposes, we assume a typical overall composting pad area is approximately 200 feet by 240 feet, consisting of eleven (11) blowers and twenty-four (24) pipe troughs (troffs). Three blowers are 20 hp (2,400 cfm), variable speed drive units capable of operating in either the positive or negative aeration mode. Blowers A, B, and C are each connected to two (2) piping troughs that run the full length of the pad. The two troughs are connected at the opposite end of the composting pad to create an "aeration pipe loop." The other eight (8) blowers are rated at 3 hp, 1,200 cfm, and are arranged one blower per 6 troughs at half-length feeding 200 cfm per trough. These blowers can be operated in the positive or negative aeration mode. The aeration piping within the six (6) pipe troughs consists of perforated

PVC plastic pipes with a six-inch inside diameter and a 1/4-inch wall thickness. The perforation holes/orifices range in size from 7/32 inch to 1/2 inch, with the diameter increasing as the distance from the blower increases.

The variable speed motor drives installed with blowers A, B, and C are controlled by five thermal probes mounted at various depths in the compost pile, and multiple parameters are fed back to the recorder. In contrast, the other eight blowers are constant speed, controlled by a timer that cycles them on and off. To ensure optimal composting operations, it is essential to verify that these thermal probes are calibrated regularly. In the constant speed system, thermal probes are installed, but all readings are taken and recorded manually.

For water and leachate drainage purposes, all aeration piping within the troughs slopes downward with the highest point at the center of the composting pad. Drain caps located at each end of the pipe length are manually removed regularly so that any buildup of debris or moisture will not interfere with the airflow.

Before the piling of the mixture on the composting pad, an 18-inch layer of wood chips is used as a base material. The primary purpose of the wood chip base is to keep the composting mixture clear of the aeration pipes, which reduces clogging of the air distribution openings in the pipes and allows for free air circulation. A secondary benefit is that the wood chips insulate the composting mixture from the pad. The compost pad is like a heat sink, and this insulating barrier improves the uniformity of heat distribution within the composting mix.

13.13.4.2 Lime Stabilization

Lime or alkaline stabilization can achieve the minimum requirements for both Class A (no detectable pathogens) and Class B (a reduced level of pathogens) biosolids concerning pathogens, depending on the amount of alkaline material added and other processes employed. Generally, alkaline stabilization meets the Class B requirements when the pH of the mixture of wastewater solids and alkaline material is at 12 or above after two hours of contact.

Class A requirements can be achieved when the pH of the mixture is maintained at or above 12 for at least 72 hours, with a temperature of 52°C maintained for at least 12 hours during this time. In one process, the mixture is air-dried to over 50% solids after 72 hours of elevated pH. Alternatively, the process may be manipulated to maintain temperatures at or above 70°F for 30 or more minutes while maintaining the pH requirement of 12. This higher temperature can be achieved by overdosing lime (i.e., adding more than is needed to reach a pH of 12), by using a supplemental heat source, or by using a combination of the two and monitoring for fecal coliforms or *Salmonella* sp. It is required before release by the generator for use.

Materials that may be used for alkaline stabilization include hydrated lime, quicklime (calcium oxide), fly ash, lime and cement kiln dust, and carbide lime. Quicklime is commonly used because it has a high heat of hydrolysis (491 British thermal units) and can significantly enhance pathogen destruction. Fly ash, lime kiln dust, or cement kiln dust are often used for alkaline stabilization because of their availability and relatively low cost.

The alkaline-stabilized product is suitable for application in many situations, such as landscaping, agriculture, and mine reclamation. The product serves as a lime substitute, a source of organic matter, and a specialty fertilizer. The addition of alkaline-stabilized biosolids results in more favorable conditions for vegetative growth by improving soil properties such as pH, texture, and water-holding capacity. Appropriate applications depend on the needs of the soil and crops that will be grown and the pathogen classification. For example, a Class B material would not be suitable for blending in a topsoil mix intended for use in home landscaping but is ideal for agriculture, mine reclamation, and landfill cover where the potential for contact is lower and access can be restricted. Class A alkaline-stabilized biosolids are beneficial in agriculture and as a top-soil blend ingredient. Alkaline-stabilized biosolids provide pH adjustment, nutrients, and organic matter, reducing reliance on other fertilizers.

Alkaline-stabilized biosolids are also useful as daily landfill cover. They meet the federal requirement that landfills must be covered with soil or soil-like material at the end of each day (40 CFR 258). In most cases, lime-stabilized biosolids are mixed with other soil to achieve the proper consistency for daily cover.

As previously mentioned, alkaline-stabilized biosolids are excellent for land reclamation in degraded areas, including acid mine spills or mine tailings. Soil conditions at such sites are very unfavorable for vegetative growth, often due to acidic content, lack of nutrients, elevated levels of heavy metals, and poor soil texture. Alkaline-stabilized biosolids help to remedy these problems, making conditions more favorable for plant growth and reducing erosion potential. In addition, once a vegetative cover is established, the quality of mine drainage improves.

13.13.4.3 Thermal Treatment

Thermal treatment, also known as wet air oxidation, involves exposing sludge to elevated temperature and pressure within a sealed reactor vessel. The high temperature and pressure rupture the cell walls of any microorganisms present in the solids and cause chemical oxidation of the organic matter. This process substantially improves dewatering and reduces the volume of material for disposal. It also produces a very high-strength waste, which must be returned to the wastewater treatment system for further treatment.

13.13.4.4 Chlorine Oxidation

Chlorine oxidation also occurs in a closed vessel. In this process, chlorine (100 to 1,000 mg/l) is mixed with a recycled solids flow. The recycled flow and process residual flow are combined in the reactor. The solids and water are separated after leaving the reactor vessel. The water is returned to the wastewater treatment system, and the treated solids are dewatered for disposal. The main advantage of chlorine oxidation is that it can be operated intermittently. The main disadvantage is the production of extremely low pH and high chlorine content in the supernatant.

13.13.4.5 Stabilization Operation

Depending on the stabilization process employed, the operational components may vary. In general, operations include pumping, observations, sampling and testing, process control calculations, maintenance, and housekeeping. Performance of the stabilization process will also vary with the type of process used. Generally, stabilization processes can produce a 40% to 60% reduction in both volatile matter (organic content) and moisture.

13.13.5 Sludge Dewatering

Digested sludge removed from the digester is still mostly liquid. The primary objective of dewatering biosolids is to reduce moisture and consequently volume to a degree that will allow for economic disposal or reuse. Epstein and Alpert (1984) make the point that if the biosolids cake is higher in solids content, it reduces the need for space, fuel, labor, equipment, and size of the receiving facility, for example, a composting facility.

Probably one of the best summarizations of the various reasons why it is essential to dewater biosolids is given by Metcalf & Eddy (1991) as follows: (1) The costs of transporting biosolids to the ultimate disposal site are significantly reduced when biosolids volume is reduced; (2) dewatered biosolids allow for easier handling; (3) dewatering biosolids (reduction in moisture content) allows for more efficient incineration; (4) if composting is the beneficial reuse choice, dewatered biosolids decrease the amount and therefore the cost of bulking agents; (5) with the USEPA's new 503 rule, dewatering biosolids may be required to render the biosolids less offensive; and (6) when landfilling is the ultimate disposal option, dewatering biosolids is required to reduce leachate production.

Again, the point being made here is that the importance of adequately dewatering biosolids for proper disposal or reuse can't be overstated.

The unit processes that are most often used for dewatering biosolids are (1) vacuum filtration, (2) pressure filtration, (3) centrifugation, and (4) drying beds. The solids content achievable by various dewatering techniques is shown in Table 13.7. The biosolids cake produced by standard dewatering processes has a consistency similar to dry, crumbly bread pudding (Spellman, 1996). This non-fluid, dewatered, dry, crumbly cake product is easily handled, non-offensive, and can be land-applied manually and by conventional agricultural spreaders (Outwater, 1994).

Dewatering processes are usually divided into natural air drying and mechanical methods (USEPA, 1997). Natural dewatering methods include those methods in which moisture is removed by evaporation and gravity or induced drainage, such as

TABLE 13.7
Solids Content of Dewatered Biosolids

Dewatering Method	Approximate Solids Content, %
Lagoons/ponds	30
Drying beds	40
Filter press	35–45
Vacuum filtration	25
Standard centrifuge	20–25
High G/high solids Centrifuge	25–40

sand beds, biosolids lagoons, paved beds, Phragmites reed beds, vacuum-assisted beds, wedgewater beds, and dewatering through freezing. These natural dewatering methods are less controllable than mechanical dewatering methods but are typically less expensive. Moreover, these natural dewatering methods require less power because they rely on solar energy, gravity, and biological processes as the source of energy for dewatering. Mechanical dewatering processes include pressure filters, vacuum filters, belt filters, and centrifuges. The aforementioned air-drying and mechanical dewatering processes will be discussed in greater detail later in this text.

13.13.5.1 Sand Drying Beds for Dewatering Sludge

Sand beds have been used successfully for years to dewater sludge. Composed of a sand bed (consisting of a gravel base, underdrains, and 8–12 inches of filter-grade sand), drying beds include an inlet pipe, splash pad containment walls, and a system to return filtrate (water) for treatment. In some cases, the sand beds are covered to protect drying solids from the elements.

In operation, solids are pumped to the sand bed and allowed to dry by first draining off excess water through the sand and then by evaporation. This is the simplest and cheapest method for dewatering sludge. Moreover, no special training or expertise is required. However, there is a downside; namely, drying beds require a great deal of manpower to clean beds; they can create odor and insect problems; and they can cause sludge buildup during inclement weather.

According to Metcalf & Eddy (1991), four types of drying are commonly used in dewatering biosolids: (1) sand, (2) paved, (3) artificial media, and (4) vacuum-assisted. In addition to these commonly used dewatering methods, several innovative natural dewatering methods will also be discussed in this section. The innovative natural dewatering methods to be addressed include experimental work on biosolids dewatering via freezing. Moreover, dewatering biosolids with aquatic plants, which has been tested and installed in several sites throughout the United States, is also discussed.

Drying beds are generally used for dewatering well-digested biosolids. Attempting to air-dry raw biosolids is usually unsuccessful and may result in odor and vector control problems. Biosolids drying beds consist of a perforated or open joint drainage system in a support medium, usually gravel, covered with a filter medium, usually sand. Still, they can consist of extruded plastic or wire mesh. Drying beds are generally separated into workable sections by wood, concrete, or other materials. Drying beds may be enclosed or open to the weather. They may rely entirely on natural drainage and evaporation processes or may use a vacuum to assist the operation (both types are discussed in the following sections).

Traditional sand drying beds *are* the oldest biosolids dewatering technique and consist of six to 12 inches of coarse sand underlain by layers of graded gravel ranging from 1/8 to 1/4 inches at the top and 3/4 to 1 1/2 inches at the bottom. The total gravel thickness is typically about 1 ft. Graded natural earth (four to six inches) usually makes up the bottom, with a web of drain tile placed on 20- to 30-ft centers. Sidewalls and partitions between bed sections are usually wooden planks or concrete and extend about 14 inches above the sand surface (McGhee, 1991).

Large open areas of land are required for sand-drying biosolids. For example, it is not unusual to have drying beds that are up to 125+ ft long and from 20 to 35 ft in width. Even at the smallest wastewater treatment plants, it is standard practice to provide at least two drying beds.

The actual dewatering process occurs as a result of two different physical processes: evaporation and drainage. The liquor, which drains off the biosolids, goes to a central sump, which pumps it back to the treatment process to undergo further treatment. The operation is very much affected by the climate. In wet climates, it may be necessary to cover the beds with a translucent material that will allow at least 85% of the sun's ultraviolet radiation to pass through.

Typical loading rates for primary biosolids in dry climates range up to 200 kg/(square meter × year) and from 60 to 125 kg/(square meter × year) for mixtures of primary and waste activated biosolids.

When a drying bed is put into operation, it is generally filled with digested biosolids to a depth ranging from 8 to 12 inches. The actual drying time is climate-sensitive; that is, drying can take from a few weeks to a few months, depending on the climate and the season. After dewatering, the biosolids' solids content will range from about 20% to 35%, and, more importantly, the volume will have been reduced to 85%. Upon completion of the drying process, the dried biosolids are generally removed from the bed with handheld forks or front-end loaders. It is important to note that in the dried biosolids removal process, a small amount of sand is lost, and the bed must be refilled and graded periodically.

The bottom line: dried solids removed from a biosolids drying bed can be either incinerated or landfilled.

Before the digested biosolids are distributed to a biosolids drying bed, it is necessary to have the sand bed prepared for the best drying results. This preparation of the bed should include the following:

1. Remove all the old biosolids as soon as convenient after they have dewatered sufficiently. Well-dewatered biosolids have a cracked, spongy appearance when squeezed and can be easily forked. The moisture content should be no more than about 70%.
2. Digested or undigested biosolids should never be added to a bed that is already covered with biosolids.
3. Remove all weeds and other vegetation that might be present. Herbicides or other types of weed killers may be used if growth is extensive. Hand weeding is also satisfactory.
4. The sand bed should be leveled and scarified by rakes, spikes, or a spring-tooth harrow. Re-leveling after scarification is recommended just before adding the digested biosolids. This will reduce surface compaction of the sand and thereby improve its filtering ability. If and when necessary, clean, coarse sand should be added to maintain approximately the same depth at all times.

After the biosolids drying bed has been adequately prepared, the digested biosolids from the digester can be applied. This withdrawal procedure should be started by opening the withdrawal valves wide open to the bed to be used. This valve should be left wide open for only a short period to allow the biosolids pipeline from the digester to the bed to be cleared of any grit or compacted material. Well-compacted material, including grit, at the inlet end of the line may make rodding or back-flushing necessary. As soon as full flow is established, the valve should be closed just enough to maintain a constant flow. This valve adjustment should prevent the lighter, watery biosolids from reaching the withdrawal pipe inlet. Additionally, it will prevent scouring of the bed surface.

Biosolids should be deposited on the bed to a depth not greater than 8 to 10 inches; however, experience will usually establish the most effective depth of biosolids application consistent with the biosolids characteristics and local weather conditions. Normally, with good drying conditions, well-digested biosolids at this depth should dewater sufficiently to be ready for removal in one to two weeks. If the biosolids are very high in solids, it may require up to three weeks or longer unless the biosolids are deposited at a lesser depth.

Another important operating feature that needs to be controlled when withdrawing biosolids from the digester is the mixing of the digester contents. It will usually be required to discontinue mixing before withdrawal. The length of time should neither be too long nor too short before withdrawal. The best time will be determined from experience.

After the biosolids have been applied to the drying bed surface, the best time for removal depends on the following factors:

1. Suitability of subsequent disposal method
2. The need to remove the next batch of biosolids from the digesters
3. The moisture content of the biosolids on the beds

Biosolids cake can be removed from the beds by shovel or fork at a moisture content of 60% to 70%. However, if it is allowed to dry until the moisture content is reduced to about 40%, it will weigh only about half to two-thirds as much and will handle better.

When the decision has been made to remove the dried biosolids, the best method should be used to minimize manual labor. When it is necessary to employ manual labor for removal, one of the best tools to use is a shovel-like fork having several tines an inch or so apart, such as an ensilage, coal, or stone fork. With a fork, biosolids can be removed with much less loss of sand than with a shovel. Nevertheless, some sand will cling to the bottom of the biosolids cake, and eventually this sand will have to be replaced, as mentioned previously.

After the operator forks or shovels the dried cake into a wheelbarrow and hauls the biosolids from the bed, it can be either hauled away to land disposal immediately or stockpiled for a year or two before land disposal. The advantage of stockpiling is that the biosolids will be more easily pulverized and therefore better for disposal on agricultural lands.

The main reason for using paved drying beds is to alleviate the problem of mechanical biosolids removal equipment damaging the underlying piping networks. The beds are paved with concrete or asphalt and are generally sloped toward the center, where a sump-like area with underlain pipes is arranged. These dewatering beds, like biosolids lagoons, depend on evaporation for dewatering the applied solids. Paved drying beds are typically rectangular in design, featuring a central drainage channel. They may incorporate buried pipes within the paved areas for heating purposes and are generally equipped with covers to protect against rain infiltration.

In this type of natural dewatering, solids contents of 45% to 50% can be achieved within 35 days in dry climates under normal conditions (McGhee, 1991). The operation of paved drying beds involves applying the biosolids to a depth of about 12 inches. The settled surface area is routinely mixed by a special vehicle-mounted machine, which is driven through the bed. Mixing is crucial because it breaks up the crust and exposes wet surfaces to the environment. Supernatant is decanted from biosolids lagoons. Biosolids loadings in relatively dry climates range from about 120 to 260 kg/m^2/year. High capital cost and larger land requirements than for sand beds are the two major disadvantages of paved drying beds.

Although the construction and operation methodologies for biosolids drying beds are well known and widely accepted, this is not to say that the wastewater industry has not attempted to incorporate further advances into their construction and operation. In an attempt to reduce the amount of dewatered biosolids that must be manually removed from drying beds, efforts have been made to construct drying beds in a specific manner. These beds can be planted with reeds, specifically, the *Phragmites communis* variety (which will be covered in greater detail later). The intent of augmenting the biosolids drying bed with reeds is to effect further desiccation. Moreover, tests have shown that the plants extend their root systems into the biosolids mass. This extended root system has the added benefit of helping to establish a rich microflora, which eventually feeds upon the organic content of the biosolids. It is interesting to note that regular plant activity works to keep the system aerobic.

13.13.5.2 Rotary Vacuum Filtration

Rotary vacuum filters have also been used for many years to dewater sludge. The vacuum filter includes filter media (belt, cloth, or metal coils), media support (drum), vacuum system, chemical feed equipment, and conveyor belt(s) to transport the dewatered solids. In operation, chemically treated solids are pumped to a vat or tank in which a rotating drum is submerged. As the drum rotates, a vacuum is applied to it. Solids collect on the media and are held there by the vacuum as the drum rotates out of the tank. The vacuum removes additional water from the captured solids. When the solids reach the discharge zone, the vacuum is released, and the dewatered solids are discharged onto a conveyor belt for disposal. The media is then washed before returning to the start of the cycle.

13.13.5.3 Pressure Filtration

Pressure filtration differs from vacuum filtration in that the liquid is forced through the filter media by positive pressure instead of a vacuum. Several types of presses are available, but the most commonly used types are plate and frame presses and belt presses. *Filter presses* include the belt or plate and frame types. The belt filter comprises two or more porous belts, rollers, and related handling systems for chemical makeup and feed, and supernatant and solids collection and transport.

The plate-and-frame filter consists of a support frame, filter plates covered with porous material, a hydraulic or mechanical mechanism for pressing plates together, and related handling systems for chemical makeup and feed, supernatant, and solids collection and transport. In the plate-and-frame filter, solids are pumped (sandwiched) between plates. Pressure (200–250 psi) is applied to the plates, and water is "squeezed" from the solids. At the end of the cycle, the pressure is released, and as the plates separate, the solids drop out onto a conveyor belt for transport to storage or disposal. Performance factors for plate and frame presses include feed sludge characteristics, type and amount of chemical conditioning, operating pressures, and the type and amount of precoat.

The belt filter uses a coagulant (polymer) mixed with the influent solids. The chemically treated solids are discharged between two moving belts. First, water drains from the solids by gravity. Then, as the two belts move between a series of rollers, pressure "squeezes" additional water out of the solids. The solids are then discharged onto a conveyor belt for transport to storage or disposal. Performance factors for the belt press include sludge feed rate, belt speed, belt tension, belt permeability, chemical dosage, and chemical selection.

Filter presses have lower operation and maintenance costs than vacuum filters or centrifuges. They typically produce a good quality cake and can be batch-operated. However, construction and installation costs are high. Moreover, chemical addition is required, and the presses must be operated by skilled personnel.

13.13.5.4 Centrifugation

Centrifuges of various types have been used in dewatering operations for at least 30 years and appear to be gaining popularity. Depending on the type of centrifuge used, in addition to centrifuge pumping equipment for solids feed and centrate removal, chemical makeup and feed equipment, and support systems for the removal of dewatered solids are required.

13.13.5.5 Sludge Incineration

Not surprisingly, incinerators produce the maximum solids and moisture reductions. The equipment required depends on whether the unit is a multiple-hearth or fluid-bed incinerator. Generally, the system will need a source of heat to reach ignition temperature, a solids feed system, and ash handling equipment. It is important to note that the system must also include all required equipment (e.g., scrubbers) to achieve compliance with air pollution control requirements. Solids are pumped to the incinerator. The solids are dried and then ignited (burned). As they burn, the organic matter is converted to carbon dioxide and water vapor, and the inorganic matter is left behind as ash or "fixed" solids. The ash is then collected for reuse or disposal.

13.13.6 Land Application of Biosolids

The purpose of land application of biosolids is to dispose of the treated biosolids in an environmentally sound manner by recycling nutrients and soil conditioners. In order to be land applied, wastewater biosolids must comply with state and federal biosolids management/disposal regulations. Biosolids must not contain materials that are dangerous to human health (i.e., toxicity, pathogenic organisms) or hazardous to the environment (i.e., toxicity, pesticides, heavy metals). Treated biosolids are land-applied by either direct injection or application and plowing in (incorporation).

13.14 PERMITS, RECORDS, AND REPORTS

Permits, records, and reports play a significant role in wastewater treatment operations. In fact, regarding the "permit," one of the first things any new operator quickly learns is the importance of "making permit" each month. In this section, we briefly cover NPDES permits and other pertinent records and reports that the wastewater operator must be familiar with.

Note: The discussion that follows is general; it does not necessarily apply to any state in particular but instead is an overview of permits, records, and reports that are an essential part of wastewater treatment plant operations. For specific guidance on requirements for your locality, refer to your state water control board or other authorized state agency for information. In this handbook, the term "Board" signifies the state-reporting agency.

Several definitions should be discussed before discussing the permit requirements for records and reporting. These definitions are listed below.

- *Average Monthly Limitation*—the highest allowable average over a calendar month, calculated by adding all of the daily values measured during the month and dividing the sum by the number of daily values measured during the month.
- *Average Weekly Limitation*—the highest allowable average over a calendar week, calculated by adding all of the daily values measured during the calendar week and dividing the sum by the number of daily values determined during the week.
- *Average Daily Limitation*—the highest allowable average over 24 hours, calculated by adding all of the values measured during the period and dividing the sum by the number of values determined during the period.
- *Average Hourly Limitation*—the highest allowable average for 60 minutes, calculated by adding all of the values measured during the period and dividing the sum by the number of values recorded during the period.
- *Daily Discharge*—means the discharge of a pollutant measured during a calendar day or any 24-hour period that reasonably represents the calendar for sampling. For pollutants with limitations expressed in units of weight, the daily discharge is calculated as the total mass of the pollutant discharged over the day. For pollutants with limitations expressed in other units, the daily discharge is calculated as the average measurement of the pollutant over the day.
- *Maximum Daily Discharge*—the highest allowable value for a daily discharge.

- *Effluent Limitation*—any restriction by the State Board on quantities, discharge rates, or concentrations of pollutants that are discharged from point sources into State waters.
- *Maximum Discharge*—the highest allowable value for any single measurement.
- *Minimum Discharge*—the lowest allowable value for any single measurement.
- *Point Source*—any discernible, defined, and discrete conveyance, including but not limited to any pipe, ditch, channel, tunnel, conduit, well, discrete fissure, container, rolling stock, vessel, or other floating craft, from which pollutants are or may be discharged. This definition does not include return flows from irrigated agricultural land.
- *Discharge Monitoring Report*—forms for use in reporting self-monitoring results of the permittee.
- *Discharge Permit*—State Pollutant Discharge Elimination System permit that specifies the terms and conditions under which a point source discharge to State waters is permitted.

13.14.1 NPDES Permits

In the United States, all treatment facilities that discharge to State waters must have a discharge permit issued by the State Water Control Board or other appropriate State agency. This permit is known on the national level as the National Pollutant Discharge Elimination System (NPDES) permit and on the state level as the (State) Pollutant Discharge Elimination System (state-PDES) permit. The permit states the specific conditions that must be met to discharge treated wastewater to State waters legally. The permit contains general requirements (applying to every discharge) and specific requirements (applying only to the point source specified in the permit). A general permit is a discharge permit that covers a specified class of dischargers. It is developed to allow dischargers within the selected category to discharge under specified conditions. All discharge permits contain general conditions that are standard for all dischargers and cover a broad range of requirements. Read the general conditions of the treatment facility's permit carefully. Permittees must retain certain records. They are as follows:

- Date, time, and exact place of sampling or measurements
- Name(s) of the individual(s) performing sampling or measurement
- Date(s) and time(s) analyses that were performed
- Name(s) of the individuals who performed the analyses
- Analytical techniques or methods used
- Observations, readings, calculations, bench data, and results
- Instrument calibration and maintenance
- Original strip chart recordings for continuous monitoring
- Information used to develop reports required by the permit
- Data used to complete the permit application

Note: All records must be kept for at least three years (longer at the request of the State Board).

Generally, reporting must be done under the following conditions/situations (requirements may vary depending on the state regulatory body with reporting authority):

- *Unusual or Extraordinary Discharge Reports*—must notify the Board by telephone within 24 hours of the occurrence and submit a written report within five (5) days. The report must include the following:
  (1) Description of the non-compliance and its cause
  (2) Non-compliance date(s), time(s), and duration
  (3) Steps planned/taken to reduce/eliminate
  (4) Steps planned/taken to prevent reoccurrence
- *Anticipated Non-Compliance*—You must notify the Board at least ten days in advance of any changes to the facility or activity that may result in non-compliance.
- *Compliance Schedules*—You must report compliance or non-compliance with any requirements contained in compliance schedules no later than 14 days following the scheduled date for completion of the requirement.
- *24-Hour Reporting*—Any non-compliance that may adversely affect State waters or endanger public health must be reported orally within 24 hours of the time the permittee becomes aware of the condition. A written report must be submitted within five days.
- *Discharge Monitoring Reports (DMRs)*—Reports self-monitoring data generated during a specified period (normally one month). When completing the DMR, remember the following:
  - More frequent monitoring must be reported.
  - All results must be included to calculate reported values.
  - Pollutants monitored by an approved method but not required by the permit must also be reported.
  - Ensure that no empty blocks on the form are left blank.
  - Averages are arithmetic unless noted otherwise.
  - Appropriate significant figures should be used.
  - All bypasses and overflows must be reported.
  - The licensed operator must sign the Report.
  - The responsible official must sign the Report.
  - The department must receive the report by the tenth of the next month.

The general requirements of the permit specify the minimum sampling and testing that must be performed on the plant discharge. Moreover, the permit will determine the frequency of sampling, sample type, and length of time for composite samples. Unless the permit requires a specific method, all sample preservation and analysis must follow the requirements outlined in the Federal Regulations *Guidelines Establishing Test Procedures for the Analysis of Pollutants Under the Clean Water Act* (40 CFR 136).

Note: All samples and measurements must be representative of the nature and quantity of the discharge.

The permit sets numerical limitations on specific parameters contained in the plant discharge. Limits may be expressed as

- Average monthly quantity (kg/day)
- Average monthly concentration (mg/l)
- Average weekly quantity (kg/day)
- Average weekly concentration (mg/l)
- Daily quantity (kg/day)
- Daily concentration (mg/l)
- Hourly average concentration (mg/l)
- Instantaneous minimum concentration (mg/l)
- Instantaneous maximum concentration (mg/l)

Suppose the facility requires additional construction or other modifications to comply with the final effluent limitations fully. In that case, the permit will contain a schedule of events to be completed to achieve full compliance.

Any special requirements or conditions set for approval of the discharge will be contained in this section. Special conditions may include the following:

- Monitoring required to determine effluent toxicity
- Pretreatment program requirements

The permit will specify, based on the treatment system complexity and the volume of flow treated, the minimum license classification required to be the designated responsible charge operator.

Several reporting systems apply to chlorination or chlorination followed by dechlorination. It is best to review this section of the specific permit for guidance. If confused, contact the appropriate State Regulatory Agency.

Failure to accurately calculate report data will result in violations of the permit.

13.15 THE BOTTOM LINE

After decades of working with water/wastewater treatment, teaching water/wastewater treatment, inspecting water/wastewater treatment plants, and writing about both, I have come to an understanding. I realize that the potable freshwater we have now is all we are going to get, and there is a growing and pressing need for clean, healthy, safe drinking water worldwide. Okay, because the supply of clean, healthy, safe drinking water is at a premium and becoming scarcer with each passing day, something needs to be done to ensure we have a supply of clean, healthy, safe drinking water now and for the future.

So, the fundamental question is, what are we going to do about our limited supply of clean, healthy, safe drinking water now and for the future?

A couple of things seem inevitable: we must protect the freshwater supplies we have today, and we must accelerate our reuse of wastewater and treat it to produce clean, healthy, safe drinking water standards.

I have concluded that to make this happen, a couple of things are necessary. First, we must make people aware that our supply of clean, healthy, safe drinking water is limited, finite, fixed, set, and restricted. After showing that clean, safe water isn't guaranteed, action is needed. Second, the first step in action is the conservation of available potable water supplies, which is a must. Third, it is necessary to treat wastewater to ensure that water released into the environment is safe and to process wastewater to meet established drinking water standards.

We can treat wastewater to drinking water standards. How? This was explained in Case Study 8.1. Maybe a reread of this one is called for.

NOTE

1 Adaptation from F.R. Spellman (2020). *Handbook of Water and Wastewater Operations*, 5th ed. Boca Raton, FL: CRC Press.

REFERENCES

Capra, F. 1982. *The Turning Point: Science, Society, and the Rising Culture*. New York: Random House.

CWNS (Clean Watersheds Needs Survey). 2000. *Report to Congress, EPA-832-R-10-002*. Washington, DC.

Daly, M. 1980. *Gyn Ecology*. Boston, MA: Beacon Press.

Epstein, E., and Alpert, J.E. 1984. Sludge dewatering and compost economics. *BioCycle*, 25(10): 31–34.

Grady, C.P.L., Jr., et al. 2011. *Biological Wastewater Treatment*, 3rd ed. Boca Raton, FL: CRC Press.

Holyningen-Huene. 1993. *Reconstructing Scientific Revolution: Thomas S. Kuhn's Philosophy of Science*. Chicago, IL: University of Chicago Press.

McGhee, T.J. 1991. *Water Supply and Sewerage*. New York: McGraw-Hill, Inc.

Metcalf & Eddy, Inc. 1991. *Wastewater Engineering: Treatment, Disposal, Reuse*, 3rd ed. New York: McGraw-Hill.

Metcalf & Eddy, Inc. 2003. *Wastewater Engineering: Treatment, Disposal, Reuse*, 4th ed. New York: McGraw-Hill.

Outwater, A.B. 1994. *Reuse of Sludge and Minor Wastewater Residuals*. Boca Raton, FL: Lewis Publishers.

Sloan Equipment. 1999. *Aeration Products*. Owings Mills, MD.

Spellman, F.R. 1996. *Stream Ecology and Self-Purification*. Boca Raton, FL: CRC Press.

Spellman, F.R. 2007. *The Science of Water*, 2nd ed. Boca Raton, FL: CRC Press.

USEPA. 1989. Technical support document for pathogen reduction in sewage sludge. *NTIS No. PB89-136618*. Springfield, VA: National Technical Information Service.

USEPA. 1997. *Innovative Uses of Compost: Disease Control for Plants and Animals*. Washington, DC: United States Environmental Protection Agency.

USEPA. 2000. *Wastewater Technology Fact Sheet Package Plants*. Washington, DC: United States Environmental Protection Agency.

USEPA. 2007a. *Wastewater Management Fact Sheet: Membrane Bioreactors*. Washington, DC: United States Environmental Protection Agency.

USEPA. 2007b. *Advanced Wastewater Treatment to Achieve Low Concentration of Phosphorus*. Washington, DC: Environmental Protection Agency.

USEPA. 2007c. *Biological Nutrient Removal Processes and Costs*. Washington, DC: United States Environmental Protection Agency.

USEPA. 2007d. *National Section 303(d) List Fact Sheet*. Accessed 07/22/2024 @ http://iaspub.epa.gov/waters/national_rept.control.

USEPA. 2011. *Principles of Design and Operations of Wastewater Treatment Pond Systems for Plant Operators, Engineers, and Managers*. Washington, DC: U.S. Environmental Protection Agency.

14 A Final Word

Water managers should take proactive steps and regularly review required measures.

14.1 AI IS A WATER HOG

Applications of artificial intelligence (AI) are universally applied in functions where they suitable. For water management personnel and end users of their product, this can be both a benefit and a harmful trend. Not only are we limited to the water supply we have today and tomorrow and maybe forever, but with population growth and pollution trends, fresh, potable, healthy drinking water is becoming harder to find (USEPA, 2023).

It has been well published that AI data centers are energy hogs, but less attention has been paid to AI data center water usage until recently, which is also beginning to catch our attention. The best way I can currently illustrate both the energy and water usage at AI data centers is to provide Case Study 14.1.

CASE STUDY 14.1 MICROSOFT AND THE CITY OF QUINCY, WASHINGTON (USEPA, 2023)[1]

In a nutshell, the Quincy Water Reuse Utility (QWRU) was constructed in a partnership between Microsoft and the city of Quincy, Washington. The Reuse Utility treats cooling water from a Microsoft data center and recirculates it to the data center for the same purpose, reducing reliance on local potable groundwater. The groundwater also has high levels of total dissolved solids (TDS), which can cause problems when used in the data center's cooling equipment. The water supplied from the QWRU is of a more suitable quality for the data center's cooling equipment than the conventionally treated, mineral-rich groundwater.

In context, the City of Quincy is in the arid eastern part of the State of Washington and receives only 7.8 inches (199 millimeters) of annual rainfall. The potable water required to service residential, commercial, and industrial needs within the city is provided by five groundwater wells. These wells in basaltic rocks reach depths of up to 1,000 feet (305 meters) into the Columbia Plateau regional aquifer system. Although the residential part of Quincy is relatively small, with about 8,100 people, its large industrial sector means its water demand is akin to that of 30,000 inhabitants. Yahoo, Sabey, Intuit, Vantage, and Microsoft have data centers along with a large food processing industry within the Quincy region.

DOI: 10.1201/9781003655848-19

So, why Quincy, Washington? Why are major companies building their data centers in Quincy? The attraction in one word is affordability, affordability, affordability. Simply put, Quincy is an attractive area to build these data centers because of the area's affordable land (as compared to land in the Silicon Valley region to the south of Washington State in California), cool temperatures, and the availability of renewable, abundant, and cheap hydropower. The surrounding area is also a hub of agricultural activity.

Let's look at the city's water needs.

Typically, every year, the city uses approximately 2.1 billion gallons (eight million liters) of groundwater. Until recently, food processors used 57% of the pumped groundwater; 33% went to residents, public, and commercial services such as schools, hotels, or restaurants; and data centers used the remaining 10% for cooling purposes.

Note that since 2007, Microsoft has operated one of the biggest data centers in Quincy on a 270-acre (110-hectare) campus. The large volume of water needed to cool high-density cloud computing server farms is making water management a growing priority for data center operators, including Microsoft. In the past, Microsoft received potable groundwater from Quincy for cooling; they would discharge the water to the sewer system for treatment at the city's municipal wastewater reclamation facility. However, after four to five cooling cycles, this water was rich in minerals, including calcium, magnesium, and silica. This was not a good situation because the high mineral content meant that long cycles in the data centers could damage heat transfer equipment or generally impair heat transfer as water evaporated during cooling. The concentration of TDS increased. To address mineral accumulation, it was necessary to perform regular blowdown or bleeding procedures—draining water from the cooling equipment—to remove mineral deposits effectively.

Complicating the TDS problem in the cooling water was the original wastewater treatment approach. The city owned and operated a municipal reclamation facility (MWRF) that uses reclaimed water to irrigate public areas and recharge shallow aquifers in the area through groundwater recharge basins. However, the treatment plant could not handle the high TDS concentrations in the discharges from the regional data centers to the MWRF. As a result, the data centers were charged an extra fee to accommodate adjustments to the facility's processes and increased maintenance needs. Another fly in the ointment or icing on the "cake," so to speak, is the high levels of TDS, calcium, magnesium, and silica that together formed a visible cake on the surface of the groundwater recharge basins in the summer. This was causing it to exceed discharge and indirect potable reuse requirements, for example, for aquifer recharge, set by the Washington State Department of Ecology.

Both the City of Quincy and Microsoft recognized the technical difficulty of using mineral-rich municipal groundwater for cooling data center equipment. Moreover, they also acknowledged the subsequent challenge of treating the data center's industrial wastewater with high levels of TDS at the city's municipal wastewater reclamation facility. The City and Microsoft formed a partnership to create the QWRU, an entirely new facility specifically designed to treat the cooling water from Microsoft's data center. Ten years of planning and construction were finally completed when QWRU officially became operational on June 30, 2001. Pipes more than 30 miles (48 kilometers) long run across Quincy's industrial area, connected to ten distinct treatment systems. To date, only Microsoft's data center campus is connected to the QWRU.

The QWRU is a closed-loop system that does not release high-mineral industrial wastewater into the environment or the city's MWRF. To prevent mineral buildup in the cooling water, salts are removed from the water before being reused at the data center and concentrated in lined brine ponds to form a solid. This sold waste is disposed of in a manner that prevents entry into state ground or surface waters. This ensures that the cooling water quality does not adversely affect cooling equipment or heat transfer within the data center. Just as importantly (maybe even more so), the QWRU substantially reduces the data center's reliance on portable groundwater for data center cooling, saving millions of gallons of water per year.

For its treatment system, the QWRU relies on a combination of the following:

- New online systems, including a lime softening facility (similar to alum treatment using calcium hydroxide), an ultrafiltration facility, a permeate tank facility [used in reverse osmosis (RO)], a blend tank facility for storage, and brine ponds.
- Previously existing infrastructure, which was part of the city's industrial wastewater treatment systems, including a high-efficiency softening (HES) facility and reverse osmosis components.

The water is cycled four to five times at the data center, and then the resulting blowdown water is transferred to a lime softening facility that adds limewater (calcium hydroxide) to remove hardness (deposits of calcium and magnesium salts). The next step is ultrafiltration.

After ultrafiltration, some of the water is transferred to a blending tank, while the rest is further softened in the HES facility and filtered at the RO plant. A pipe transfers the HES calcium and magnesium-laden brine and the RO brine silica to the brine ponds. Nature takes over, and the brine undergoes maximum evaporation, and the residuals are disposed of separately. Note that Quincy expects it will need to do this every two or three years.

The treated wastewater continues to flow from the RO system and is stored in a permeate tank for approximately eight hours, acting as a buffer whenever the system is down for maintenance of the HES or RO systems. The ultrafiltration and RO water are blended at various ratios throughout the day, depending on the conductivity levels (ideally 300–350 microsiemens). Finally, the treated industrial wastewater is returned to the data center for use as cooling water.

In operation, the QWRU system must be topped off or topped up often because water evaporates during the cooling process, and a small volume of highly concentrated brine needs disposal. Because the data centers do not need potable water for cooling, the City of Quincy and Microsoft have been able to use make-up water from untreated surface water sources by accessing rights to water drawn from the Columbia Basin Project in the U.S. Bureau of Reclamation and the Quincy Irrigation District canal.

DID YOU KNOW?

According to the U.S. Bureau of Reclamation (USBR, 2023), more than 670,000 acres in east central Washington serve the Columbia Basin Project. The main facilities of the project include Grand Coulee Dam, Franklin D. Roosevelt Lake, three power plants, four switchyards, and a pump-generating plant. Primary irrigation facilities include the Feeder Canal, Banks Lake, the Main, West, East High, and East Low canals, O'Sullivan Dam, Potholes Reservoir, and Potholes Canal. There are over 300 miles of main canals, and 2,000 miles of laterals, and 3,500 miles of drains and wasteways on the project. In addition to supplying water for irrigation, producing electricity, controlling floods, providing recreation, and regulating streamflow, the Columbia Basin Project also provides water for cities, industries, navigation, and endangered species.

During the summer, cooling demand in data centers is at its highest. The QWRU uses almost 300 million gallons per year of Columbia Basin Project water from the irrigation canal. This is a positive practice because it substantially lowers the need to use groundwater for cooling and reduces the city's potable water demand and treatment costs, since there is no need to soften the water before sending it back to the data center. Note that the irrigation canal, from which the water is extracted, runs dry in winter. However, this is not a significant issue, since cooling demand is comparatively lower in the winter months.

Five percent of the makeup water, on an annual basis, is still potable groundwater, which is mixed and softened before entering the cooling systems. Forging ahead, the City of Quincy aims to purchase more water from the Bureau of Reclamation and get permission from the Quincy Irrigation District to withdraw more water from the canal because it requires minimal treatment for the data centers and could eliminate the data center's reliance on potable groundwater supplies.

14.2 THE BOTTOM LINE

AI data centers require vast amounts of electricity for operation and water for cooling. The future need for more electricity and greater access to water for powering the exponential growth of AI is a given. The problem is that predictions of future energy and water demand are weighed down with uncertainties due to a lack of visibility into proprietary private sector planning for new model training. This is the case even though there is some visible movement toward increasing the amount of nuclear power as a source of electricity. There is also a concern about procuring electrical components such as transformers, switching equipment, advanced transmission lines, and generation equipment in the near- and longer-term. Access to water supplies for data centers is a significant issue, and, in line with the need for more electrical power sources, the current movement toward water reuse may ameliorate the problem of finding adequate supplies of cooling water.

Oh, there is a tight rein on humans in everything we do or want to do, and in some cases, not what we want, but it happens anyway because we make mistakes, and in some cases, these mistakes are not checked until a limit enters the scene, so to speak. In the case of expanding AI, two limiting factors are the need for large amounts of electricity and water, and the need to address the sources found or developed.

Some would say, including this author, that the recent advances in AI have been hailed as potentially transformative for a long list of industries and other sectors, including national security. The key to using AI in water management, or in any other enterprise, is to make sure we have a connected mindset with "normal" human operators and to ensure that the operators stay away from bias or any other nefarious goal or use.

The question is, what is the limit?

NOTE

1 From F.R. Spellman (2024). *The Science of Water Reuse*. Boca Raton, FL: CRC Press.

REFERENCES

USBR. 2023. *Reclamation Managing Water in the West*. Washington, DC: U.S. Department of Interior.

USEPA. 2023. *Case Studies That Demonstrate the Benefits of Water Reuse*. Accessed 12/22/2024 @ https://epa.gov/waterreuse/forms/contact-us--about-water-reuse-and-recycling.

Glossary

A

Absorption: any process by which one substance penetrates the interior of another substance.

Acid: has a pH of water less than 5.5; pH modifier used in the U.S. Fish and Wildlife Service wetland classification system; in common usage, acidic water has a pH less than 7.

Acidic deposition: the transfer of acidic or acidifying substances from the atmosphere to the surface of the Earth or to objects on its surface. Transfer can be either by wet-deposition processes (rain, snow, dew, fog, frost, hail) or by dry deposition (gases, aerosols, or fine to coarse particles).

Acid Rain: precipitation with higher-than-normal acidity, caused primarily by sulfur and nitrogen dioxide air pollution.

Acre-foot (acre-ft.): the volume of water needed to cover an acre of land to a depth of one foot; equivalent to 43,560 cubic feet or 32,851 gallons.

Activated carbon: a very porous material that after being subjected to intense heat to drive off impurities can be used to adsorb pollutants from water.

Adsorption: the process by which one substance is attracted to and adheres to the surface of another substance, without actually penetrating its internal structure.

Aeration: a physical treatment method that promotes the biological degradation of organic matter. The process may be passive (when waste is exposed to air) or active (when a mixing or bubbling device introduces the air).

Aerobic Bacteria: a type of bacteria that requires free oxygen to carry out metabolic function.

Algae: chlorophyll-bearing nonvascular, primarily aquatic species that have no true roots, stems, or leaves; most algae are microscopic, but some species can be as large as vascular plants.

Algal bloom: the rapid proliferation of passively floating, simple plant life, such as blue-green algae, in and on a body of water.

Alkaline: has a pH greater than 7; pH modifier in the U.S. Fish and Wildlife Service wetland classification system; in common usage, a pH of water greater than 7.4.

Alluvial aquifer: a water-bearing deposit of unconsolidated material (sand and gravel) left behind by a river or other flowing water.

Alluvium: a general term for sediments of gravel, sand, silt clay, or other particulate rock material deposited by flowing water, usually in the beds of rivers and streams, on a flood plain, on a delta, or at the base of a mountain.

Alpine snow glade: a marshy clearing between slopes above the timberline in mountains.

Amalgamation: the process of dissolving or blending a metal (commonly gold and silver) in mercury to separate it from its parent material.

Ammonia: a compound of nitrogen and hydrogen (NH_3) that is a common by-product of animal waste. Ammonia readily converts to nitrate in soils and streams.

Anadromous fish: migratory species that are born in freshwater, live mostly in estuaries and ocean water, and return to freshwater to spawn.

Anaerobic: pertaining to, taking place in, or caused by the absence of oxygen.

Anomalies: as related to fish, externally visible skin or subcutaneous disorders, including deformities, eroded fins, lesions, and tumors.

Anthropogenic: having to do with or caused by humans.

Anticline: a fold in the Earth's crust, convex upward, whose core contains stratigraphically older rocks.

Aquaculture: the science of faming organisms that live in water, such as fish, shellfish, and algae.

Aquatic: living or growing in or on water.

Aquatic guidelines: specific levels of water quality that, if reached, may adversely affect aquatic life. These are non-enforceable guidelines issued by a governmental agency or other institution.

Aquifer: a geologic formation, group of formations, or part of a formation that contains sufficient saturated permeable material to yield significant quantities of water to springs and wells.

Arroyo: a small, deep, flat-floored channel or gully of an ephemeral or intermittent stream, usually with nearly vertical banks cut, into unconsolidated material.

Artificial recharge: augmentation of natural replenishment of groundwater storage by some method of construction, spreading of water, or by pumping water directly into an aquifer.

Atmospheric deposition: the transfer of substances from the air to the surface of the Earth, either in wet form (rain, fog, snow, dew, frost, hail) or in dry form (gases, aerosols, particles).

Atmospheric pressure: the pressure exerted by the atmosphere on any surface beneath or within it; equal to 14.7 pounds per square inch at sea level.

Average discharge: as used by the U.S. Geological Survey, the arithmetic average of all complete water years of record of surface water discharge, whether consecutive or not. The term "average" is generally reserved for the average of records, and "mean" is used for averages of shorter periods, namely, daily, monthly, or annual mean discharges.

B

Background concentration: a concentration of a substance in a particular environment that is indicative of minimal influence by human (anthropogenic) sources.

Backwater: a body of water in which the flow is slowed or turned back by an obstruction such as a bridge or dam, an opposing current, or the movement of the tide.

Bacteria: single-cell microscopic organisms.

Bank: the sloping ground that borders a stream and confines the water in the natural channel when the water level, or flow, is normal.

Bank storage: the change in the amount of water stored in an aquifer adjacent to a surface-water body resulting from a change in the stage of the surface-water body.

Barrier bar: an elongate offshore ridge, submerged at least at high tide, built up by the action of waves or currents.

Base flow: the sustained low flow of a stream, usually groundwater inflow to the stream channel.

Basic: the opposite of acidic; water that has a pH greater than 7.

Basin and Range physiography: a region characterized by a series of generally north-trending mountain ranges separated by alluvial valleys.

Bed material: sediment comprising the streambed.

Bed sediment: the material that is temporarily stationary in the bottom of a stream or other watercourse.

Bedload: sediment that moves on or near the streambed and is in almost continuous contact with the bed.

Bedrock: a general term used for solid rock that underlies soils or other unconsolidated material.

Benthic invertebrates: insects, mollusk, crustaceans, worms and other organisms without a backbone that live in, on, or near the bottom of lakes, streams, or oceans.

Benthic organism: a form of aquatic life that lives on or near the bottom of streams, lakes, or oceans.

Bioaccumulation: the biological sequestering of a substance at a higher concentration than that at which it occurs in the surrounding environment or medium. Also, the process whereby a substance enters organisms through the gills, epithelial tissues, dietary, or other sources.

Bioavailability: the capacity of a chemical constituent to be taken up by living organisms either through physical contact or by ingestion.

Biochemical: refers to chemical processes that occur inside or are mediated by living organisms.

Biochemical process: a process characterized by, produced by, or involving chemical reactions in living organism.

Biochemical Oxygen Demand (BOD): the amount of oxygen required by bacteria to stabilize decomposable organic matter under aerobic conditions.

Biodegradation: the transformation of a substance into new compounds through biochemical reactions or the actions of microorganisms such as bacteria.

Biological Treatment: a process that uses living organisms to bring about chemical changes.

Biomass: the amount of living matter in the form of organisms present in a particular habitat, usually expressed as weight per unit area.

Biota: all living organisms of an area.

Bog: a rain-fed, acidic wetland with spongy sphagnum moss and thick peat, nutrient-poor, and typically without inflow or outflow.

Brackish water: water with a salinity intermediate between seawater and freshwater (containing from 1,000 to 10,000 milligrams per liter of dissolved solids).

Breakdown product: a compound derived by chemical, biological, or physical action upon a pesticide. The breakdown is a natural process that may result in a more toxic or less toxic compound and a more persistent or less persistent compound.

Breakpoint chlorination: the addition of chlorine to water until the chlorine demand has been satisfied and free chlorine residual is available for disinfection.

C

Calcareous: a rock or substance formed of calcium carbonate or magnesium carbonate by biological deposition or inorganic precipitation, or containing those minerals in sufficient quantities to effervesce when treated with cold hydrochloric acid.

Capillary fringe: the zone above the water table in which water is held by surface tension. Water in the capillary fringe is under a pressure less than atmospheric.

Carbonate rocks: rocks (such as limestone or dolostone) that are composed primarily of minerals (such as calcite and dolomite) containing a carbonate ion.

Center pivot irrigation: an automated sprinkler system involving a rotating pipe or boom that supplies water to a circular area of an agricultural field through sprinkler heads or nozzles.

Channel scour: erosion caused by flowing water and sediment on a stream channel; results in the removal of mud, silt, and sand on the outside curve of a stream bend and the bed material of a stream channel.

Channelization: the straightening and deepening of a stream channel to allow the water to move faster or to drain a wet area for farming.

Chemical treatment: a process that results in the formation of a new substance or substances. The most common chemical water treatment processes include coagulation, disinfection, water softening, and filtration.

Chlordane: octachlor-4,7-methanotetrahydroindane. An organochlorine insecticide no longer registered for use in the United States. Technical chlordane is a mixture that primarily contains cis- and trans-chlordane, cis- and trans-nonachlor, and heptachlor.

Chlorinated solvent: a volatile organic compound containing chlorine. Some common solvents are trichloroethylene, tetrachloroethylene, and carbon tetrachloride.

Chlorofluorocarbons: a class of volatile compounds consisting of carbon, chlorine, and fluorine. Commonly called freons, they have been used in refrigeration mechanisms, as blowing agents in the fabrication of flexible and rigid foams, and, until banned from use several years ago, as propellants in spray cans.

Chlorination: the process of adding chlorine to water to kill disease-causing organisms or to act as an oxidizing agent.

Chlorine Demand: a measure of the amount of chlorine that will combine with impurities, and is therefore unavailable to act as a disinfectant.

Ciénaga: a marshy area where the ground is wet due to the presence of seepage of springs.

Clean Water Act (CWA): federal law dating to 1972 (with several amendments) with the objective to restore and maintain the chemical, physical, and biological integrity of the nation's waters. Its long-range goal is to eliminate the discharge of pollutants into navigable waters, and to make national waters fishable and swimmable.

Climate: the sum total of the meteorological elements that characterize the average and extreme conditions of the atmosphere over a long period of time at any one place or region of the Earth's surface.

Coagulants: chemicals that cause small particles to stick together to form larger particles.

Coagulation: a chemical water treatment process that clusters tiny suspended particles into larger ones. This is accomplished by the addition of a coagulant that neutralizes the electrostatic charges that cause particles to repel each other.

Coliform bacteria: a group of bacteria not only inhabiting the intestines of humans or animals but also occasionally found elsewhere. The presence of bacteria in water is used as an indication of fecal contamination (contamination by animal or human wastes).

Color: a physical characteristic of water. Color is most commonly tan or brown from oxidized iron, but contaminants may cause other colors, such as green or blue. Color differs from turbidity, which is water's cloudiness.

Combined sewer overflow: a discharge of untreated sewage and stormwater to a stream when the capacity of a combined storm/sanitary sewer system is exceeded by storm runoff.

Communicable diseases: Usually caused by *microbes*—microscopic organisms including bacteria, protozoa, and viruses. Most microbes are essential components of our environment and do not cause disease. Those that do are called pathogenic organisms, or simply *pathogens*.

Community: in ecology, the species that interact in a common area.

Community Water System: a public water system serving at least 15 year-round connections or 25 year-round residents.

Composite sample: a series of individual or grab samples taken at different times from the same sampling point and mixed together.

Concentration: the ratio of the quantity of any substance present in a sample of a given volume or a given weight compared to the volume or weight of the sample.

Cone of depression: the depression of heads around a pumping well caused by the withdrawal of water.

Confined aquifer (artesian aquifer): an aquifer that is completely filled with water under pressure and is overlain by material that restricts the movement of water.

Confining layer: a body of impermeable or distinctly less permeable material stratigraphically adjacent to one or more aquifers that restricts the movement of water into and out of the aquifers.

Confluence: the meeting point where two or more streams merge; the location where a tributary connects with the mainstream.

Conglomerate: a coarse-grained sedimentary rock composed of fragments larger than two millimeters in diameter.

Constituent: a chemical or biological substance in water, sediment, or biota that can be measured by an analytical method.

Consumptive use: the quantity of water that is not available for immediate use because it has been evaporated, transpired, or incorporated into products, plant tissue, or animal tissue.

Contact recreation: recreational activities, such as swimming and kayaking, in which contact with water is prolonged or intimate, and in which there is a likelihood of ingesting water.

Contaminant: a toxic material found as an unwanted residue in or on a substance.

Contamination: degradation of water quality compared to original or natural conditions due to human activity.

Contributing area: the area in a drainage basin that contributes water to streamflow or recharge to an aquifer.

Core sample: a sample of rock, soil, or other material obtained by driving a hollow tube into the undisturbed medium and withdrawing it with its contained sample.

Criterion: a standard rule or test on which a judgment or decision can be based.

Cross Connection: any connection between safe drinking water and a nonpotable water or fluid.

CxT Value: the product of the residual disinfectant concentration **C,** in milligrams per liter, and the corresponding disinfectant contact time **T,** in minutes. Minimum **CxT** values are specified by the Surface Water Treatment Rule as a means of ensuring adequate kill or inactivation of pathogenic microorganisms in water.

D

Datum plane: a horizontal plane to which ground elevations or water surface elevations are referenced.

Deepwater habitat: permanently flooded lands lying below the deepwater boundary of wetlands.

Degradation products: compounds resulting from the transformation of an organic substance through chemical, photochemical, and/or biochemical reactions.

Denitrification: a process by which oxidized forms of nitrogen, such as nitrate, are reduced to form nitrites, nitrogen oxides, ammonia, or free nitrogen, commonly brought about by the action of denitrifying bacteria and usually resulting in the escape of nitrogen to the air.

Detection limit: the concentration of a constituent or analyte below which a particular analytical method cannot determine the concentration with a high degree of certainty.

Diatoms: single-celled, colonial, or filamentous algae with siliceous cell walls constructed of two overlapping parts.

Direct runoff: the runoff entering stream channels promptly after rainfall or snowmelt.

Discharge: the volume of fluid passing a point per unit of time, commonly expressed in cubic feet per second, million gallons per day, gallons per minute, or seconds per minute per day.

Discharge area (groundwater): an area where subsurface water is discharged to the land surface, surface water, or the atmosphere.

Disinfectants-disinfection by-products (D-DBPs): a term used in connection with state and federal regulations designed to protect public health by limiting the concentration of either disinfectants or the by-products formed by the reaction of disinfectants with other substances in the water (such as trihalomethanes—THMs).

Disinfection: a chemical treatment method involving the addition of a substance (e.g., chlorine, ozone, or hydrogen peroxide) that destroys or inactivates harmful microorganisms, or inhibits their activity.

Dispersion: the extent to which a liquid substance introduced into a groundwater system spreads as it moves through the system.

Dissociate: the process of ion separation that occurs when an ionic solid is dissolved in water.

Dissolved constituent: operationally defined as a constituent that passes through a 0.45-μm filter.

Dissolved Oxygen (DO): the oxygen dissolved in water is usually expressed in milligrams per liter, parts per million, or percent of saturation.

Dissolved solids: any material that can dissolve in water and be recovered by evaporating the water after filtering the suspended material.

Diversion: a turning aside or alteration of the natural course of a flow of water, normally considered to physically leave the natural channel. In some states, this can be a consumptive use from another stream, such as by livestock watering. In other States, a diversion must consist of actions such as taking water through a canal, pipe, or conduit.

Dolomite: a sedimentary rock consisting chiefly of magnesium carbonate.

Domestic withdrawals: water used for normal household purposes, such as drinking, food preparation, bathing, washing clothes and dishes, flushing toilets, and watering lawns and gardens. The water may be obtained from a public supplier or may be self-supplied and is also called residential water use.

Drainage area: the drainage area of a stream at a specified location is the area, measured in a horizontal plane, enclosed by a drainage divide.

Drainage basin: the land area drained by a river or stream.

Drainage divide: the boundary between adjoining drainage basins.

Drawdown: the difference between the water level in a well before pumping and the water level in the well during pumping. Also, for flowing wells, the reduction of the pressure head results from the discharge of water.

Drinking Water Standards: water quality standards measured in terms of suspended solids, unpleasant taste, and harmful microbes to human health. Drinking water standards are included in state water quality rules.

Drinking Water Supply: any raw or finished water source that is or may be used as a public water system or as drinking water by one or more individuals.

Drip irrigation: an irrigation system in which water is applied directly to the root zone of plants by means of applicators (orifices, emitters, porous tubing, or perforated pipe) operated under low pressure. The applicators can be placed on or below the surface of the ground or can be suspended from support.

Drought: a prolonged period of less-than-normal precipitation such that the lack of water causes a severe hydrologic imbalance.

E

Ecoregion: an area of similar climate, landform, soil, potential natural vegetation, hydrology, or other ecologically relevant variables.

Ecosystem: a community of organisms considered together with the nonliving factors of its environment.

Effluent: outflow from a particular source, such as a stream that flows from a lake or liquid waste that flows from a factory or sewage treatment plant.

Effluent Limitations: standards developed by the EPA to define the levels of pollutants that could be discharged into surface waters.

Electrodialysis: the process of separating substances in a solution by dialysis, using an electric field as the driving force.

Electronegativity: the tendency for atoms that do not have a complete octet of electrons in their outer shell to become negatively charged.

Emergent plants: erect, rooted, herbaceous plants that may be temporarily or permanently flooded at the base but do not tolerate prolonged inundation of the entire plant.

Enhanced Surface Water Treatment Rule (ESWTR): a revision of the original Surface Water Treatment Rule that includes new technology and requirements to deal with newly identified problems.

Environment: the sum of all conditions and influences affecting the life of organisms.

Environmental sample: a water sample collected from an aquifer or stream for chemical, physical, or biological characterization of the sample resource.

Environmental setting: a land area characterized by a unique combination of natural and human-related factors, such as row-crop cultivation or glacial-till soils.

Ephemeral stream: a stream or part of a stream that flows only in direct response to precipitation; it receives little or no water from springs, melting snow, or other sources; its channel is at all times above the water table.

EPT richness index: an index based on the sum of the number of taxa in three insect orders, Ephemeroptera (mayflies), Plecoptera (stoneflies), and Trichoptera (caddisflies), that are primarily composed of species considered to be relatively intolerant to environmental alterations.

Erosion: the process whereby materials of the Earth's crust are loosened, dissolved, or worn away and simultaneously moved from one place to another.

Eutrophication: the process by which water becomes enriched with plant nutrients, most commonly phosphorus and nitrogen.

Evaporite minerals (deposits): minerals or deposits of minerals formed by the evaporation of water containing salts. These deposits are common in arid climates.

Evaporites: a class of sedimentary rocks composed primarily of minerals precipitated from a saline solution as a result of extensive or total evaporation of water.

Evapotranspiration: the process by which water is discharged to the atmosphere as a result of evaporation from the soil and surface-water bodies and transpiration by plants.

F

Facultative Bacteria: a type of anaerobic bacteria that can metabolize its food either aerobically or anaerobically.

Fall line: an imaginary line marking the boundary between the ancient, resistant crystalline rocks of the Piedmont province of the Appalachian Mountains, and the younger, softer sediments of the Atlantic Coastal Plain province in the Eastern United States. Along rivers, this line is commonly reflected by waterfalls.

Fecal bacteria: microscopic single-celled organisms (primarily fecal coliforms and fecal streptococci) found in the wastes of warm-blooded animals. Their presence in water is used to assess the sanitary quality of water for body-contact recreation or consumption. Their presence indicates contamination by the waste of warm-blooded animals and the possible presence of pathogenic (disease-producing) organisms.

Federal Water Pollution Control Act (1972): The Act outlines the objective "to restore and maintain the chemical, physical, and biological integrity of the nation's waters." The 1972 act and subsequent Clean Water Act amendments are the most far-reaching water pollution control legislation ever enacted. They established broad water pollution controls, standard laws, and interstate cooperation. They provided grants for research, investigations, training, and information on national programs on surveillance, the effects of pollutants, pollution control, and the identification and measurement of pollutants. Additionally, they offer a lot of grants and loans for the construction of treatment works. The Act established national discharge standards with enforcement provisions.

The Federal Water Pollution Control Act established several milestone achievement dates. It required secondary treatment of domestic waste by publicly owned treatment works (POTWs) and the application of the "best practicable" water pollution control technology by industry by 1977. Virtually all industrial sources have achieved compliance (Due to economic difficulties and cumbersome federal requirements, certain POTWs obtained an extension to July 1, 1988, for compliance). The Act also called for new levels of technology to be imposed during the 1980s and 1990s, particularly for controlling toxic pollutants.

The Act mandates a strong pretreatment program to control toxic pollutants discharged by the industry into POTWs. The 1987 amendments require that stormwater from industrial activity be regulated.

Fertilizer: any of a large number of natural or synthetic materials, including manure and nitrogen, phosphorus, and potassium compounds, spread on or worked into the soil to increase its fertility.

Filtrate: liquid that has passed through a filter.

Filtration: a physical treatment method for removing solid (particulate) matter from water by passing the water through porous media such as sand or a man-made filter.

Flocculation: the water treatment process following coagulation, which uses gentle stirring to bring suspended particles together so that they will form larger, more settleable clumps called floc.

Flood: any relatively high streamflow that overflows the natural or artificial banks of a stream.

Flood attenuation: a weakening or reduction in the force or intensity of a flood.

Flood irrigation: the application of irrigation water whereby the entire surface of the soil is covered by ponded water.

Flood plain: a strip of relatively flat land bordering a stream channel that is inundated at times of high water.

Flow line: the idealized path followed by particles of water.

Flow path: an underground route for groundwater movement, extending from a recharge (intake) zone to a discharge (output) zone, such as a shallow stream.

Fluvial: related to a river or stream.

Freshwater: water that contains less than 1,000 milligrams per liter of dissolved solids.

Freshwater chronic criteria: the highest concentration of a contaminant that freshwater aquatic organisms can be exposed to for an extended period (4 days) without adverse effects.

G

Grab sample: a single water sample collected at one time from a single point.

Groundwater: the fresh water found under the Earth's surface, usually in aquifers. Groundwater is a major source of drinking water, and a source of growing concern in areas where leaching agricultural or industrial pollutants, or substances from leaking underground storage tanks, are contaminating groundwater.

H

Habitat: the part of the physical environment in which a plant or animal lives.

Hardness: a characteristic of water caused primarily by the salts of calcium and magnesium. It causes deposition of scale in boilers, damage in some industrial processes, and sometimes an objectionable taste.

Headwaters: the source and upper part of a stream.

Hydraulic conductivity: the capacity of a rock to transmit water. It is expressed as the volume of water at the existing kinematic viscosity that will move in unit time under a unit hydraulic gradient through a unit area measured at right angles to the direction of flow.

Hydraulic gradient: the change in hydraulic head per unit of distance in a given direction.

Hydrogen Bonding: the term used to describe the weak but effective attraction that occurs between polar covalent molecules.

Hydrograph: a graph showing the variation of water elevation, velocity, streamflow, or other properties of water with respect to time.

Hydrologic Cycle: literally the water-earth cycle. The movement of water in all three physical forms through various environmental media (air, water, biota, and soil).

Hydrology: the science that deals with water as it occurs in the atmosphere, on the surface of the ground, and underground.

Hydrostatic pressure: the pressure exerted by the water at any given point in the body of water at rest.

Hygroscopic: a substance that readily absorbs moisture.

I

Impermeability: the incapacity of a rock to transmit a fluid.

Index of Biotic Integrity (IBI): an aggregated number, or index, based on several attributes or metrics of a fish community that provides an assessment of biological conditions.

Indicator sites: stream sampling sites located at outlets of drainage basins with relatively homogeneous land use and physiographic conditions; most indicator-site basins have drainage areas ranging from 20 to 200 square miles.

Infiltration: the downward movement of water from the atmosphere into soil or porous rock.

Influent: Water flowing into a reservoir, basin, or treatment plant.

Inorganic: containing no carbon; matter other than plant or animal.

Inorganic chemical: a chemical substance of mineral origin not having carbon in its molecular structure.

Inorganic soil: soil with less than 20% organic matter in the upper 16 inches.

Ionic Bond: the attractive forces between oppositely charged ions; for example, the forces between the sodium and chloride ions in a sodium chloride crystal.

Instantaneous discharge: the volume of water that passes a point at a particular instant in time.

Instream use: water use that occurs within the stream channel for purposes such as hydroelectric power generation, navigation, water-quality improvement, fish propagation, and recreation. Sometimes called no withdrawal use or in-channel use.

Intermittent stream: a stream that flows only when it receives water from rainfall, runoff, springs, or some surface source such as melting snow.

Internal drainage: surface drainage whereby the water does not reach the ocean, such as drainage toward the lowermost or central part of an interior basin or closed depression.

Intertidal: alternately flooded and exposed by tides.

Intolerant organisms: organisms that are not adaptable to human alterations to the environment and thus decline in numbers where alterations occur.

Invertebrate: an animal having no backbone or spinal column.

Ion: a positively or negatively charged atom or group of atoms.

Irrigation: controlled application of water to arable land to supply the requirements of crops not satisfied by rainfall.

Irrigation return flow: the part of irrigation water applied to the surface that is not consumed by evapotranspiration or uptake by plants and that migrates to an aquifer or surface-water body.

Irrigation withdrawals: withdrawals of water for application on land to assist in the growing of crops and pastures or to maintain recreational lands.

K

Kill: Dutch term for stream or creek.

L

Lacustrine: relating, produced by, or formed in a lake.

Leachate: a liquid that has percolated through soil containing soluble substances and contains certain amounts of these substances in solution.

Leaching: the removal of materials in solution from soil or rock; also refers to the movement of pesticides or nutrients from the land surface to groundwater.

Limnetic: the deepwater zone (greater than two meters deep).

Littoral: the shallow-water zone (less than two meters deep).

Load: material that is moved or carried by streams, reported as the weight of material transported during a specified time period, such as tons per year.

M

Main stem: the principal trunk of a river or a stream.

Marsh: a water-saturated, poorly drained area, intermittently or permanently water-covered, with aquatic and grasslike vegetation.

Maturity (stream): the stage in the development of a stream at which it has reached its maximum efficiency, when velocity is just sufficient to carry the sediment delivered to it by tributaries; characterized by a broad, open, flat-floored valley with a moderate gradient and gentle slope.

Maximum Contaminant Level (MCL): the maximum allowable concentration of a contaminant in drinking water, as established by state and/or federal regulations. Primary MCLs are health-related and mandatory. Secondary MCLs are related to the aesthetics of the water and are highly recommended but not required.

Mean discharge : the arithmetic mean of individual daily mean discharges of a stream during a specific period, usually daily, monthly, or annually.

Membrane filter method: a laboratory method used for coliform testing. The procedure uses an ultra-thin filter with a uniform pore size smaller than bacteria (less than a micron). After water is forced through the filter, the filter is incubated in a special medium that promotes the growth of coliform bacteria. Bacterial colonies with a green-gold sheen indicate the presence of coliform bacteria.

Method detection limit: the minimum concentration of a substance that can be accurately identified and measured with current lab technologies.

Midge: a small fly in the family Chironomidae. The larval (juvenile) life stages are aquatic.

Minimum reporting level (MRL): the smallest measured concentration of a constituent that may be reliably reported using a given analytical method. In many cases, the MRL is used when documentation for the method detection limit is not available.

Mitigation : actions taken to avoid, reduce, or compensate for the effects of human-induced environmental damage.

Modes of transmission of disease: the ways in which diseases spread from one person to another.

Monitoring: repeated observation, measurement, or sampling at a site, on a scheduled or event basis, for a particular purpose.

Monitoring well: a well designed for measuring water levels and testing groundwater quality.

Multiple-tube fermentation method: a laboratory method used for coliform testing, which uses a nutrient broth placed in a culture tube. Gas production indicates the presence of coliform bacteria.

N

National Primary Drinking Water Regulations (NPDWRs): regulations developed under the Safe Drinking Water Act, which establish maximum contaminant levels, monitoring requirements, and reporting procedures for contaminants in drinking water that endanger human health.

National Pollutant Discharge Elimination System (NPDES): a requirement of the CWA that discharges meet certain requirements prior to discharging waste to any water body. It sets the highest permissible effluent limits, by permit, prior to making any discharge.

Near Coastal Water Initiative: This initiative was developed in 1985 to provide for the management of specific problems in waters near coastlines that are not dealt with in other programs.

Nitrate: an ion consisting of nitrogen and oxygen (NO_3). Nitrate is a plant nutrient and is very mobile in soils.

Nonbiodegradable: substances that do not break down easily in the environment.

Nonpoint source: a source (of any water-carried material) from a broad area, rather than from discrete points.

Nonpoint-source contaminant: a substance that pollutes or degrades water that comes from lawn or cropland runoff, the atmosphere, roadways, and other diffuse sources.

Nonpoint-source water pollution: water contamination that originates from a broad area (such as leaching of agricultural chemicals from crop land) and enters the water resource diffusely over a large area.

Nonpolar Covalently bonded: a molecule composed of atoms that share their electrons equally, resulting in a molecule that does not have polarity.

Nutrient: any inorganic or organic compound needed to sustain plant life.

O

Organic: containing carbon, but possibly also containing hydrogen, oxygen, chlorine, nitrogen, and other elements.

Organic chemical: a chemical substance of animal or vegetable origin having carbon in its molecular structure.

Organic detritus: any loose organic material in streams—such as leaves, bark, or twigs—removed and transported by mechanical means, such a disintegration or abrasion.

Organic soil: soil that contains more than 20% organic matter in the upper 16 inches.

Organochlorine compound: synthetic organic compounds containing chlorine. As generally used, the term refers to compounds containing mostly or exclusively carbon, hydrogen, and chlorine.

Outwash: soil material washed down a hillside by rainwater and deposited upon more gently sloping land.

Overland flow: the flow of rainwater or snow melt over the land surface toward stream channels.

Oxidation: when a substance either gains oxygen or loses hydrogen or electrons in a chemical reaction, it is a type of chemical treatment.

Oxidizer: a substance that oxidizes another substance.

P

Parts per million: the number of weight or volume units of a constituent present in each one million units of the solution or mixture. Formerly used to express the results of most water and wastewater analyses, **ppm** is being replaced by milligrams per liter **m/l**. For drinking water analyses, concentrations in parts per million and milligrams per liter are equivalent. A single PPM can be compared to a shot glass full of water inside a swimming pool.

Pathogens: types of microorganisms that can cause disease.

Perched groundwater: unconfined groundwater separated from an underlying main body of groundwater by an unsaturated zone.

Percolation: the movement, under hydrostatic pressure, of water through interstices of a rock or soil (except the movement through large openings such as caves).

Perennial stream: a stream that usually has water in its channel at all times.

Periphyton: microorganisms that coat rocks, plants, and other surfaces on lake bottoms.

Permeability: the capacity of a rock to transmit a fluid; a measure of the relative ease with which a porous medium can transmit a liquid.

pH: a measure of the acidity (less than 7) or alkalinity (greater than 7) of a solution; a pH of 7 is considered neutral.

Phosphorus: a nutrient essential for growth that can play a key role in stimulating aquatic growth in lakes and streams.

Photosynthesis: the synthesis of compounds with the aid of light.

Physical treatment: any process that does not produce a new substance (e.g., screening, adsorption, aeration, sedimentation, and filtration).

Point source: originating at any discrete source.

Polar covalent bond: the shared pair of electrons between two atoms is not equally held. Thus, one of the atoms becomes slightly positively charged, and the other atom becomes slightly negatively charged.

Polar covalent molecule: (water) one or more polar covalent bonds result in a molecule that is polar covalent. Polar covalent molecules exhibit partial positive and negative poles, causing them to behave like tiny magnets. Water is the most common polar covalent substance.

Pollutant: any substance introduced into the environment that adversely affects the usefulness of the resource.

Pollution: the presence of matter or energy whose nature, location, or quantity produces undesired environmental effects. Under the Clean Water Act, for example, the term is defined as a man-made or man-induced alteration of the physical, biological, and radiological integrity of water.

Polychlorinated biphenyls (PCBs): a mixture of chlorinated derivatives of biphenyl, marketed under the trade name Aroclor with a number designating the chlorine content (such as Aroclor 1260). PCBs were used in transformers and capacitors for insulating purposes and in gas pipeline systems as a lubricant. Further sale or new use was banned by law in 1979.

Polycyclic aromatic hydrocarbon (PAH): a class of organic compounds with a fused-ring aromatic structure. PAHs result from the incomplete combustion of organic carbon (including wood), municipal solid waste, and fossil fuels, as well as from the natural or anthropogenic introduction of uncombusted coal and oil. PAHs include benzo(a) pyrene, fluoranthene, and pyrene.

Population: a collection of individuals of one species or mixed species making up the residents of a prescribed area.

Porosity: the ratio of the volume of voids in a rock or soil to the total volume.

Potable water: water that is safe and palatable for human consumption.

Precipitation: any or all forms of water particles that fall from the atmosphere, such as rain, snow, hail, and sleet. The act or process of producing a solid phase within a liquid medium.

Pretreatment: any physical, chemical, or mechanical process used before the main water treatment processes. It can include screening, presedimentation, and chemical addition.

Primary Drinking Water Standards: regulations on drinking water quality (under SDWA) are considered essential for the preservation of public health.

Primary Treatment: the first step of treatment at a municipal wastewater treatment plant. It typically involves screening and sedimentation to remove materials that float or settle.

Public-supply withdrawals: water withdrawn by public and private water suppliers for use within a general community. Water is used for a variety of purposes, such as domestic, commercial, industrial, and public water use.

Public water system: as defined by the Safe Drinking Water Act, any system, publicly or privately owned, that serves at least 15 service connections for 60 days out of the year or serves an average of 25 people for at least 60 days out of the year.

Publicly Owned Treatment Works (POTW): a waste treatment works owned by a state, local government unit, or Indian tribe, usually designed to treat domestic wastewaters.

R

Rain shadow: a dry region on the lee side of a topographic obstacle, usually a mountain range, where rainfall is noticeably less than on the windward side.

Reach: a continuous part of a stream between two specified points.

Reaeration: the replenishment of oxygen in water from which oxygen has been removed.

Receiving waters: a river, lake, ocean, stream, or other water source into which wastewater or treated effluent is discharged.

Recharge: the process by which water is added to a zone of saturation, usually by percolation from the soil surface.

Recharge area (groundwater): an area within which water infiltrates the ground and reaches the zone of saturation.

Reference Dose (RfD): an estimate of the amount of a chemical that a person can be exposed to daily that is not anticipated to cause adverse systemic health effects over the person's lifetime.

Representative Sample: a sample containing all the constituents present in the water from which it was taken.

Return flow: that part of irrigation water that is not consumed by evapotranspiration and that returns to its source or another body of water.

Reverse Osmosis (RO): a process where a semi-permeable membrane separates solutions with different ion concentrations. Typically, water flows from the chamber with a lower ion concentration into the chamber with a greater ion concentration, creating hydrostatic or osmotic pressure. In RO, enough external pressure is applied to overcome this hydrostatic pressure, thus reversing the flow of water. This results in the water on the other side of the membrane becoming depleted in ions and demineralized.

Riffle: a shallow part of the stream where water flows swiftly over completely or partially submerged obstructions to produce surface agitation.

Riparian: relating to or situated on the bank of a natural body of flowing water.

Riparian rights: a concept of water law under which authorization to use water in a stream is based on ownership of the land adjacent to the stream.

Riparian zone: relating to or located on the bank of a body of water, especially a stream.

Runoff: that part of precipitation or snowmelt that appears in streams or surface-water bodies.

Rural withdrawals: water used in suburban or farm areas for domestic and livestock needs. The water generally is self-supplied and includes domestic use, drinking water for livestock, and other uses such as dairy sanitation, evaporation from stock-watering ponds, and cleaning and waste disposal.

S

Safe Drinking Water Act (SDWA): a federal law passed in 1974 to establish federal standards for drinking water quality, protect underground sources of water, and set up a system of state and federal cooperation to ensure compliance with the law.

Saline water: water that is considered unsuitable for human consumption or for irrigation because of its high content of dissolved solids; generally expressed as milligrams per liter (mg/l) of dissolved solids. Seawater is usually considered to contain more than 35,000 mg/l of dissolved solids. A general salinity scale is:

Concentration of Dissolved Solids in mg/l	
Slightly saline	1,000–3,000
Moderately saline	3,000–10,000
Very saline	10,000–35,000
Brine	More than 35,000

Saturated zone: a subsurface zone in which all the interstices or voids are filled with water under pressure greater than that of the atmosphere.

Screening: a pretreatment method that uses coarse screens to remove large debris from the water to prevent clogging of pipes or channels leading to the treatment plant.

Secondary Drinking Water Standards: regulations developed under the Safe Drinking Water Act that establish maximum levels of substances affecting the aesthetic characteristics (taste, color, or odor) of drinking water.

Secondary maximum contaminant level (SMCL): the maximum level of a contaminant or undesirable constituent in public water systems that, in the judgment of the USEPA, is required to protect public welfare. SMCLs are secondary (non-enforceable) drinking water regulations established by the US EPA for contaminants that may adversely affect the odor or appearance of the water.

Secondary treatment: the second step of treatment at a municipal wastewater treatment plant. This step uses a growing numbers of microorganisms to digest organic matter and reduce the amount of organic waste. Water leaving this process is chlorinated to destroy any disease-causing microorganisms before its release.

Sedimentation: a physical treatment method that involves reducing the velocity of water in basins so that the suspended material can settle out by gravity.

Seep: a small area where water percolates slowly to the land surface.

Seiche: a sudden oscillation of the water in a moderate-sized bay, caused by wind.

Sinuosity: the ratio of the channel length between two points on a channel to the straight-line distance between the same two points; a measure of meandering.

Soil horizon: a layer of soil that is distinguishable from adjacent layers by characteristic physical and chemical properties.

Soil moisture: water occurring in the pore spaces between the soil particles in the unsaturated zone from which water is discharged by the transpiration of plants or by evaporation from the soil.

Solution: formed when a solid, gas, or another liquid in contact with a liquid becomes dispersed homogeneously throughout the liquid. The substance, called a solute, is said to dissolve. The liquid is called the solvent.

Solvated: when either a positive or negative ion becomes surrounded by polar solvent molecules.

Sorb: to take up and hold either by absorption or by adsorption.

Sorption: a general term for the interaction (binding or association) of a solute ion or molecule with a solid.

Specific yield: the ratio of the volume of water that will drain under the influence of gravity to the volume of saturated rock.

Spring: a place where a concentrated discharge of groundwater flows at the ground surface.

Surface runoff: runoff that travels over the land surface to the nearest stream channel.

Surface tension: the attractive forces exerted by the molecules below the surface upon those at the surface, resulting in them crowding together and forming a higher density.

Surface water: all water naturally open to the atmosphere, including all springs, wells, or other collectors that are directly influenced by surface water.

Surface Water Treatment Rule (SWTR): a federal regulation established by the USEPA under the Safe Drinking Water Act that imposes specific monitoring and treatment requirements on all public drinking water systems drawing water from a surface water source.

Suspended sediment: sediment transported in suspension by a stream.

Suspended solids: different from suspended sediment only in the way that the sample is collected and analyzed.

Synthetic organic chemicals (SOCs) are generally applied to manufactured chemicals that are less volatile than volatile organic compounds. Included are herbicides, pesticides, and chemicals widely used in industries.

T

Total head: the height above a datum plane of a column of water in a groundwater system. It is composed of the elevation head and the pressure head.

Total Suspended Solids (TSS): solids present in wastewater.

Transpiration: the process by which water passes through living organisms, primarily plants, into the atmosphere.

Trihalomethanes (THMs): a group of compounds formed when natural organic compounds from decaying vegetation and soil (such as humic and fulvic acids) react with chlorine.

Turbidity: a measure of the cloudiness of water caused by the presence of suspended matter, which shelters harmful microorganisms and reduces the effectiveness of disinfecting compounds.

U

Unconfined aquifer: an aquifer whose upper surface is a water table free to fluctuate under atmospheric pressure.

Unsaturated zone: a subsurface zone above the water table in which the pore spaces may contain a combination of air and water.

V

Vehicle of disease transmission: any nonliving object or substance contaminated with pathogens.

Vernal pool : a small lake or pond that is filled with water for only a short time during the spring.

W

Wastewater: the spent or used water from individual homes, a community, a farm, or an industry that contains dissolved or suspended matter.

Water budget: an accounting of the inflow to, outflow from, and storage changes of water in a hydrologic unit.

Water column: an imaginary column extending through a water body from its floor to its surface.

Water demand: water requirements for a particular purpose, such as irrigation, power, municipal supply, plant transpiration, or storage.

Water Table: the top water surface of an unconfined aquifer at atmospheric pressure.

Waterborne disease: water is a potential vehicle of disease transmission, and waterborne disease is possibly one of the most preventable types of communicable illness. The application of basic sanitary principles and technology has virtually eliminated severe outbreaks of waterborne diseases in developed countries. The most prevalent waterborne diseases include *typhoid fever, dysentery, cholera, infectious hepatitis*, and *gastroenteritis*.

Water Softening: a chemical treatment method that uses either chemicals to precipitate or a zeolite to remove those metal ions (typically Ca^{2+}, Mg^{2+}, Fe^{3+}) responsible for hard water.

Watershed: the land area that drains into a river, river system, or other body of water.

Wellhead Protection: the protection of the surface and subsurface areas surrounding a water well or well field supplying a public water system from contamination by human activity.

Y

Yield: the mass of material or constituents transported by a river in a specified period divided by the drainage area of the river basin.

Index

Note: Page numbers in *italics* indicate a figure and page numbers in **bold** indicate a table on the corresponding page.

E

T

U

Y

Z